Designing and Building Fuel Cells

Colleen Spiegel

New York Chicago San Francisco Lisbon London Madrid
Mexico City Milan New Delhi San Juan Seoul
Singapore Sydney Toronto

The McGraw·Hill Companies

Library of Congress Cataloging-in-Publication Data

Spiegel, Colleen.
   Designing and building fuel cells / Colleen Spiegel.—1st ed.
     p.   cm.
   ISBN 0-07-148977-0 (alk. paper)
   1. Fuel cells—Design and construction. I. Title.
TK2931.S65   2007
621.31'2429—dc22

                                            2007007508

McGraw-Hill books are available at special quantity discounts to use as premiums and sales promotions, or for use in corporate training programs. For more information, please write to the Director of Special Sales, Professional Publishing, McGraw-Hill, Two Penn Plaza, New York, NY 10121-2298. Or contact your local bookstore.

Designing and Building Fuel Cells

Copyright © 2007 by The McGraw-Hill Companies. All rights reserved. Printed in the United States of America. Except as permitted under the Copyright Act of 1976, no part of this publication may be reproduced or distributed in any form or by any means, or stored in a database or retrieval system, without the prior written permission of publisher.

3 4 5 6 7 8 9 10   IBT/IBT   1 9 8 7 6 5 4 3 2 1

ISBN-13: 978-0-07-148977-5
ISBN-10:   0-07-148977-0

Sponsoring Editors
Kenneth P. McCoombs
Larry Hagar

Editorial Supervisor
Jody McKenzie

Project Manager
Vastavikta Sharma, International Typesetting and Composition

Acquisitions Coordinator
Laura Hahn

Copy Editor
Michael McGee

Proofreader
Mona Mehra

Indexer
Steve Ingle

Production Supervisor
George Anderson

Composition
International Typesetting and Composition

Illustration
International Typesetting and Composition

Art Director, Cover
Jeff Weeks

Cover Designer
Libby Pisacreta

Information has been obtained by McGraw-Hill from sources believed to be reliable. However, because of the possibility of human or mechanical error by our sources, McGraw-Hill, or others, McGraw-Hill does not guarantee the accuracy, adequacy, or completeness of any information and is not responsible for any errors or omissions or the results obtained from the use of such information.

*To my husband, Brian,
who inspires me to be a better person every day,
encourages me to pursue all of my dreams,
and has the patience and endurance to stand by me
while I work at them.*

*To my unborn son, Howard,
who had to endure endless sleepless nights with me
while completing this book.*

ABOUT THE AUTHOR

Colleen S. Spiegel is the founder of Clean Fuel Cell Energy, LLC. She has been an R&D manager and chemical engineer for more than seven years, and has expertise in the areas of fuel cell design and modeling. Ms. Spiegel is a member of the American Institute of Chemical Engineers (AICHE) and the Institute of Electrical & Electronic Engineers (IEEE).

Contents

Foreword   xii

Chapter 1. An Introduction to Fuel Cells   1

   1.1 What Is a Fuel Cell?   3
        1.1.1 Comparison with batteries   4
        1.1.2 Comparison with heat engine   5
   1.2 Why Do We Need Fuel Cells?   6
        1.2.1 Portable sector   7
        1.2.2 Transportation sector   7
        1.2.3 Stationary sector   7
   1.3 History of Fuel Cells   7
        1.3.1 PEM fuel cells   9
        1.3.2 Solid oxide fuel cells   10
        1.3.3 Molten carbonate fuel cells   10
        1.3.4 Phosphoric acid fuel cells   11
        1.3.5 Alkali fuel cells   11
   1.4 How Do Fuel Cells Work?   12
   Chapter Summary   13
   Problems   13
   Bibliography   13

Chapter 2. Fuel Cells and the Hydrogen Economy   15

   2.1 Characteristics of Hydrogen   16
        2.1.1 Safety aspects of hydrogen as a fuel   18
   2.2 World Energy Demand   19
   2.3 Development of the Hydrogen Economy   21
   2.4 Hydrogen Production, Distribution, and Storage   22
        2.4.1 Technologies for hydrogen production   22
        2.4.2 Technologies for hydrogen storage   24
        2.4.3 Worldwide hydrogen refueling stations   25

Contents

2.5	Investment of Hydrogen Infrastructure	26
	2.5.1 Government support	28
	2.5.2 Long-term projections of hydrogen use	28
	2.5.3 Key players in hydrogen R&D	29
Chapter Summary		33
Problems		33
Bibliography		34

Chapter 3. Fuel Cell Types — 35

3.1	Polymer Electrolyte Membrane Fuel Cells (PEMFCs)	36
3.2	Alkaline Fuel Cells (AFCs)	39
3.3	Phosphoric Acid Fuel Cells (PAFCs)	40
3.4	Solid Oxide Fuel Cells (SOFCs)	42
3.5	Molten-Carbonate Fuel Cells (MCFCs)	43
3.6	Direct Methanol Fuel Cells (DMFCs)	44
3.7	Zinc Air Fuel Cells (ZAFCs)	46
3.8	Protonic Ceramic Fuel Cells (PCFCs)	47
3.9	Biological Fuel Cells (BFCs)	49
Chapter Summary		50
Problems		51
Bibliography		51

Chapter 4. Fuel Cell Applications — 53

4.1	Portable Power	53
4.2	Backup Power	55
	4.2.1 Basic electrolyzer calculations	56
4.3	Transportation Applications	57
	4.3.1 Automobiles	58
	4.3.2 Buses	66
	4.3.3 Utility vehicles	66
	4.3.4 Scooters and bicycles	69
4.4	Stationary Power Applications	72
Chapter Summary		83
Problems		84
Bibliography		84

Chapter 5. Basic Fuel Cell Thermodynamics — 87

5.1	Basic Thermodynamic Concepts	87
5.2	Fuel Cell Reversible and Net Output Voltage	92
5.3	Theoretical Fuel Cell Efficiency	99
	5.3.1 Energy efficiency	100
5.4	Fuel Cell Temperature	101
5.5	Fuel Cell Pressure	102
Chapter Summary		104
Problems		104
Bibliography		105

Chapter 6. Fuel Cell Electrochemistry — 107

6.1	Electrode Kinetics	110
6.2	Voltage Losses	112

6.3	Internal Currents and Crossover Currents	116
6.4	Improving Kinetic Performance	117
	Chapter Summary	118
	Problems	118
	Bibliography	119

Chapter 7. Fuel Cell Charge Transport — 121

7.1	Voltage Loss Due to Charge Transport	121
7.2	Microscopic Conductivity in Metals	126
7.3	Ionic Conductivity in Aqueous Electrolytes	126
7.4	Ionic Conductivity of Polymer Electrolytes	127
7.5	Ionic Conduction in Ceramic Electrolytes	130
	Chapter Summary	131
	Problems	132
	Bibliography	132

Chapter 8. Fuel Cell Mass Transport — 133

8.1	Convective Mass Transport from Flow Channels to Electrode	134
8.2	Diffusive Mass Transport in Fuel Cell Electrodes	135
8.3	Convective Mass Transport in Flow Structures	139
	8.3.1 Mass transport in flow channels	139
	8.3.2 Pressure drop in flow channels	144
	Chapter Summary	149
	Problems	149
	Bibliography	150

Chapter 9. Heat Transfer — 151

9.1	Fuel Cell Energy Balance	152
	9.1.1 General energy balance procedure	152
	9.1.2 Energy balance of fuel cell stack	154
	9.1.3 General energy balance for fuel cell	154
	9.1.4 Energy balance for fuel cell components and gases	155
9.2	Heat Generation and Flux in Fuel Cell Layers	158
9.3	Heat Conduction	158
9.4	Heat Dissipation Through Natural Convection and Radiation	159
9.5	Fuel Cell Heat Management	160
	9.5.1 Heat exchanger model	162
	9.5.2 Air cooling	163
	9.5.3 Edge cooling	166
	Chapter Summary	167
	Problems	168
	Bibliography	168

Chapter 10. Fuel Cell Modeling — 171

10.1	Conservation of Mass	175
10.2	Conservation of Momentum	175

10.3 Conservation of Energy	176
10.4 Conservation of Species	177
10.5 Conservation of Charge	178
10.6 The Electrodes	178
10.6.1 Mass transport	179
10.6.2 Electrochemical behavior	181
10.6.3 Ion/electron transport	183
10.6.4 Heat transport in the electrodes	184
10.7 The Electrolyte	185
Chapter Summary	186
Problems	187
Bibliography	187

Chapter 11. Fuel Cell Materials   189

11.1 Electrolyte Layer	190
11.1.1 PEMFCs and DMFCs	192
11.1.2 PAFCs	195
11.1.3 AFCs	196
11.1.4 MCFCs	196
11.1.5 SOFCs	198
11.2 Fuel Cell Electrode Layers	199
11.2.1 PEMFC, DMFC, and PAFC catalysts	201
11.2.2 PEMFC, DMFC, and PAFC gas diffusion layers	205
11.2.3 AFC electrodes	207
11.2.4 MCFC electrodes	208
11.2.5 SOFC electrodes	208
11.3 Low-Temperature Fuel Cell Processing Techniques	210
11.4 SOFC manufacturing method	212
11.5 Method for Building a Fuel Cell	213
11.5.1 Preparing the polymer electrolyte membrane	213
11.5.2 Catalyst/electrode layer material	214
11.5.3 Hot-pressing the MEA	215
Chapter Summary	217
Problems	217
Bibliography	218

Chapter 12. Fuel Cell Stack Components and Materials   221

12.1 Bipolar Plates	221
12.1.1 Bipolar plate materials for low and medium temperature fuel cells	223
12.1.2 Coated metallic plates	224
12.1.3 Composite plates	226
12.2 Flow-Field Design	227
12.3 Materials for SOFCs	232
12.4 Materials for MCFCs	233
12.5 PAFC Materials and Design	234
12.6 Channel Shape, Dimensions, and Spacing	235
12.7 Bipolar Plate Manufacturing	235
12.7.1 Nonporous graphite plate fabrication	235
12.7.2 Coated metallic plate fabrication	236
12.7.3 Composite plate fabrication	237

| | | Contents | ix |

12.8	Gaskets and Spacers	237
	12.8.1 PEMFCs/DMFCs/AFCs	238
	12.8.2 SOFC Seals	238
12.9	End Plates	240
12.10	Constructing the Fuel Cell Bipolar Plates, Gaskets, End Plates, and Current Collectors	241
	12.10.1 Bipolar plate design	241
	12.10.2 Gasket selection	242
	12.10.3 End plates	243
	12.10.4 Current collectors	244
	Chapter Summary	244
	Problems	245
	Bibliography	245

Chapter 13. Fuel Cell Stack Design — 247

13.1	Fuel Cell Stack Sizing	249
13.2	Number of Cells	254
13.3	Stack Configuration	255
13.4	Distribution of Fuel and Oxidants to the Cells	260
13.5	Cell Interconnection	262
	13.5.1 SOFCs	263
	13.5.2 AFCs	264
13.6	Stack Clamping	264
13.7	Water Management for PEMFCs	265
	13.7.1 Water management methods	266
13.8	Putting the fuel cell stack together	267
	Chapter Summary	268
	Problems	269
	Bibliography	269

Chapter 14. Fuel Cell System Design — 273

14.1	Fuel Subsystem	276
	14.1.1 Humidification systems	276
	14.1.2 Fans and Blowers	281
	14.1.3 Compressors	284
	14.1.4 Turbines	289
	14.1.5 Fuel cell pumps	291
14.2	Electrical Subsystem	296
	14.2.1 Power diodes	297
	14.2.2 Switching devices	298
	14.2.3 Switching regulators	300
	14.2.4 Inverters	303
	14.2.5 Supercapacitors	304
	14.2.6 Power electronics for cellular phones	305
	14.2.7 DC-DC converters for automotive applications	306
	14.2.8 Multilevel converters for larger applications	307
14.3	Fuel Cell Hybrid Power Systems	308
14.4	System Efficiency	308
	Chapter Summary	310
	Problems	311
	Bibliography	311

x   Contents

Chapter 15. Fuel Types, Delivery, and Processing   313

15.1 Hydrogen   315
    15.1.1 Gas   315
    15.1.2 Liquid   320
    15.1.3 Carbon nanofibers   322
15.2 Other Common Fuel Types   324
    15.2.1 Methanol   324
    15.2.2 Ethanol   325
    15.2.3 Metal hydrides   325
    15.2.4 Chemical hydrides   328
    15.2.5 Ammonia   331
    15.2.6 Natural gas   333
    15.2.7 Propane   333
    15.2.8 Gasoline and other petroleum-based fuels   333
    15.2.9 Bio-fuels   334
15.3 Fuel Processing   334
    15.3.1 Desulfurization   336
    15.3.2 Steam reforming   337
    15.3.3 Carbon formation   339
    15.3.4 Internal reforming   339
    15.3.5 Direct hydrocarbon oxidation   341
    15.3.6 Partial oxidation   341
    15.3.7 Pyrolysis   343
    15.3.8 Methanol reforming   343
15.4 Bioproduction of Hydrogen   344
    15.4.1 Photosynthesis   344
    15.4.2 Digestion processes   345
15.5 Electrolyzers   346
    15.5.1 Electrolyzer efficiency   347
    15.5.2 High pressure in electrolyzers   347
Chapter Summary   348
Problems   349
Bibliography   349

Chapter 16. Fuel Cell Operating Conditions   353

16.1 Operating Pressure   353
16.2 Operating Temperature   356
16.3 Flow Rates of Reactants   357
16.4 Humidity of Reactants   361
16.5 Fuel Cell Mass Balance   363
Chapter Summary   370
Problems   370
Bibliography   371

Chapter 17. Fuel Cell Characterization   373

17.1 Fuel Cell Testing Setup   373
17.2 Verification of the Assembly   376
17.3 Fuel Cell Conditioning   376
17.4 Baseline Test Conditions and Operating Parameters   377
    17.4.1 Temperature   378
    17.4.2 Pressure   378

	17.4.3	Flow rate	378
	17.4.4	Compression force	378
17.5	Polarization Curves		378
17.6	Fuel Cell Resistance		380
	17.6.1	Current interrupt	381
	17.6.2	The AC resistance method	382
	17.6.3	The high-frequency resistance (HFR) method	382
	17.6.4	Electrochemical (EIS) impedance spectroscopy	383
	17.6.5	Stoichiometry (utilization) sweeps	385
	17.6.6	Limiting current	386
	17.6.7	Cyclic voltammetry	387
17.7	Current Density Mapping		388
17.8	Neutron Imaging		388
17.9	Characterization of Fuel Cell Layers		388
	17.9.1	Porosity determination	389
	17.9.2	BET surface area determination	389
	17.9.3	Transmission electron microscopy (TEM)	390
	17.9.4	Scanning electron microscopy (SEM)	390
	17.9.5	X-ray diffraction (XRD)	390
	17.9.6	Energy dispersive spectroscopy (EDS)	390
	17.9.7	X-ray fluorescence (XRF)	391
	17.9.8	Inductively coupled plasma mass spectroscopy (ICP-MS)	391
Chapter Summary			391
Problems			392
Bibliography			392

Appendix A. Useful Constants and Conversions	395
Appendix B. Thermodynamic Properties of Selected Substances	397
Appendix C. Molecular Weight, Gas Constant and Specific Heat for Selected Substances	399
Appendix D. Gas Specific Heats at Various Temperatures	401
Appendix E. Specific Heat for Saturated Liquid Water at Various Temperatures	403
Appendix F. Thermodynamic Data for Selected Fuel Cell Reactants at Various Temperatures	405
Appendix G. Binary Diffusion Coefficients for Selected Fuel Cell Substances	413
Appendix H. Product Design Specifications	415
Appendix I. Fuel Cell Design Requirements and Parameters	417

Index   423

Foreword

The creation and consumption of energy is a necessity in today's world. It has drastically increased the quality of life of modern society and has allowed the rapid advancement of modern technology. The majority of energy required to power our homes, offices, and automobiles is created from fossil fuels, and though they are essential to our society, use of fossil fuels has also resulted in increased health risks and global warming. Their diminishing supply is also causing international tension and contributing to high inflation. Thankfully, fuel cell technology has the potential to meet the energy needs of our ever-growing population, resolve many of the conflicts fossil fuels are causing, and will accomplish these goals in an environmentally friendly way.

During the last decade, fuel cells have been researched and developed more quickly than at any time since their discovery in 1839. Interest in them has increased tremendously as international tensions have intensified due to increasing fossil fuel prices, the scarcity of fossil fuels, the power of countries with national fossil fuel resources, and the threat of global warming. The study of fuel cells crosses many academic disciplines, such as mechanical, chemical, environmental, and electrical engineering, as well as chemistry and material science. The writing of this book was motivated by the need to have a practical, informational fuel cell reference that could aid someone who has never built a fuel cell before, yet provide enough theory to help accurately design a state-of-the-art fuel cell.

The primary audience of this textbook is professional engineers and scientists who need an all-encompassing fuel cell guide. It is also intended for the professional or student who wants to design and build commercial-grade fuel cells; therefore, it incorporates a unique balance of theory, practical knowledge, and design. This book is also meant to be an engineering or science text, and it includes examples and problems to help students bridge the gap between the various interdisciplinary sciences needed to accurately design and build fuel cells.

The text is organized into 17 chapters. The first four chapters provide an overview of fuel cell technology, types, applications, and offers a peek at the coming hydrogen economy. The first chapter describes the basics of fuel cells and compares them with batteries and heat engines, outlining fuel cell markets, and the history and fundamentals of how fuel cells work. Chapter 2 describes fuel cells and the hydrogen economy. It compares hydrogen with other fuel types, and describes how the hydrogen economy can be developed. Hydrogen production, storage, and distribution are discussed, as well as those countries working on building a hydrogen economy. Chapter 3 describes different fuel cell types. It also explores a few fuel cell types sometimes not considered fuel cells, or that are in mostly the R&D stage. Chapter 4 discusses the main applications of commercial and R&D fuel cells such as portable power, backup power, transportation, and stationary power applications.

Chapters 5 through 9 provide an introduction to the necessary science involved in predicting fuel cell performance. Chapter 5 covers basic fuel cell thermodynamics and demonstrates the calculation of theoretical fuel cell voltage, theoretical fuel cell efficiency, fuel cell temperature, and pressure. Chapter 6 describes the basics of fuel cell electrochemistry and introduces electrode kinetics, voltage losses, internal and crossover currents, and improved kinetic performance. Chapter 7 explores fuel cell charge transport especially in fuel cell electrolytes, while Chapter 8 covers fuel cell mass transport and specifically discusses diffusive transport in fuel cell electrodes, convective transport in flow channels, and concentration polarization. Chapter 9 is devoted to heat transfer, and covers the topics of energy balance, heat conduction, active heat removal, stack heat dissipation, and stack cooling options. Chapter 10 describes fuel cell modeling and uses the concepts outlined in Chapters 5 through 9 to model the fuel cell layers.

Chapters 11 through 14 discuss the basic materials and plant subsystems used for building a fuel cell system. Chapter 11 details state-of-the art fuel cell materials for the electrolyte, catalyst, and electrode layers, and explores processing techniques for creating the fuel cell. Chapter 12 tackles fuel cell stack components and materials, and specifically discusses bipolar plate materials, flow-field design, channel shape, and dimensions. It also delves into bipolar plate manufacturing techniques, gaskets, and end plates. Chapter 13 is about fuel cell design and explores stack sizing, the number of cells, stack configuration, cell interconnection, stack clamping, and how to put a fuel cell stack together. Chapter 14 covers fuel cell system design and introduces fuel cell plant components and the electrical parts of the system.

The remaining chapters (15 to 17) explain fuel types, fuel cell operating conditions, and testing. Chapter 15 compares fuel types, storage, and processing. It specifically covers hydrogen, methanol, ethanol, metal hydrides, chemical hydrides, ammonia, propane, and petroleum-based fuels. Chapter 16 shows how to calculate fuel cell operating conditions based upon operating temperature, pressure, flow rates, humidity, and mass balances. Finally, Chapter 17 summarizes popular methods of characterizing fuel cells and describes common fuel cell test methods.

I would like to acknowledge all of the teachers, engineers, and scientists that have contributed to my learning and knowledge of this subject. I also want to express my heartfelt gratitude to my husband for his continued patience, understanding, and encouragement throughout the writing of this text, and for being a loving partner as we move on toward life's other goals together.

Colleen Spiegel
Clearwater, Florida, USA

Chapter 1

An Introduction to Fuel Cells

The current movement towards environmentally friendlier and more efficient power production has caused an increased interest in alternative fuels and power sources. Fuel cells are one of the older energy conversion technologies, but only within the last decade have they been extensively studied for commercial use. The reliance upon the combustion of fossil fuels has resulted in severe air pollution, and extensive mining of the world's oil resources. In addition to being hazardous to the health of many species (including our own), the pollution is indirectly causing the atmosphere of the world to change (global warming). This global warming trend will become worse due to an increase in the combustion of fossil fuels for electricity because of the large increase in world population. In addition to health and environmental concerns, the world's fossil fuel reserves are decreasing rapidly. The world needs a power source that has low pollutant emissions, is energy efficient, and has an unlimited supply of fuel for a growing world population. Fuel cells have been identified as one of the most promising technologies to accomplish these goals.

Many other alternative energy technologies have been researched and developed. These include solar, wind, hydroelectric power, bioenergy, geothermal energy, and many others. Each of these alternative energy sources have their advantages and disadvantages, and are in varying stages of development. In addition, most of these energy technologies cannot be used for transportation or portable electronics. Other portable power technologies, such as batteries and supercapacitors also are not suitable for transportation technologies, military applications, and the long-term needs of future electronics.

The ideal option for a wide variety of applications is using a hydrogen fuel cell combined with solar or hydroelectric power. Compared to other fuels, hydrogen does not produce any carbon monoxide or other pollutants. When it is fed into a fuel cell, the only by-products are oxygen and heat. The oxygen is recombined with hydrogen to form water when power is needed.

Fuel cells can utilize a variety of fuels to generate power—from hydrogen, methanol, and fossil fuels to biomass-derived materials. Using fossil fuels to generate hydrogen is regarded as an intermediate method of producing hydrogen, methane, methanol, or ethanol for utilization in a fuel cell before the hydrogen infrastructure has been set up. Fuels can also be derived from many sources of biomass, including methane from municipal wastes, sewage sludge, forestry residues, landfill sites, and agricultural and animal waste.

Fuel cells can also help provide electricity by working with large power plants to become more decentralized and increase efficiency. Most electricity produced by large fossil-fuel burning power plants are distributed through high voltage transmission wires over long distances [1]. These power plants seem to be highly efficient because of their large size; however, a 7 to 8 percent electric energy loss in Europe, and a 10 percent energy loss in the United States occurs during long distance transmission [1]. One of the main issues with these transmission lines is that they do not function properly all the time. It would be safer for the population if electricity generation did not occur in several large plants, but is generated where the energy is needed. Fuel cells can be used wherever energy is required without the use of large transmission lines.

Fossil fuels are limited in supply, and are located in select regions throughout the world. This leads to regional conflicts and wars which threaten peace. The limited supply and large demand dries up the cost of fossil fuels tremendously. The end of low-cost oil is rapidly approaching. Other types of alternative energy technology such as fuel cells, can last indefinitely when non-fossil fuel–based hydrogen is used.

This chapter discusses the basics of fuel cells:

- What is a fuel cell?
- Why do we need fuel cells?
- The history of fuel cells
- How do fuel cells work?

By discussing the fuel cell basics, one can appreciate the relevance and significance of fuel cells in addressing environmental and industrial problems, as well as the physical and chemical mechanisms that underlie fuel cell operation.

1.1 What Is a Fuel Cell?

Fuel cells are electrochemical devices that convert chemical energy of the reactants directly into electricity and heat with high efficiency. Generally speaking, a fuel cell is simply an energy conversion device for power generation. The basic physical structure of a fuel cell consists of an electrolyte layer in contact with a porous anode and cathode on either side. A schematic representation of a fuel cell with reactant/product gases and the ion conduction flow directions through the cell is shown in Figure 1-1.

In a typical fuel cell, gaseous fuels are fed continuously to the anode (negative electrode), while an oxidant (oxygen from the air) is fed continuously to the cathode (positive electrode). Electrochemical reactions take place at the electrodes to produce an electric current.

Some of the advantages of fuel cell systems include:

- Fuel cells have the potential for a high operating efficiency that is not a strong function of system size.
- Fuel cells have a highly scalable design.
- Numerous types of potential fuel sources are available.
- Fuel cells produce zero or near-zero greenhouse emissions.

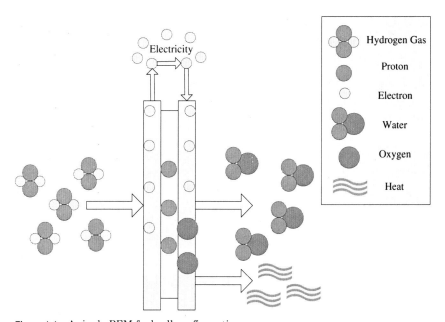

Figure 1-1 A single PEM fuel cell configuration.

- Fuel cells have no moving parts (other than pumps or compressors in some fuel cell plant subsystems). This provides for stealthy, vibration-free, highly reliable operation.
- Fuel cells provide nearly instantaneous recharge capability when compared to batteries.

Limitations common to all fuel cell systems include the following:

- Cost-effective, mass produced pure hydrogen storage and delivery technology.
- If pure fuel is not used, fuel reformation technology needs to be taken into account.
- If fuels other than pure hydrogen are used, then fuel cell performance gradually decreases over time due to catalyst degradation and electrolyte poisoning.

1.1.1 Comparison with batteries

A fuel cell has many similar characteristics with batteries, but also differs in many respects. Both are electrochemical devices that produce energy directly from an electrochemical reaction between the fuel and the oxidant. The battery is an energy storage device. The maximum energy available is determined by the amount of chemical reactant stored in the battery itself. A battery has the fuel and oxidant reactants built into itself (onboard storage), in addition to being an energy conversion device. In a secondary battery, recharging regenerates the reactants. This involves putting energy into the battery from an external source. The fuel cell is an *energy conversion* device that theoretically has the capability of producing electrical energy for as long as the fuel and oxidant are supplied to the electrodes [7]. Figure 1-2 shows a comparison of a fuel cell and battery.

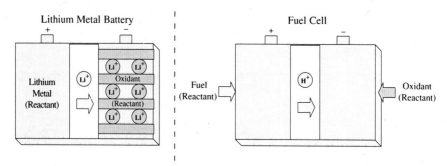

Figure 1-2 Comparison of a fuel cell and a battery.

The lifetime of a primary battery is limited because when the amount of chemical reactants stored in a battery runs out, the battery stops producing electricity. In addition, when a battery is not being used, a very slow electrochemical reaction takes place that limits the lifetime of the battery. The electrode of a battery is also used in the process; therefore, the lifetime of the battery is dependent on the lifetime of the electrode.

In comparison, a fuel cell is an energy conversion device where the reactants are supplied. The fuels are stored outside the fuel cell. A fuel cell can supply electrical energy as long as fuel and oxidant are supplied [1]. The amount of energy that can be produced is theoretically unlimited as long as the fuel and oxidant are supplied. Also, no "leakage" occurs in a fuel cell, and no corrosion of cell components occurs when the system is not in use.

1.1.2 Comparison with heat engine

A heat engine also converts chemical energy into electric energy, but through intermediate steps. The chemical energy is first converted into thermal energy through combustions, then thermal energy is converted into mechanical energy by the heat engine, and finally the mechanical energy is converted into electric energy by an electric generator [1]. This multistep energy process requires several devices in order to obtain electricity. The maximum efficiency is limited by Carnot's law because the conversion process is based upon a heat engine, which operates between a low and high temperature [1]. The process also involves moving parts, which implies that they wear over time. Regular maintenance of moving components is required for proper operation of the mechanical components. Figure 1-3 shows a comparison between a fuel cell and a heat engine/electrical generator.

Since fuel cells are free of moving parts during operation, they can work reliably and with less noise. This results in lower maintenance costs, which make them especially advantageous for space and underwater missions. Electrochemical processes in fuel cells are not governed

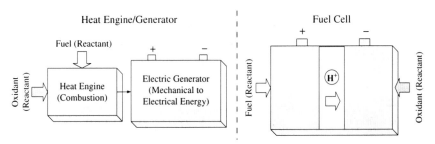

Figure 1-3 Comparison of a fuel cell to a heat generator.

by Carnot's law, therefore high operating temperatures are not necessary for achieving high efficiency. In addition, the efficiency of fuel cells is not strongly dependent on operating power. It is their inherent high efficiency that makes fuel cells an attractive option for a wide range of applications, including road vehicle power sources, distributed electricity and heat production, and portable systems [6].

1.2 Why Do We Need Fuel Cells?

Conventional power generation relies upon fossil fuels, which produce a significant amount of pollutants, and there is a limited supply. Many alternative energy approaches have been proposed, such as biofuel, hydroelectric power, batteries, wind, solar, bioenergy, and geothermal energy. All of these sources can provide energy, but every method has advantages and disadvantages.

Fuel cells are needed because they provide electric power in applications that are currently energy-limited. For example, one of the most annoying things about a laptop computer is that the battery gives out after a couple of hours! Table 1-1 compares the weight, energy and

TABLE 1-1 General Fuel Cell Comparison with Other Power Sources

	Weight	Energy	Volume
Fuel Cell	9.5 lbs	2190 Whr	4.0 L
Zinc-Air Cell	18.5 lbs	2620 Whr	9.0 L
Battery	24 lbs	2200 Whr	9.5 L

volume of batteries with a typical PEM fuel cell. Each market needs fuel cells for varying reasons as described in the next few paragraphs.

1.2.1 Portable sector

In coming years, portable devices—such as laptops, cell phones, video recorders, and others—will need greater amounts of power for longer periods of time. Fuel cells are very scalable and have easy recharging capabilities compared to batteries. Cell phone technology is advancing rapidly, but the limiting factor for the new technology is the power. More power is required to provide consumers with all of the functions in devices they require and want. The military also has a need for long-term portable power for new soldier's equipment. In addition, fuel cells operate silently, and have low heat signatures, which are clear advantages for the military.

1.2.2 Transportation sector

Many factors are contributing to the fuel cell push in the automotive market. The availability of fossil fuels is limited, and due to this, an inevitable price increase will occur. In addition, legislation is becoming stricter about controlling environmental emissions in many countries all over the world. One of the new pieces of legislation that will help introduce the fuel cell automobile market in the United States is the Californian zero emission vehicle (ZEV) mandate, which requires that a certain number of vehicles be sold annually in California. Fuel cell vehicles also have the ability to be more fuel efficient than vehicles powered by other fuels. This power technology allows a new range of power use in small two-wheeled and four-wheeled vehicles, boats, scooters, unmanned vehicles, and other utility vehicles.

1.2.3 Stationary sector

Stationary fuel cells can produce enough electricity and heat to power an entire house or business, which can result in significant savings. These fuel cells may even make enough power to sell some of it back to the grid. Fuel cells can also power residences and businesses where no electricity is available. Sometimes it can be extremely expensive for a house not on the grid to have the grid connected to it. Fuel cells are also more reliable than other commercial generators used to power houses and businesses. This can benefit many companies, given how much money they can lose if the power goes down for even a short time.

1.3 History of Fuel Cells

Fuel cells have been known to science for about 150 years [4]. They were minimally explored in the 1800s and extensively researched in the second half of the twentieth century. Initial design concepts for fuel cells were explored in 1800, and William Grove is credited with inventing the first fuel cell in 1839 [4]. Various fuel cell theories were contemplated throughout the nineteenth century, and these concepts were studied for their practical uses during the twentieth century. Extensive fuel cell research was started by NASA in the 1960s, and much has been done since then. During the last decade, fuel cells were extensively researched, and are finally nearing commercialization. A summary of fuel cell history is shown in Figure 1-4.

In 1800, William Nicholson and Anthony Carlisle described the process of using electricity to break water into hydrogen and oxygen [4]. William Grove is credited with the first known demonstration of the fuel cell in 1839. Grove saw notes from Nicholson and Carlisle and thought he might "recompose water" by combining electrodes in a series circuit, and soon accomplished this with a device called a "gas battery." It operated with separate platinum electrodes in oxygen and hydrogen submerged in a dilute sulfuric acid electrolyte solution. The sealed containers contained water and gases, and it was observed that the water level rose in both tubes as the current flowed. The "Grove cell," as it came to be called, used a platinum electrode immersed in nitric acid and a zinc electrode in zinc sulfate to generate about 12 amps of current at about 1.8 volts [4].

Friedrich Wilhelm Ostwald (1853–1932), one of the founders of physical chemistry, provided a large portion of the theoretical understanding of how fuel cells operate. In 1893, Ostwald experimentally determined the roles of many fuel cell components [4].

Ludwig Mond (1839–1909) was a chemist that spent most of his career developing soda manufacturing and nickel refining. In 1889, Mond and his assistant Carl Langer performed numerous experiments using a

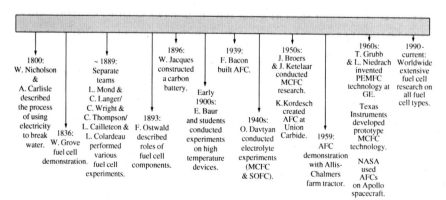

Figure 1-4 The history of fuel cells.

coal-derived gas. They used electrodes made of thin, perforated platinum, and had many difficulties with liquid electrolytes. They achieved 6 amps per square foot (the area of the electrode) at 0.73 volts [4].

Charles R. Alder Wright (1844–1894) and C. Thompson developed a similar fuel cell around the same time. They had difficulties in preventing gases from leaking from one chamber to another. This and other causes prevented the battery from reaching voltages as high as 1 volt. They felt that if they had more funding, they could create a better, robust cell that could provide adequate electricity for many applications [4].

The French team of Louis Paul Cailleteton (1832–1913) and Louis Joseph Colardeau came to a similar conclusion, but thought the process was not practical due to needing "precious metals." In addition, many papers were published during this time saying that coal was so inexpensive that a new system with a higher efficiency would not decrease the prices of electricity drastically [4].

William W. Jacques (1855–1932), an electrical engineer and chemist, did not pay attention to these critiques, and startled the scientific world by constructing a "carbon battery" in 1896 [4]. Air was injected into an alkali electrolyte to react with a carbon electrode. He thought he was achieving an efficiency of 82 percent, but actually obtained only an 8-percent efficiency [4].

Emil Baur (1873–1944) of Switzerland and several of his students conducted many experiments on different types of fuel cells during the early 1900s. His work included high-temperature devices, and a unit that used a solid electrolyte of clay and metal oxides [4].

O. K. Davtyan of the Soviet Union did many experiments to increase the conductivity and mechanical strength of the electrolyte in the 1940s. Many of the designs did not yield the desired results, but Davtyan's and Baur's work contributed to the necessary preliminary research for today's current molten carbonate and solid oxide fuel cell devices.

1.3.1 PEM fuel cells

Thomas Grubb and Leonard Niedrach invented PEM fuel cell technology at General Electric in the early 1960s. GE developed a small fuel cell for the U.S. Navy's Bureau of Ships (Electronics Division) and the U.S. Army Signal Corps [4]. The fuel cell was fueled by hydrogen generated by mixing water and lithium hydride. It was compact, but the platinum catalysts were expensive.

NASA initially researched PEM fuel cell technology for Project Gemini in the early U.S. space program. Batteries were used for the preceding Project Mercury missions, but Project Apollo required a power source that would last a longer amount of time. Unfortunately, the first PEM cells developed had repeated difficulties with the internal cell contamination and leakage of oxygen through the membrane. GE redesigned their fuel cell, and the new model performed adequately for the rest of the Gemini flights. The designers of Project Apollo and the Space Shuttle ultimately chose to use alkali fuel cells [4].

GE continued to work on PEM fuel cells in the 1970s, and designed PEM water electrolysis technology, which lead to the U.S. Navy Oxygen Generating Plant. The British Royal Navy used PEM fuel cells in the early 1980s for their submarine fleet, and during the last decade, PEM fuel cells have been researched extensively by commercial companies for transportation, stationary, and portable power markets.

1.3.2 Solid oxide fuel cells

Emil Baur and H. Preis experimented with solid oxide electrolytes during the late 1930s. The initial designs were not as electrically conductive as they hoped, and many unwanted chemical reactions were reported. Solid oxide and molten carbonate fuel cells had a similar history up until the 1950s. O. K. Davtyan of Russia during the 1940s also performed experiments with electrolytes, but experienced unwanted chemical reactions and short life ratings [4].

Solid oxide research later began to accelerate at the Central Technical Institute in The Hague, Netherlands, the Consolidation Coal Company in Pennsylvania, and General Electric in Schenectady, New York [4]. Several issues with solid oxide arose, such as high internal electrical resistance, melting, and short-circuiting due to semiconductivity. Due to these problems, many researchers began to believe that molten carbonate fuel cells showed more short-term promise. Recently, increasing energy prices and advances in materials has reinvigorated work on SOFC, and about 40 companies are currently researching this technology [4].

1.3.3 Molten carbonate fuel cells

Emil Baur and H. Preis in Switzerland experimented with high temperature solid oxide electrolytes in the 1930s. Many problems were encountered with the electrical conductivity and unwanted chemical reactions. O. K. Davtyan researched this further during the 1940s, but had little success. By the late 1950s, Dutch scientists G. H. J. Broers and J. A. A. Ketelaar built upon previous work, but were also unsuccessful. They then decided to focus on the electrolytes of fused (molten) carbonate salts instead, and successfully created a fuel cell that ran for six months. However, they found that the molten electrolyte was partially lost through reactions with gasket materials [4].

The U.S. Army's Mobility Equipment Research and Development Center (MERDC) at Ft. Belvoir tested many molten carbonate fuel cells made by Texas Instruments in the 1960s. These fuel cells ranged from 100–1000 Watts and were designed to run on hydrogen from a gasoline reformer.

1.3.4 Phosphoric acid fuel cells

Phosphoric acid fuel cells were slower to develop than other fuel cells. G. V. Elmore and H. A. Tanner conducted experiments with electrolytes that were 35 percent phosphoric acid, and 65 percent silica powder. They presented their experimental results in a paper "Intermediate Temperature Fuel Cells," in which they describe an acid cell that operated for six months with a current density of 90 mA/cm^2 and 0.25 volts without any deterioration [4].

During the 1960s and '70s, advances in electrode materials and issues with other fuel cell types created renewed interest in PAFCs. The U.S. Army explored PAFCS that ran with common fuels in the 1960s.

Karl Kordesch and R. F. Scarr of Union Carbide created a thin electrode made of carbon paper as a substrate, and a Teflon-bonded carbon layer as a catalyst carrier. An industry partnership known as TARGET was primarily led by Pratt & Whitney and the American Gas Association, and produced significant improvements in fuel cell power plants, increasing the power from 15 kW in 1969 to 5 mW in 1983 [4].

1.3.5 Alkali fuel cells

Francis Thomas Bacon (1904–1992) built an alkali electrolyte fuel cell that used nickel gauze electrodes in 1939 [4]. He employed potassium hydroxide for the electrolyte instead of the acid electrolytes, and porous "gas-diffusion electrodes" rather than the solid electrodes used since Grove. He also used pressurized gases to keep the electrolyte from "flooding" in the electrodes [4]. During World War II, he thought the alkali electrolyte fuel cells might provide a good source of power for the Royal Navy submarines in place of dangerous storage batteries in use at the time. In the following 20 years, he created large-scale demonstrations with his alkali cells using potassium hydroxide as the electrolyte, instead of the acid electrolytes used since Grove's time. One of the first demonstrations was a 1959 Allis–Chalmers farm tractor, which was powered by a stack of 1008 cells [4]. This tractor was able to pull a weight of 3000 pounds [4]. Allis–Chalmers continued fuel cell research for many years, and also demonstrated a fuel cell–powered golf cart, submersible, and fork lift.

In the late 1950s and 1960s, Union Carbide also experimented with alkali cells. Karl Kordesch and his colleagues designed alkali cells with carbon gas–diffusion electrodes based upon the work of G. W. Heise and E. A. Schumacher in the 1930s. They demonstrated a fuel cell–powered mobile radar set for the U.S. Army, as well as a fuel cell–powered motorbike. Eduard Justi of Germany designed gas-diffusion electrodes around the same time [4].

Pratt & Whitney licensed the Bacon patents in the early 1960s, and won the National Aeronautics and Space Administration (NASA) contract to power the Apollo spacecraft with alkali cells [4].

Based upon the research, development, and advances made during the last century, technical barriers are being resolved by a world network of scientists. Fuel cells have been used for over 20 years in the space program, and the commercialization of fuel cell technology is rapidly approaching.

1.4 How Do Fuel Cells Work?

A fuel cell consists of a negatively charged electrode (anode), a positively charged electrode (cathode), and an electrolyte membrane. Hydrogen is oxidized on the anode and oxygen is reduced on the cathode. Protons are transported from the anode to the cathode through the electrolyte membrane, and the electrons are carried to the cathode over the external circuit. On the cathode, oxygen reacts with protons and electrons, forming water and producing heat. Both the anode and cathode contain a catalyst to speed up the electrochemical processes. Figure 1-5 shows a schematic of a single fuel cell.

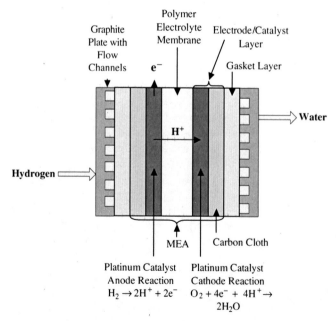

Figure 1-5 Generalized schematic of a single fuel cell.

Figure 1-5 shows an example of a typical fuel cell (proton exchange membrane fuel cell) with the following reactions:

Anode: $H_2 (g) \rightarrow 2H^+ (aq) + 2e^-$
Cathode: $\frac{1}{2}O_2 (g) + 2H^+ (aq) + 2e^- \rightarrow H_2O (l)$
Overall: $H_2 (g) + \frac{1}{2}O_2 (g) \rightarrow H_2O (l)$ + electric energy + waste heat

Reactants are transported by diffusion and/or convection to the catalyzed electrode surfaces where the electrochemical reactions take place. The half cell reactions will be different for each other fuel cell type (see Chapter 3), but the overall cell reaction will be similar as the overall reaction listed previously. The water and waste heat generated by the fuel cell must be continuously removed, and may present critical issues for the operation of certain fuel cells.

Chapter Summary

While the fuel cell is a unique and fascinating system, accurate system selection, design, and modeling for prediction of performance is needed to obtain optimal performance and design. In order to make strides in performance, cost, and reliability, one must possess an interdisciplinary understanding of electrochemistry, materials, manufacturing, and mass and heat transfer. The remaining chapters in this book will provide the necessary basis in these areas in order to design and build a fuel cell.

Problems

1. Describe the differences between a fuel cell and a battery, and a fuel cell and a heat engine.

2. William Grove is usually credited with the invention of the fuel cell. In what way does his gaseous voltaic battery represent the first fuel cell?

3. Describe the functions of the electrodes, and the electrolyte and catalyst layers of the fuel cell.

4. Describe the history of each of the fuel cell types briefly.

5. Why does society need fuel cells, and what can they be used for?

Bibliography

[1] Li, Xianguo. *Principles of Fuel Cells*. 2006. New York: Taylor & Francis Group.
[2] O'Hayre, Ryan, Suk-Won Cha, Whitney Colella, and Fritz B. Prinz. *Fuel Cell Fundamentals*. 2006. New York: John Wiley & Sons.
[3] Barbir, Frano. *PEM Fuel Cells: Theory and Practice*. 2005. Burlington, MA: Elsevier Academic Press.

[4] *Collecting the History of Fuel Cells.* http://americanhistory.si.edu/fuelcells/index.htm. Smithsonian Institution, 2006. Last accessed September 15, 2006. Last updated December 2005.

[5] Stone, C. and A.E. Morrison. "From Curiosity to Power to Change the World." *Solid State Ionics*, 152–153, pp. 1–13.

[6] "Fuel Cells." Energy Research Center of the Netherlands (ECN). http://www.ecn.nl/en/h2sf/additional/fuel-cells-explained/. Last updated September 18, 2006.

[7] Mikkola, Mikko. "Experimental Studies on Polymer Electrolyte Membrane Fuel Cell Stacks." Helsinki University of Technology, Department of Engineering Physics and Mathematics, Masters Thesis, 2001.

[8] Applyby, A., and Foulkes.F. *Fuel Cell Handbook*. 1989. New York: Van Nostrand Reinhold.

Chapter 2

Fuel Cells and the Hydrogen Economy

Many policy makers, energy analysts, and scientists believe that hydrogen is the fuel of the future. It is commonly known that fossil fuels can only sustain the world energy demand for a limited amount of time. The increase in oil prices has lead to extensive research and development of hydrogen-based power sources and other types of alternative energy technologies. Fuel cells are viewed to have the most potential compared with other alternative power technologies. Hydrogen is commercially available in small quantities, but the structure for hydrogen refueling, transport, and storage on a large scale is not in place. Hydrogen can be generated from a variety of energy sources, and stored and transported in many ways, and for conversion into usable energy. One of the biggest obstacles that needs to be overcome in order for fuel cells to be commercially viable is creating a structure for distributing and storing hydrogen. This infrastructure is practically nonexistent at the current time.

This chapter covers the current status of the hydrogen economy, and its future. The topics to be specifically covered include the following:

- The Characteristics of Hydrogen
- World Energy Demand
- Development of the Hydrogen Economy
- Hydrogen Production, Distribution, and Storage
- Investment in the Hydrogen Infrastructure

These concepts will be discussed in detail in this chapter, and in Chapter 15.

2.1 Characteristics of Hydrogen

Hydrogen has many unusual characteristics compared with other elements. It is the lightest and most abundant element, and it can burn with oxygen to release large amounts of energy. Hydrogen has a high energy content by weight. Many pollutants are formed when hydrogen is burned in air because of the high nitrogen content of the air. Hydrogen is highly flammable, and burns when it makes up 4 to 75% of air by volume [4]. It has a low energy density by volume at standard temperature and atmospheric pressure. The volumetric density can drastically be lowered by storing compressed hydrogen under pressure, or converting it to liquid hydrogen. Table 2-1 compares the relevant properties of hydrogen, methane, methanol, ethanol, propane, and gasoline—all of which can be used as fuel for fuel cells. Hydrogen does not exist in its natural form on earth; therefore, it must be manufactured through one of various ways: the steam reforming of natural gas, the gasification of coal, electrolysis, or the reforming/oxidation of other hydrocarbons or biomass.

TABLE 2-1 Hydrogen Compared with Other Fuels (Compiled from [14] and [15])

Property	Hydrogen	Methane	Methanol	Ethanol	Propane	Gasoline
Molecular Weight (g/mol)	2.016	16.043	32.04	46.0634	44.10	~107.0
Density (kg/m^3) 20°C and 1 atm	0.08375	0.6682	791	789	1.865	751
Normal Boiling point (°C)	−252.8	−161.5	64.5	78.5	−42.1	27–225
Flash Point (°C)	<−253	−188	11	13	−104	−43
Flammability Limits in Air (Volume %)	4.0–75.0	5.0–15.0	6.7–36.0	3.3–19	2.1–10.1	1.0–7.6
CO_2 Production per Energy Unit	0	1.00	1.50			1.80
Autoignition Temperature in Air (°C)	585	540	385	423	490	230–480
Higher Heating Value (MJ/kg)	142.0	55.5	22.9	29.8	50.2	47.3
Lower Heating Value (MJ/kg)	120.0	50.0	20.1	27.0	46.3	44.0

Hydrogen is a good choice for a future energy source for many reasons. Some of these reasons are [3]:

- Hydrogen can be made from various sources. It is completely renewable. The most abundant and cleanest precursor for hydrogen is water.
- Hydrogen can be stored in many forms, from gaseous to liquid to solid. It can also be stored in various chemicals and substances such as methanol, ethanol, and metal hydrides.
- It can be produced from, and converted to, electricity with high efficiencies.
- It can be transported and stored as safely as any other fuel.

Hydrogen and other fuels for fuel cells can be made cheaply and easily by processing hydrocarbons. Hydrogen has the potential to provide energy to all parts of the economy: industry, residences, transportation, and mobile applications. It can eventually replace petroleum-based fuels used for automobiles, and may provide an attractive solution for remote communities that cannot obtain electricity through the grid. One of the main attractions for using hydrogen is its environmental advantage over fossil fuels—but the hydrogen is only as clean as the technologies used to produce it. The production of hydrogen can be pollutant-free if it is produced by one of three methods:

- Through electrolysis using electricity derived solely from nuclear power or renewable energy sources.
- Through steam reforming of fossil fuels combined with new carbon capture and storage technologies.
- Through thermochemical or biological techniques based on renewable biomass [4].

A major disadvantage of processing hydrocarbons is the pollution and carbon dioxide, which eliminates one of the main reasons for using hydrogen in the first place. The best low-pollution alternative for creating hydrogen is the electrolysis of water. This method creates no carbon dioxide, or nitrous or sulfurous oxides, but it is more costly compared to other methods if electricity from electrical plants is used to break the water. The commercial worldwide production of hydrogen is approximately 40 million tonnes [4]. This is used primarily to make other chemicals such as ammonia and urea, for the cracking of petrochemicals, and a feedstock for the food, electronics, and metallurgical processing industries.

2.1.1 Safety aspects of hydrogen as a fuel

One of the critical issues with the use of hydrogen is safety. Contrary to popular perception, hydrogen is less flammable than gasoline and other fossil fuels. Like any other fuel, hydrogen has risks if not properly stored or transported. However, the transportation of hydrogen at very high pressure, or very low temperature, brings other hazards. These hazards can be controlled through the use of proper handling and controls.

The reputation of hydrogen as unsafe has been unfairly tainted by the Hindenberg incident and the hydrogen bomb. Investigation into the Hindenberg incident proved that the aluminum power-filled paint varnish that coated the ship started the fire—not the hydrogen. Since hydrogen is a small molecule, it has a tendency to escape through small openings more so than other gaseous or liquid fuels. Hydrogen will tend to leak through holes or joints of low pressure fuel lines only 1.26 to 2.8 times faster than a natural gas leak through the same hole [3]. Natural gas has an energy density three times greater than hydrogen, so a natural gas leak results in a greater energy release than a hydrogen leak. If a hydrogen leak occurs, hydrogen disperses much more quickly than other fuels. Hydrogen is lighter and more diffusive than gasoline, propane, or natural gas. If an explosion occurred, hydrogen has the lowest explosive energy per unit of stored fuel.

Liquid hydrogen has a different set of safety issues. If liquid hydrogen spills, there could be burns associated with it. Liquid hydrogen has characteristics similar to an oil spill, but it dissipates rapidly. If a pressure valve fails, an explosion could occur because of the liquid rapidly turning into vapor. This can be prevented by putting the correct controls and valves into the hydrogen system.

When hydrogen is aboard a vehicle, the potential dangers include explosion and toxicity. Hydrogen as a source of fire can occur from fuel storage or the fuel cell itself. The fuel cell has a lower hazard than the hydrogen fuel. Small amounts of hydrogen and oxygen do exist in the fuel cell, but they are separated by a thin electrolyte layer. A fuel cell system has controls with detailed safety algorithms that will disconnect the fuel supply lines in case of fuel mixing, fuel leaks or other safety hazards. When designing hydrogen systems, many safety factors need to be considered:

- A catastrophic rupture due to a defect in the tank or puncture by a sharp object.
- Mixture of fuel cell reactants in the cell.
- A large leak due to faulty controls or a puncture.
- A slow leak due to stress cracks in the tank liner, or pressure relief valve.

In addition to having a robust control system with detailed safety algorithms, these failure modes can be prevented in many ways. They include the following:

- Leak prevention through thorough testing of tanks and equipment.
- Installing more than one valve.
- Designing equipment for shocks, vibrations, and wide temperature ranges.
- Adding several hydrogen or oxygen sensors or leak detectors.
- Ignition prevention by illuminating a source of electrical sparks.
- Designing fuel cell supply lines that are physically separated from other equipment.

In order for hydrogen to become widely accepted, international regulations codes and standards need to be developed for construction, maintenance, and operation of hydrogen facilities and equipment. Uniformity of safety requirements will increase customer confidence in using hydrogen.

2.2 World Energy Demand

In order to obtain an idea of the necessity and potential of fuel cells, the world energy demand can be examined. Figure 2-1 shows the current and projected energy consumption from 1980–2030. International energy consumption is estimated to increase by 2.0 percent per year from

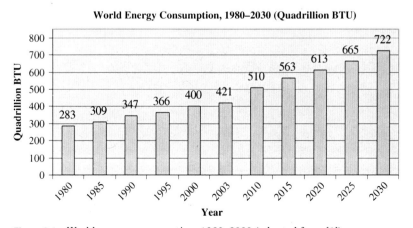

Figure 2-1 World energy consumption, 1980–2030 (adapted from [1]).

2003 to 2030 [1]. Total worldwide energy use grows from 421 quadrillion British thermal units (BTU) in 2003 to 563 quadrillion BTU in 2015 and 722 quadrillion BTU in 2030 [1]. The most rapid growth in energy demand from 2003 to 2030 is projected for Asia, including China and India, Central and South America, Africa, the Middle East, and Eurasia. The energy requirements for these nations are increasing by 5.0 percent per year on average [1].

Figure 2-2 shows the world energy consumption by end-use sector. Trends in the end use sector depend upon the pace of economic development in a given region. On a worldwide basis, energy demand in the industrial sector grows most rapidly, at an average rate of 2.4 percent per year as shown in Figure 2-2 [1]. Residential energy use rises an average of 1.7 percent per year, and commercial energy use increases at a rate of 1.8 percent per year from 2003 to 2030 (see Figure 2-2) [1]. The slowest growth is expected in the transportation end-use sector at 1.4 percent per year [1]. The slower growth in the transportation sector is due to higher world oil prices. Again, the most rapid growth in energy demand from 2003 to 2030 is projected for Asia, including China and India, Central and South America, Africa, the Middle East, and Eurasia. Slower growth (0.6 percent per year) is expected for the United States, Canada, most of Europe, and Australia [1].

Since the need for energy is increasing rapidly (as shown in Figures 2-1 and 2-2), it is critical to obtain a low-emission power technology for the future.

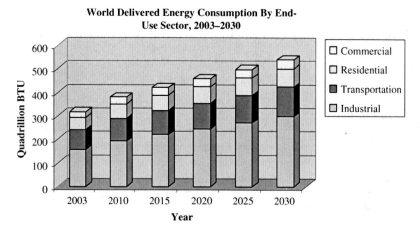

Figure 2-2 World delivered energy consumption by end use sector, 2003–2030 (adapted from [1]).

2.3 Development of the Hydrogen Economy

Most of the current energy needs in the world is being met by fossil fuels. These fuels are easily obtained, stored, and transported because large amounts of money have been used to create, build, and maintain the system. Because of the current fuel distribution system, technology has advanced at a faster pace during the last two centuries than in all of recorded history. Despite all the advantages it has provided for our society, it has also had negative effects on the environment, some of which likely have yet to be seen. Some of these harmful effects include air pollution due to NO_x and SO_x emissions, water and soil pollution due to spills and leaks, and carbon dioxide accumulation in the atmosphere. These pollutants have the potential to warm the global atmosphere and kill many species [3].

In addition to the negative environmental consequences of using these fuels, there is a finite supply of fossil fuels that will inevitably force the use of another form of energy. The demand for energy will also continuously increase due to the constant increase of the global population.

Technologies to produce, store, transport, and convert hydrogen have already been in use for many years. Other technologies, such as those designed to capture carbon dioxide and other pollutants produced during the processing of hydrocarbons, have also been demonstrated. Fuel cells have been extensively researched by most large car manufacturers, and prototypes are currently being driven around the country. Fuel cells that supply power to remote communities are ready to come to market. Portable fuel cells have been demonstrated and are close to commercialization. Many governments along with the major oil and coal companies have been researching and developing carbon capture and storage for more than a decade [4].

Several technological and cost breakthroughs are necessary for the hydrogen economy to become a reality. The cost to supply hydrogen using existing technologies is costly compared to conventional technologies. Many technological problems also need to be addressed. Hydrogen production from renewables, carbon capture and storage, and a hydrogen distribution and storage infrastructure needs further development for fuel cells to become a low-cost, safe, and viable energy solution.

The amount of time it will take for hydrogen to become a cost-effective, viable alternative energy is hard to predict. Hydrogen will not take the place of existing systems overnight. In order for a "hydrogen economy" to exist, an interim solution must occur where fossil fuels are first meeting the demand for hydrogen. This transition would be gradual, possibly taking several decades. Due to this, carbon capture and storage technology will be an important element as long as fossil fuels remain the world's primary energy source. Strong government incentives would certainly aid in the transition process.

2.4 Hydrogen Production, Distribution, and Storage

The governments in many countries all over the world have set goals for hydrogen use, production and costs. For example, the U.S. Department of Energy set a production cost target of $2 to $3 per gallon of hydrogen by 2015 [4]. In order to meet this target, the current cost of hydrogen needs to be reduced by half. Steam reforming of natural gas will most likely be the cheapest method of producing hydrogen. However it is produced, it will require a large and costly infrastructure to store, transport, and dispense the hydrogen. As a side note, many studies have been conducted comparing the costs of a hydrogen infrastructure with other types of fuel infrastructures (such as methanol), and it was found that the hydrogen infrastructure did not cost anymore than other types of fuel infrastructures. In order to store hydrogen, it must be compressed and stored in pressurized containers, or converted to the liquid form and stored in a cryogenic liquid hydrogen tank. These techniques are in use today, but it is costly to process and build the necessary tanks because a small number are made. Hydrogen can be transported by pipeline, but only over short distances. A few hydrogen-pipeline systems exist today in the United States and Europe, but none are greater than 200 kilometers in length [4]. It is easier to transport hydrogen by rail, or trucks, for long distances. Bringing the costs down will make hydrogen a viable alternative energy solution [4].

2.4.1 Technologies for hydrogen production

Currently, hydrogen is produced from various fossil fuels such as oil, natural gas, and coal. Some of the current uses of hydrogen include hydrotreating and hydrocracking, which are processes used in refineries to upgrade crude oil. It is used in the chemical industry to make various chemical compounds, such as ammonia and methanol, and in metallurgical processes. Some of the technologies to produce hydrogen include steam reforming of natural gas, partial oxidation of hydrocarbons, and coal gasification. However, these technologies will not help to decrease the dependence on fossil fuels.

The electrolysis of water is a mature technology that was developed for hydrogen production. It is efficient, but requires large amounts of electricity. This can be solved, however, by using solar energy to produce the electricity required to break the hydrogen. This technology is mature enough to be used on a large scale for electricity and hydrogen generation. Other options for generating hydrogen include hydropower, nuclear plants during off-peak hours, direct thermal decomposition, thermolysis, thermochemical cycles, and photolysis. Many of these technologies are at various stages of development, and a few have been abandoned. The most common methods of producing hydrogen are described in the next few paragraphs. More details about some of these technologies can be found in Chapter 15.

Steam reforming. The cheapest method of producing hydrogen on a large scale is through steam reforming of fossil fuels. The current methods use a nickel catalyst. Methane first reacts with steam to produce carbon monoxide and hydrogen. The carbon monoxide passes over the catalyst, then reacts with the steam to produce carbon dioxide and hydrogen according to the following reaction [4]:

$$CH_4 + H_2O \rightarrow CO + 3H_2$$
$$CO + H_2O \rightarrow CO_2 + H_2$$

Natural gas is the cheapest feedstock for producing hydrogen from steam reforming, but this cost is still two to three times higher than producing gasoline from crude oil. Currently, a lot of research is being conducted on how to improve the efficiency and lower production costs of steam reforming [4].

Partial oxidation. Another method used to produce hydrogen is partial oxidation. This process involves reacting the membrane with oxygen to produce hydrogen and carbon monoxide. The conversion efficiency is lower than steam reforming, which is why that process still dominates commercial hydrogen production.

Coal gasification. The gasification of coal is one of the oldest methods for producing hydrogen. It was used to produce "town gas" before natural gas became available. The coal is heated to a gaseous state, then mixed with steam in the presence of a catalyst to produce synthesis gas. This gas can be processed to extract hydrogen and other chemicals, or burned to produce electricity. Current R&D on coal gasification is focusing on the lowering of pollutants, such as nitrogen and sulfur oxides, mercury, and carbon monoxide.

Biomass. Hydrogen can be produced from many types of biomass such as agricultural and animal residues using pyrolysis and gasification processes. These produce a carbon-rich synthesis gas. Using biomass instead of fossil fuels produce no carbon dioxide emissions. Unfortunately, biomass hydrogen production costs are much higher than hydrogen production costs from fossil fuels. Biological processes for producing hydrogen from biomass includes fermentation, anaerobic digestion, and metabolic processing techniques, but these are far from being competitive with traditional hydrogen-producing techniques.

Water electrolysis. Hydrogen production using water electrolysis is seldom used due to the cost of electricity. This technique is employed when extremely pure hydrogen is needed. The environmental benefits of using electrolysis depend on what method is used to produce the electricity to break the water. If the electricity were produced from nuclear or renewable energy sources such as wind, solar, and biomass, it would produce pollutant-free hydrogen. If electrolysis is performed with electricity or nuclear power, a large reduction in cost is required to compete with conventional energy sources [3]. There are several hydrogen gas stations for vehicles that produce hydrogen at the site via electrolysis using solar panels.

Metal hydrides applications. Hydrogen's ability to form metal hydrides can also be used for energy conversion. In order for hydrogen to be released, heat must be applied. The heat generated during the charging and discharging process are functions of the hydrating substance, hydrogen pressure, and the temperature at which the heat is supplied or extracted.

2.4.2 Technologies for hydrogen storage

Many commercially available technologies exist for storing hydrogen. The most common storage method is the pressurized storage tank, which is available in many sizes and pressure ranges. Other storage methods that may be considered for various applications are described in this section. More details about hydrogen storage methods can be found in Chapter 15.

Large underground storage. Hydrogen can be stored underground in caverns, aquifers, and depleted petroleum and gas fields. These large underground storage systems will be of the same type and content as natural gas systems, but will be approximately three times more expensive. Not many technical problems are expected for underground hydrogen storage. There are already several instances of hydrogen being stored underground. The city of Kiel stores, Germany town gas underground. Gaz de France, the French gas company, stores natural gas underground. Imperial Chemical Industries of Great Britain stores hydrogen in salt mines in Teeside, United Kingdom.

Vehicular pressurized hydrogen tanks. These tanks are made of ultralight composite materials that allow pressures in excess of 20 bars. They are used in prototype fuel cell automobiles and buses.

Liquid hydrogen storage. Large amounts of hydrogen can be stored in the liquid form compared with compressed gas storage. The production of liquid hydrogen fuel requires large amounts of energy. It also requires a system for keeping the hydrogen at extremely cold temperatures. Therefore, the method of hydrogen storage can sometimes be costly and inconvenient compared with compressed gas storage systems. However, some prototype hydrogen-powered automobiles also use specially developed liquid hydrogen.

Metal hydride storage. Certain metals can store hydrogen in spaces in the lattice of suitable metals or alloys. A storage is created similar to liquid hydrogen. When the mass of the metal or alloy is taken into account, the metal hydride is comparable to pressurized hydrogen. During the storage process, heat is released which must be removed. During the hydrogen release process, hydrogen must be transferred to the storage tank.

Novel hydrogen storage methods. Many novel hydrogen methods are currently being investigated that offer the potential for higher energy density than conventional methods. These include hydrogen storage in activated carbon at cryogenic temperatures, in carbon nanotubes, and in glass microspheres. These hydrogen technologies are still in the R&D stages, and it is unknown when they may be commercialized.

2.4.3 Worldwide hydrogen refueling stations

The number of hydrogen refueling stations around the world is listed in Table 2-2. There has been a slow buildup of refueling stations in the countries listed since the 1990s. Certain countries, such as Germany, will begin rapidly introducing hydrogen refueling stations during the next five years. The more rapid introduction is due to several reasons [16]:

- Many fuel cell vehicle manufacturers will not begin vehicle production for a niche market.
- Both the automotive industry and fuel suppliers need to mass produce a certain number of units in order to reduce cost.

TABLE 2-2 Worldwide Fueling Stations
(Compiled from *Fuel Cells 2000* [5])

Location	No. of refueling stations
United States	60
Australia	1
Austria	1
Belgium	2
Canada	9
China	6
Denmark	9
France	2
Germany	20
Greece	1
Iceland	1
India	1
Italy	6
Japan	18
Luxemburg	1
Netherlands	1
Portugal	2
Singapore	2
South Korea	2
Spain	3
Sweden	3
Switzerland	1
United Kingdom	1
Norway	10

- Consumers will not accept a fuel that is not "widely available."
- There is an alternative fuel and vehicle market in Brazil and Argentina. Brazil uses ethanol, and Argentina uses natural gas. Lessons learned from studying the introduction of alternative fuel vehicles and fuel stations implies that a rapid introduction is a better way to gain customers and acceptance of the new technology [16].

The United States has the most hydrogen refueling stations—primarily in California. Germany has the next highest number of refueling stations, followed by Japan. This correlates with the countries that have the highest funding for fuel cell and hydrogen infrastructure research.

2.5 Investment of Hydrogen Infrastructure

As mentioned previously, in order to successfully convert the oil-based infrastructure to a hydrogen one, a radical transformation of the global energy-supply system is required. A large infrastructure to store, distribute, and transport hydrogen would need to be built, and consumers need to be interested in purchasing fuel cell vehicles and other fuel cell products. The total cost of building a hydrogen infrastructure would depend on when and how fast it is built. Even if hydrogen only replaced petroleum-based fuels for automobiles, the cost of converting this structure would probably be in the trillions of dollars. The cost of replacing all vehicles currently used (~800 million vehicles) with fuel cell vehicles is estimated to be $2 trillion, if the average vehicle is estimated to cost $2500 more than a conventional vehicle [4]. The cost of developing the rest of the necessary hydrogen infrastructure would be about $8 trillion [4].

The main barrier to converting the economy to hydrogen is cost. In order to begin the conversion to the hydrogen economy, the hydrogen infrastructure needs to be established where fuel costs are extremely high and/or where there is public concern for the environment. These factors, combined with government incentives, will help the hydrogen market to develop. As the supply and demand increases for hydrogen, mass production of hydrogen equipment and fuel cells will lower the costs and make using hydrogen a more popular option.

A major factor in implementing the hydrogen economy is a strong, long-term commitment by governments. This will encourage fuel suppliers and equipment manufacturers, and increase consumer's confidence in the new energy source. Figure 2-3 shows a diagram of how current and future technologies can create hydrogen and other fuels

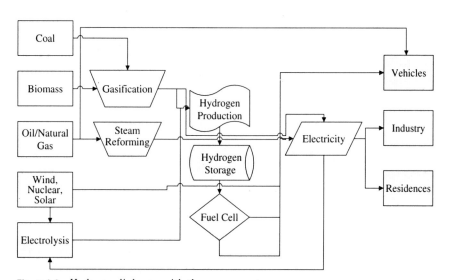

Figure 2-3 Hydrogen linkages with the energy system.

simultaneously. The best option would be for hydrogen to be manufactured using traditional processes at first. Natural gas and coal would initially be used to create the majority of hydrogen, and would continue to produce electricity and transitional fuels.

Hydrogen would compete against traditional energy technologies such as electricity, gas, and oil. It could initially be a good method of handling other alternative electricity fluctuations for wind or solar power. Fuel cells could also enhance the security of the existing power supply by providing backup in case of power station or transmission system failure.

2.5.1 Government support

In order to begin the transition to the hydrogen economy, there is a need for government help in two major areas [4]:

- The research and development of hydrogen fuel cells
- Incentive to encourage investment in the hydrogen infrastructure

Government, public, and private funding for fuel cell research has increased tremendously during the last decade. Many governments are expecting the transition to start occurring within the next 10 to 20 years and are working to speed up the process through international and joint private–public sector programs. The International Energy Agency estimates that public hydrogen spending is about $1 billion per year [17]. This sounds like a lot, but it is modest in comparison to what other governments are spending on other energy technologies per year. Only 15 percent of R&D spending on energy technologies is spent on hydrogen and fuel cell technology [4].

The largest hydrogen programs are in the United States, Japan, and the European Union. The total amount of hydrogen funding spent in these countries is about two-thirds of the total hydrogen funding for the world. The U.S. increased funding significantly in 2003 with a five-year $1.7 billion hydrogen program [17]. Japan has allocated $320 million to its hydrogen research activities in 2005 [4]. Total European funding is expected to reach €2.8 billion until 2011 [4]. China, Brazil, and India have also launched their own programs.

2.5.2 Long-term projections of hydrogen use

It is difficult to predict how soon the transition to a hydrogen economy will occur. The transition will undoubtedly take several decades. The construction, operation, and decommissioning of the current structure will take at least 10 to 20 years. Automobiles typically last about a decade or two, but power stations and pipelines are built to last for decades. The U.S. Energy Information Administrations' International Energy Outlook 2005 and the IEA's World Energy Outlook 2005 project that hydrogen will play a small role in fulfilling the world's total energy needs unless major breakthroughs are created.

2.5.3 Key players in hydrogen R&D

Many countries have active fuel cell programs with worldwide public funding at $1 billion per year. Fuels cells make up half of this amount, and the rest is for production, transportation, and storage, with small amounts going toward gas turbines and internal combustion engines.

Spending has increased a lot during the last few years. The biggest increases in spending occurred in the United States and the European Union. Some countries have focused their funding specifically on fuel cells, the hydrogen supply, or the new hydrogen infrastructure. A country's focus depends upon what they traditionally manufacture or what type of natural resources they have. For example, Germany's strength is in manufacturing vehicles; therefore, most of their research has been directed towards fuel cells for vehicles.

The United States. The U.S. government has hydrogen and fuel cell research programs primarily under the Hydrogen, Fuel Cells, and Infrastructure Technologies Program run by the Department of Energy (DOE) [4]. The U.S. government's strategy is to focus on high-risk applied R&D. The administration increased funding in 2003 with a $1.2 billion program known as the Hydrogen Fuel Initiative [4]. The DOE has identified four phases in the transition to the hydrogen economy [4]:

1. The government and private sector will conduct R&D and develop industry standards. In 2015, the administration will make a decision on whether the technologies will be commercialized in the near term, and how much R&D should be continued [4].
2. As early as 2010, the government will begin modification of the existing infrastructure to support hydrogen applications.
3. The government will begin implementing a large-scale infrastructure for manufacturing fuel cells and distributing hydrogen.
4. The realization of the hydrogen economy is projected to begin in 2025.

Figure 2-4 shows an illustration of the U.S. hydrogen economy timeline.

Japan. Japan was the first country to have a large-scale hydrogen R&D program. The first program was a ten-year $165 million program that

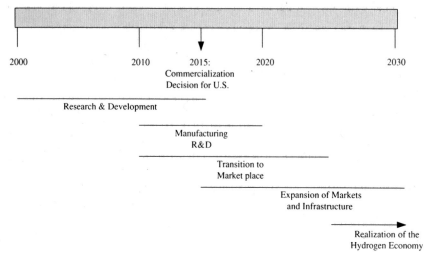

Figure 2-4 The U.S. hydrogen economy timeline (adapted from [4] and [13]).

was completed in 2002 [4]. In 2003, the New Hydrogen Project (NEP) focused on commercialization, and then raised $320 million in 2005 [4]. The Japanese government believes fuel cells will be competitive within the next two decades. The NEP sets target dates of 2010 and 2020 for the introduction of fuel cell vehicles, refueling stations, and stationary power [4]. The NEP has identified three stages in the transition to the hydrogen economy:

1. The initial stage was completed in 2005 and focused on fuel cell technology development and standards.
2. The induction stage will end in 2010 and involves the acceleration of vehicle sales while constructing the refueling infrastructure.
3. The diffusion stage will be from 2011 to 2020 and will continue to set up the infrastructure started in the second stage.

Figure 2-5 shows an illustration of Japan's hydrogen economy timeline.

The European Union. European funding for hydrogen and fuel cell R&D is provided under the Renewable Energy Sixth Framework Programme, which ran from 2002 to 2006 [4]. The program has $120 million in funds, which was matched by private investment. Total funding is expected to reach €2.8 billion by 2011 [4]. Production projects will account for €1.3 billion and end-use projects in communities will account for €1.5 billion [4]. All of the hydrogen projects funded by the European Union are in

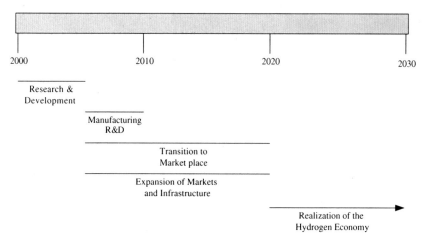

Figure 2-5 The Japan hydrogen economy timeline.

support of the Quick Start initiative. This program aims to attract private investment and accelerate the commercialization of fuel cells. Other production-related projects are aimed at building large demonstration plants that can produce electricity and hydrogen, as well as separate and store the CO_2 created in the process. Projects have also been designated that will create hydrogen communities with a centralized and decentralized hydrogen infrastructure, hydrogen vehicles, and a refueling structure. Research will also be conducted for different hydrogen production processes, including wind and biomass.

Programs in other countries. Many programs exist in other countries, including Australia, Canada, France, Germany, Italy, Korea, China, India, Russia, and Brazil. Projects are usually conducted in collaboration with private organizations.

- **Australia:** Australia's program is aimed at reducing carbon dioxide in the energy supply and creating low-emission ways of utilizing its large fossil fuel resources. One main focus is the production of hydrogen through coal gasification in the COAL21 program.
- **Canada:** Canada's hydrogen R&D focuses on production from renewable energy and fuel cells. Some successes include Ballard's demonstration of a fuel cell bus in 1993, and the Hydrogenics alkaline water electrolyzer. Public funding has been about $25 million per year, and total spending has exceeded $200 million[4].
- **France:** R&D areas include PEM and solid oxide fuel cells, coal gasification technology with carbon capture, solar and nuclear energy,

and biomass and fossil fuel reforming and storage. Total R&D funding has to date been about $48 million [4].

- **Germany:** Germany is a world leader in hydrogen and fuel cell development. Fuel cells are a main focus of the country's R&D efforts. About 75 percent of the fuel cell research conducted in Europe is in Germany. Total funding is estimated to be about $41 million per year [4].
- **Italy:** Funding for hydrogen and fuel cells has averaged about €30 million per year since 2000, with about 60 percent used for hydrogen production and fuel cells [4].
- **Korea:** Korea started funding hydrogen and fuel cell development in 1998, but is already becoming a major contributor of fuel cell knowledge and technology. A program was launched to run from 2004 to 2011 with a budget of $586 million [4]. The goals for 2012 include the development of hydrogen production systems using electrolysis-based renewables, a stationary 370 MW fuel cell, and the introduction of 10,000 fuel cell vehicles [4].
- **China:** China's R&D efforts are motivated by severe pollution and worries about energy security. Annual public funding is estimated to be in the tens of millions of dollars, with a larger amount being spent by private organizations. The Shanghai government also aims to introduce 1000 fuel-cell vehicles by 2010 [4].
- **India:** India has budgeted 2.5 billion rupees ($58 million) for hydrogen and fuel cell research in universities and national laboratories until 2007. One large project plans to blend small amounts of hydrogen into diesel fuel for use in 50 buses in New Delhi. The goal of India's government is to introduce 1000 hydrogen-powered vehicles by 2010. Automobile makers are expected to contribute at least 5 billion rupees ($116 million) to the development and demonstration of fuel-cell vehicles over the next five years [4].
- **Russia:** Russia has been researching fuel cells for many years. A national program is currently being discussed. Hydrogen research increased in 2003 with an agreement between the Russian Academy of Science and the Norilsk Nikel Company on a fuel cell development program. Total funding is $120 million; $30 million was budgeted in 2005 [4].
- **Brazil:** The goal of Brazil's R&D program is to commercialize fuel cells for automobiles and off-grid energy. The focus of the research is on storage technologies, hydrogen production from electrolysis, and the reforming of natural gas, ethanol, and other biofuels.

Private industry. Private sector funding for hydrogen and fuel cell R&D is thought to be much larger than public budgets. The International

Energy Agency [17] estimates that the R&D funding is between $3 to 4 billion per year, which is approximately four times the amount spent by the governments. The main sources of funding are oil and gas companies, car manufacturers, electricity, and gas utility and power plant companies. One of the largest projects involving private firms is the FreedomCAR and Fuel Partnership, which was set up in 2002 by the U.S. Department of Energy, General Motors, Ford, and DaimlerChrysler. The target of this program is to develop vehicles that run on hydrogen or other energy sources besides petroleum-based fuels. Five additional energy companies joined the program in 2003: BP America, ChevronTexaco Corporation, ConocoPhillips, Exxon, Mobil Corporation, and Shell Hydrogen [17]. The U.S. government is also searching for private funding of FutureGen, a project that will be the world's first zero-emission hydrogen-production power plant. The total cost of the project is $950 million, and $250 million will come from the private sector [4]. Another example of a successful collaborative private/public sector partnership is the California Fuel Cell Partnership, which involves car manufacturers, energy companies, fuel cell developers, and government agencies. The goal of this partnership is to develop fuel-cell vehicles and the refueling hydrogen infrastructure.

Chapter Summary

A transition from our traditional fossil-fuel based economy to a hydrogen economy seems inevitable. To obtain this goal, one or more intermediate steps may be necessary, such as using propane or methanol as fuels, and/or initially processing hydrogen from fossil-based fuels and coal. If the switch is made to a hydrogen economy sooner rather than later, the proper infrastructure can be put into place sooner. Although hydrogen will most definitely be made initially from fossil fuels, many non-fossil fuel–based methods can be used in the future to obtain hydrogen, such as nuclear, biological, wind, and, ultimately, solar power. Hydrogen technologies for production, utilization, and storage are already in use, and although many improvements could still be made, there is no major technological obstacle for utilizing hydrogen. More research and development needs to be conducted on solar hydrogen for future applications since hydrogen can be produced at lost cost with no pollution after the infrastructure has been set up.

Problems

1. Why is hydrogen a better fuel choice compared with other fuels?

2. What are the advantages/disadvantages of a hydrogen economy?

3. Do you think safety concerns regarding hydrogen are overrated?

4. If you were required to design a plan for the transition from the current economy to a hydrogen-based economy, what technologies would you choose for hydrogen production, storage, and delivery? (Chapter 15 may help provide additional information to answer this question.)

5. What government seems to be doing the best job in preparing for a transition to a hydrogen economy?

Bibliography

[1] Energy Information Administration. Official Energy Statistics from the U.S. Government. http://www.eia.doe.gov/. Last accessed September 15, 2006.
[2] O'Hayre, Ryan, Suk-Won Cha, Whitney Colella, and Fritz B. Prinz. *Fuel Cell Fundamentals*. 2006. New York: John Wiley & Sons.
[3] Barbir, Frano. *PEM Fuel Cells: Theory and Practice*. 2005. Burlington, MA: Elsevier Academic Press.
[4] *The Hydrogen Economy: A Non-Technical Review*. 2006. United Nations Environment Program E.
[5] "World Wide Hydrogen Fueling Stations." *Fuel Cells 2000*. http://www.fuelcells.org/info/charts/h2fuelingstations.pdf. Last accessed September 18, 2006.
[6] Barber, David B. "Nuclear Energy and the Future: The Hydrogen Economy or the Electricity Economy?" March 24, 2005. http://www.iags.org/barber.pdf. Last accessed September 12, 2006.
[7] Barreto, L., A. Makihira, and K. Riahi. "The Hydrogen Economy in the 21st Century: A Sustainable Development Scenario." *Florida International Journal of Hydrogen Energy*. 2002.
[8] Brown, Lester R. *Eco-Economy: Building an Economy for Earth*. "Chapter 5: Building the Solar/Hydrogen Economy." 2001. New York: W.W. Norton & Co.
[9] Crabtree, George W., Mildred S. Dresselhaus, and Michelle V. Buchanan. "The Hydrogen Economy." *Physics Today*. American Institute of Physics. December 2004.
[10] Bossel, Ulf. "Does a Hydrogen Economy Make Sense?" European Fuel Cell Forum. http://www.efcf.com. April 7, 2005. Last Accessed September 11, 2006.
[11] Rifkin, Jeremy. *The Hydrogen Economy*. 2002. Penguin Putnam.
[12] Yacobucci, Brent D. and Aimee E. Curtright. *A Hydrogen Economy and Fuel Cells: An Overview*. CRS Report for Congress. January 14, 2004.
[13] "Roadmap on Manufacturing R&D for the Hydrogen Economy." Based on the Results of the Workshop on Manufacturing R&D for the Hydrogen Economy, Washington D.C. July 13–14, 2005. December 2005.
[14] U.S. Department of Energy Hydrogen Analysis Resource Center. "Comparative Properties of Hydrogen and Fuels." http://hydrogen.pnl.gov.
[15] *Perry's Chemical Engineers' Handbook*, 7th ed. 1997. McGraw-Hill.
[16] Wurster, Reynold. "Pathways to a Hydrogen Refueling Infrastructure Between Today and 2020." L-B-Systemtechnik GmbH. http://www.hyweb.de/. Last accessed September 13, 2006.
[17] "Hydrogen & Fuel Cells: A Review of National R&D Programs." International Energy Agency. OECD/IEA. 2004.

Chapter 3
Fuel Cell Types

Many types of fuel cells are currently being researched. The nine main types are the subject of this chapter and are differentiated from one another on the basis of the electrolytes and/or fuel used with that particular fuel cell type. The most common fuel cell types include the following:

- Polymer electrolyte membrane fuel cells (PEMFCs)
- Alkaline fuel cells (AFCs)
- Phosphoric acid fuel cells (PAFCs)
- Solid oxide fuel cells (SOFCs)
- Molten carbonate fuel cells (MCFCs)
- Direct methanol fuel cells (DMFCs)
- Zinc air fuel cells (ZAFCs)
- Protonic ceramic fuel cells (PCFCs)
- Biological fuel cells (BFCs)

Polymer electrolyte membrane fuel cells (PEMFCs—sometimes called proton exchange membrane fuel cells) and DMFCs are very similar except for a few minor differences in the fuel cell construction and the fuel used. When conducting research on fuel cell types, one may also come across the names of other fuel cells, such as:

- Direct borohydride fuel cells
- Metal hydride fuel cells
- Formic acid fuel cells
- Direct ethanol fuel cells

- Regenerative fuel cells
- Microbial fuel cells
- Enzymatic fuel cells

Direct borohydride fuel cells and metal hydride fuel cells are subcategories of alkaline fuel cells (AFCs), which use a solution of sodium borohydride and metal hydrides for their fuel, respectively. Formic acid fuel cells are almost identical to PEMFC, but formic acid is used as the fuel, and the catalyst is palladium instead of platinum. Direct ethanol fuel cells (DEFCs) are very similar to PEMFCs and DMFCs, but are given the new name because ethanol is used as the fuel in the fuel cell. Regenerative fuel cells are a PEMFC that acts as a regular PEMFC and a "reversible" PEMFC (like an electrolyzer). The electrolyzer can be integrated into a single stack of cells, or two separate stacks can be used. PEMFC is the most common type of "regenerative fuel cell," but other fuel cell types can also be used. Several types of biological fuel cells also exist. Two common types (listed earlier) are microbial fuel cells and enzymatic fuel cells. Biological fuel cells (BFCs), protonic ceramic fuel cells (PCFCs), direct borohydride fuel cells, direct ethanol fuel cells, and formic acid fuel cells are still in the research stages. The other fuel cell types are on the verge of commercialization. Details of each fuel cell type described in this chapter are summarized in Table 3-1.

3.1 Polymer Electrolyte Membrane Fuel Cells (PEMFCs)

The polymer electrolyte (also called proton exchange membrane or PEM) fuel cell delivers high-power density while providing low weight, cost, and volume. A PEM fuel cell consists of a negatively charged electrode (anode), a positively charged electrode (cathode), and an electrolyte membrane, as shown in Figure 3-1. Hydrogen is oxidized on the anode and oxygen is reduced on the cathode. Protons are transported from the anode to the cathode through the electrolyte membrane and the electrons are carried over an external circuit load. On the cathode, oxygen reacts with protons and electrons forming water and producing heat.

In the PEM fuel cell, transport from the fuel flow channels to the electrode takes place through an electrically conductive carbon paper, which covers the electrolyte on both sides. These backing layers typically have a porosity of 0.3 to 0.8 and serve the purpose of transporting the reactants and products to and from the bipolar plates to the reaction site [8]. An electrochemical oxidation reaction at the anode produces electrons that flow through the bipolar plate/cell interconnect to the external circuit, while the ions pass through the electrolyte to the opposing

TABLE 3-1 Fuel Cell Types

Fuel cell system	Proton exchange membrane fuel cell (PEMFC)	Direct methanol fuel cell (DMFC)	Solid oxide fuel cell (SOFC)	Alkaline fuel cell (AFC)	Phosphoric acid fuel cell (PAFC)	Liquid molten carbonate fuel cell (MCFC)	Zinc–air fuel cells (ZAFC)	Protonic ceramic fuel cell (PCFC)	Biological fuel cell (BFC)
Fuel	H_2	$CH_3OH + H_2O$	CO, H_2	H_2	H_2	$H_2/CO/$ reformate	Zinc Oxide	CO, H_2	carbohydrate and hydrocarbon
Oxidizer	O_2, air	O_2, air	O_2, air	O_2, air	O_2, air	CO_2, O_2, air	O_2, air	O_2, air	O_2, air
Most Common Electrolyte	Perflourosulfonic acid membrane (Nafion by DuPont)	Perflourosulfonic acid membrane (Nafion by DuPont)	Yttria stabilized zirconia (YSZ)	Potassium hydroxide	Liquid phosphoric acid	Lithium, sodium and/or potassium carbonate soaked in a matrix	Potassium hydroxide	10% yttrium barium cerate (BCY10) protonic ceramic	Phosphate solution
Electrolyte Thickness	~50–175 μm	~50–175 μm	~25–250 μm	N/A	N/A	0.5 to 1 mm	N/A	~460 μm	Varies
Ion Transferred	$H+$	$H+$	O^{2-}	OH^-	$H+$	CO_3^{2-}	OH^-	O^{2-}	$H+$
Most Common Anode Catalyst	Pt	Pt/Ru	Nickel/YSZ	Pt or Ni	Pt	Nickel	Zinc	Platinum, nickel coating	Microorganism/enzyme
Anode Catalyst Layer Thickness	~10 to 30 μm	~10 to 30 μm	~25 to 150 μm	N/A	~10 to 30 μm	0.20 to 1.5 mm	N/A	Varies	Varies
Bipolar-Plate/Interconnect Material	Graphite, titanium, stainless steel, and doped polymers	Graphite, titanium, stainless steel and doped polymers	Doped $LaCrO_3$, $YcrO_3$, Iconel alloys	N/A	Graphite, titanium, stainless steel, and doped polymers	Stainless steel	N/A	Stainless steel	Graphite, electrically conductive polymer stainless steel

(*Continued*)

TABLE 3-1 Fuel Cell Types (Continued)

Fuel cell system	Proton exchange membrane fuel cell (PEMFC)	Direct methanol fuel cell (DMFC)	Solid oxide fuel cell (SOFC)	Alkaline fuel cell (AFC)	Phosphoric acid fuel cell (PAFC)	Liquid molten carbonate fuel cell (MCFC)	Zinc–air fuel cells (ZAFC)	Protonic ceramic fuel cell (PCFC)	Biological fuel cell (BFC)
Operating Temperature	Room temperature to 100°C	Room temperature to 100°C	600 to 1000°C	Room Temperature–250°C	150 to 220°C	620 to 660°C	700°C	500 to 700°C	Room temperature
Operating Pressure (atm)	1 to 3	1 (anode), 1 to 3 cathode	1 atm	1 to 4	3 to 10	1 to 10	1 atm	1 atm	1 atm
Major Contaminants	CO < 100 ppm, sulfur, dust	CO < 100 ppm, sulfur, dust	<100 ppm sulfur	CO_2	CO <100 ppm, sulfur, dust, NH_3	H_2S, HCl, As, H_2Se, NH_3, AsH_3, dust	N/A	CO < 100 ppm, sulfur, dust	N/A
Maximum Fuel Cell Efficiency (current)	~58%	~40%	~65%	~64%	~42%	~50%	N/A	~55 to 65%	~40%
Primary Applications	Stationary, portable, and vehicular	Portable electronics or as an APU	Stationary and distributed power, or as an APU	Space programs, portable power	Stationary power	Stationary power	Portable and vehicular power	Stationary and distributed power, or as an APU	Many utility Applications

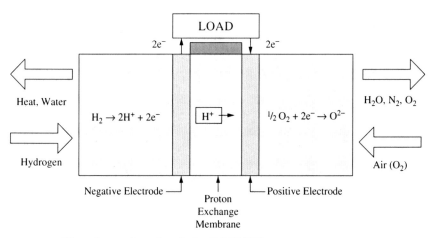

Figure 3-1 The polymer electrolyte fuel cell (PEMFC).

electrode. The electrons return from the external circuit, while the ions pass through the electrolyte to the opposing electrode. The electrons return from the external circuit to participate in the electrochemical reduction reaction at the cathode. The reactions at the electrode are

Anode: $H_2 (g) \rightarrow 2H^+ (aq) + 2e^-$
Cathode: $\frac{1}{2}O_2 (g) + 2H^+ (aq) + 2e^- \rightarrow H_2O (l)$
Overall: $H_2 (g) + \frac{1}{2}O_2 (g) \rightarrow H_2O (l)$

The standard electrolyte material currently used in PEM fuel cells is a fully fluorinated Teflon-based material produced by DuPont for space applications in the 1960s. The DuPont electrolytes have the generic brand name Nafion, and the types used most frequently are 1135, 115, and 117. The Nafion membranes are fully fluorinated polymers that have very high chemical and thermal stability. The electrodes are thin films that are bonded to the membrane. Electrodes with low platinum loading perform as well or better than high-platinum-loaded electrodes. To improve the utilization of platinum, a soluble form of the polymer is incorporated into the porosity of the carbon support structure. This increases the interface between the electrocatalyst and the solid polymer electrolyte [9].

3.2 Alkaline Fuel Cells (AFCs)

Alkaline fuel cells (AFCs) have been used by NASA on space missions and can achieve power-generating efficiencies of up to 70 percent. The operating temperature of these cells range between room temperature to 250°C [11]. The electrolyte is aqueous solution of alkaline potassium hydroxide soaked in a matrix. (This is advantageous because the cathode

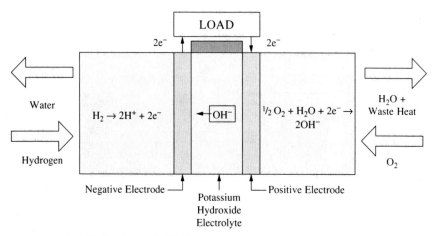

Figure 3-2 An alkaline fuel cell (AFC).

reaction is faster in the alkaline electrolyte, which means higher performance). Several companies are examining ways to reduce costs and improve operating flexibility. AFCs typically have a cell output from 300 watts to 5 kW [10]. The chemical reactions that occur in this cell are as follows:

Anode: $2H_2 (g) + 4(OH)^- (aq) \rightarrow 4H_2O (l) + 4e^-$

Cathode: $O_2 (g) + 2H_2O (l) + 4e^- \rightarrow 4(OH)^- (aq)$

Overall: $2H_2 (g) + O_2 (g) \rightarrow 2H_2O (l)$

A diagram of the alkaline fuel cell is shown in Figure 3-2. Another advantage of AFCs are the materials such as the electrolyte and catalyst used are low-cost. The catalyst layer can use either platinum or non-precious metal catalysts such as nickel. A disadvantage of AFCs is that pure hydrogen and oxygen have to be fed into the fuel cell because it cannot tolerate the small amount of carbon dioxide from the atmosphere. Over time, carbon dioxide degrades the KOH electrolyte which can lead to significant issues. Two commonly used solutions are refreshing the KOH electrolyte or carbon dioxide scrubbers. Due to these limitations, AFCs are not used for many power applications.

3.3 Phosphoric Acid Fuel Cells (PAFCs)

The phosphoric acid fuel cell (PAFC) is one of the few commercially available fuel cells. Several hundred fuel cell systems have been tested, and these fuel cells have been installed all over the world. Most of the PAFC plants that have been built are in the 50 to 200 kW range, but large plants of 1 MW and 5 MW also have been built. The largest plant operated to date achieved 11 MW of grid-quality alternating current (AC) power [11].

PAFCs are very efficient fuel cells, generating electricity at more than 40 percent efficiency. About 85 percent of the steam produced by the PAFC is used for cogeneration. This efficiency may be compared to about 35 percent for the utility power grid in the United States. Operating temperatures are in the range of 150 to 220°C. At lower temperatures, PAFC is a poor ionic conductor, and carbon monoxide (CO) poisoning of the platinum catalyst in the anode can become severe [11].

Two main advantages of the phosphoric acid fuel cell include a cogeneration efficiency of nearly 85 percent, and its ability to use impure hydrogen as fuel. PAFCs can tolerate a carbon monoxide concentration of about 1.5 percent, which increases the number of fuel types that can be used. Disadvantages of PAFCs include their use of platinum as a catalyst (like most other fuel cells) and their large size and weight. PAFCs also generate low current and power comparable to other types of fuel cells [12].

Phosphoric acid fuel cells are the most mature fuel cell technology. The commercialization of these cells was brought about through the Department of Energy (DOE) and ONSI (which is now United Technologies Company (UTC) Fuel Cells) and organizational linkages with Gas Research Institute (GRI), electronic utilities, energy service companies, and user groups [54]. The chemical reactions for PAFCs are as follows:

Anode: $H_2 \text{ (g)} \rightarrow 2H^+ \text{ (aq)} + 2e^-$

Cathode: $\frac{1}{2} O_2 \text{ (g)} + 2H^+ \text{ (aq)} + 2e^- \rightarrow H_2O \text{ (l)}$

Overall: $H_2 \text{ (g)} + \frac{1}{2}O_2 \text{ (g)} + CO_2 \rightarrow H_2O \text{ (l)} + CO_2$

A diagram of the phosphoric fuel cell is shown in Figure 3-3.

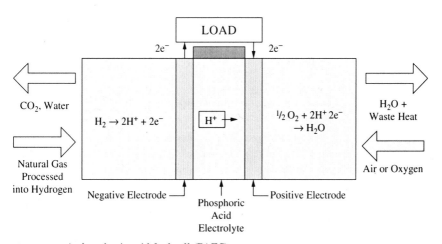

Figure 3-3 A phosphoric acid fuel cell (PAFC).

3.4 Solid Oxide Fuel Cells (SOFCs)

Solid oxide fuel cells (SOFCs) seem promising for large, high-power applications such as industrial and large-scale central electricity generating stations. Some developers also see potential SOFC use in motor vehicles and are developing fuel cell auxiliary power units (APUs). The chemistry of the solid oxide fuel cells consists of an electrolyte that is a non-porous solid, such as Y_2O_3 stabilized ZrO_2 with conductivity based oxygen ions (O_2). The anode is usually made of a $Co\text{-}ZrO_2$ or $Ni\text{-}ZrO_2$ cement, while the cathode is made of Sr-doped $LaMnO_3$. Cells are being constructed in three main configurations: tubular, bipolar and a planar configuration adopted more recently by many other developers as shown in Figure 3-4. The operating temperatures can reach 1000°C. Power-generating efficiencies could reach 60 to 85 percent with cogeneration and when cell output is up to 100 kW. The anode, cathode, and overall cell reactions are

Anode: $H_2 (g) + O^{2-} \rightarrow H_2O (g) + 2e^-$
Cathode: $\frac{1}{2}O_2 (g) + 2e^- \rightarrow O^{2-}$
Overall: $H_2 (g) + \frac{1}{2}O_2 (g) \rightarrow H_2O (g)$

SOFCs can be constructed in many ways. Tubular SOFC designs are closer to commercialization and are being produced by several companies around the world. Tubular SOFC technology has produced as much as 220 kW. Japan has two 25-kW units online, and a 100-kW plant is being tested in Europe [10].

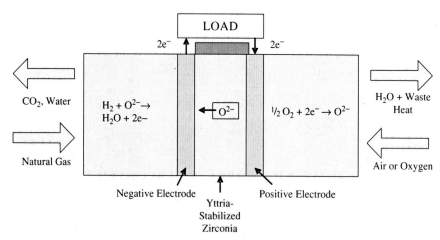

Figure 3-4 A solid oxide fuel cell (SOFC).

SOFCs that employ a ceramic, solid-state electrolyte (zirconium oxide stabilized with yttrium oxide) may be the only fuel cell technology with the potential to span market-competitive applications from residential loads as small as 2 kW to wholesale distributed generation units of 10 to 25 MW according to the Electric Power Research Institute (EPRI). Even though SOFCs operate at higher temperatures than MCFCs, their simple efficiency is theoretically not as good as that of MCFCs. The waste heat that SOFCs produce (between 850 to 1000°C) is extremely beneficial when used for cogeneration or for driving an integrated gas turbine because it can boost overall system energy efficiency to very attractive levels. SOFCs are able to operate at a high enough temperature to incorporate an internal fuel reformer that uses heat from the fuel cell. The recycled steam and a catalyst can convert the natural gas directly into a hydrogen-rich fuel cell.

SOFCs coupled with small gas turbines are high-efficiency systems that have a combined rating in the range of 250 kW to 25 MW, and are expected to fit into grid support or industrial onsite generation markets. These fuel cells could potentially compete head-on with wholesale power rates [11].

3.5 Molten-Carbonate Fuel Cells (MCFCs)

Two major corporations are pursuing the commercialization of molten carbonate fuel cells (MCFCs) in the United States: FuelCell Energy and M-C Power Corporation. The electrolyte in the molten-carbonate fuel cell uses a liquid solution of lithium, sodium, and/or potassium carbonates, soaked in a matrix. MCFCs have high fuel-to-electricity efficiencies ranging from 60 to 85 percent with cogeneration, and operate at about 620–660°C [11]. The high operating temperature is a big advantage because it enables a higher efficiency and the flexibility to use more types of fuels and inexpensive catalysts. This high operating temperature is needed to achieve sufficient conductivity of the electrolyte.

Molten carbonate fuel cells can use hydrogen, carbon monoxide, natural gas, propane, landfill gas, marine diesel, and coal gasification products as the fuel. MCFCs producing 10-kW to 2-MW MCFCs have been tested with a variety of fuels and are primarily targeted to electric utility applications. MCFCs for stationary applications have been successfully demonstrated in several locations throughout the world. A disadvantage of MCFCs is that high temperatures enhance corrosion and the breakdown of cell components. The reactions at the anode, cathode, and the overall reaction for the MCFC are [29]:

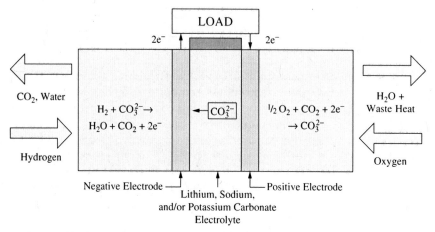

Figure 3-5 A molten carbonate fuel cell (MCFC).

Anode: $H_2\ (g) + CO_3^{2-} \rightarrow H_2O\ (g) + CO_2\ (g) + 2e^-$

Cathode: $\frac{1}{2}O_2\ (g) + CO_2\ (g) + 2e^- \rightarrow CO_3^{2-}$

Overall: $H_2\ (g) + \frac{1}{2}O_2\ (g) + CO_2\ (g) \rightarrow H_2O\ (g) + CO_2\ (g)$

A diagram of a molten carbonate fuel cell is shown in Figure 3-5.

3.6 Direct Methanol Fuel Cells (DMFCs)

The large potential market for fuel cell portable applications has generated a strong interest in a fuel cell that can run directly on methanol. Operating on liquid fuel would expedite the introduction of fuel cell technology into the commercial marketplace because it has the perception of being a "safer" fuel than hydrogen, and it allows the fuel cell to be refueled easily [12].

The direct methanol fuel cell (DMFC) uses the same polymer electrolyte membrane as the PEM fuel cell. The fuel for the DMFC, however, is methanol instead of hydrogen. Methanol flows through the anode as fuel and is broken down into protons, electrons, and water. Advantages of methanol include its wide availability and its ability to be easily reformed from gasoline or biomass. Although it only has a fifth of the energy density of hydrogen by weight, it offers more than four times the energy per volume when compared to hydrogen at 250 atmospheres since it is liquid [9]. The chemical reactions for this fuel cell are as follows:

Anode: $CH_3OH\ (l) + H_2O\ (l) \rightarrow CO_2 + 6H^+ + 6e^-$
Cathode: $6H^+ + {}^3/_2 O_2 + 6e^- \rightarrow 3H_2O(l)$
Overall: $CH_3OH\ (l) + {}^3/_2 O_2\ (g) \rightarrow CO_2\ (g) + 2H_2O\ (l)$

For the lower heating value reaction, E is 1.170 V, which is slightly less than hydrogen at 1.185 V. For each mole of methanol, six electrons are transferred. This seems far superior to the two electrons of a hydrogen PEMFC. However, the comparison is not with one mole of hydrogen, but one mole of methanol reformed into hydrogen. Theoretically, three moles of hydrogen molecules can be reformed so that the total number of electrons is equal in each case.

A major issue with the DMFC is that the oxidation of methanol produces intermediate hydrocarbon species, which poison the electrode [12]. Another limitation is that the oxidation of methanol at the anode becomes as slow as the oxygen electrode reaction, and large overpotentials are required for high power output. There is also the problem of high crossover of the methanol through the electrolyte (the fuel molecules diffuse directly through the electrolyte to the oxygen electrode). Thus, power is severely compromised since as much as 30 percent of methanol can be lost this way [12].

The performance of a direct methanol fuel cell is on the order of 180 to 250 mW/cm^2 with air. This is insufficient for current applications except with very small portable fuel cells. Platinum loadings are 4 mg/cm^2, almost a factor of ten times greater than hydrogen–air PEM fuel cells [9].

Many companies and laboratories are researching direct methanol fuel cells. Some of the goals of this research are as follows:

1. Improve membranes to have higher selectivity (ratio of conductivity to crossover rate).
2. Improve the anode for improved methanol oxidation catalysis.
3. Increase cathode performance to methanol-tolerant oxygen reduction reaction (ORR) catalysts and improve air electrodes for lower loading.
4. Implement new activity related to high temperature membranes.
5. Increase MEA durability.
6. Establish reliable MEA fabrication approaches.
7. Improve understanding of factors influencing electro-osmosis.

Efficiencies of about 40 percent are expected with the direct methanol fuel cell. The fuel cell would typically operate at a temperature between room temperature to 100°C. This is a relatively low temperature range, which makes this fuel cell attractive for tiny to mid-sized applications. Many companies are working on DMFC prototypes used by the military

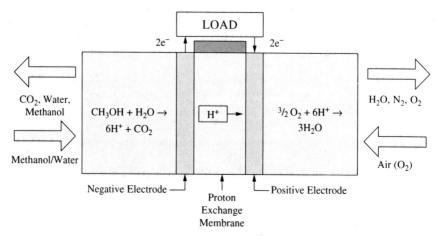

Figure 3-6 A direct methanol liquid-feed fuel cell (DMFC).

for powering electronic equipment in the field. A diagram of the direct methanol fuel cell is shown in Figure 3-6.

3.7 Zinc Air Fuel Cells (ZAFCs)

In zinc air fuel cells (ZAFC), there is a gas diffusion electrode (GDE), a zinc anode separated by electrolyte, and some form of mechanical separators, as shown in Figure 3-7. Like other fuel cell types, the GDE is a permeable membrane that allows the oxidant to pass through it. Zinc oxide is created by hydroxyl ions and water (from oxygen), which react

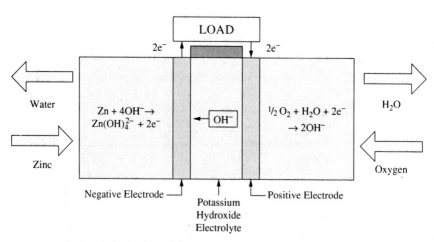

Figure 3-7 A zinc air fuel cell (ZAFC).

with the zinc at the anode, thus creating an electrical potential. Zinc–air fuel cells can be connected together to obtain the desired power requirements. This electrochemical process is very similar to that of a PEM fuel cell, but the refueling process shares characteristics with batteries [7].

ZAFCs contain a zinc "fuel tank" that automatically regenerates the fuel. Zinc fuel is in the form of small pellets, and is consumed, and releases electrons to drive a load. Oxygen from ambient air accepts electrons from the load, and potassium zincate is produced from this process. Electrolysis is used to reprocess potassium zincate back into zinc pellets and oxygen. This regeneration process is powered by an outside source (such as solar cells) and can be repeated indefinitely [7].

ZAFCa are often referred to as "regenerative" fuel cells. A regenerative fuel cell system is a system where no fuel is added, and none of the products or by-products is wasted. The oxygen used by the fuel cell is released back into the air by the electrolyzer. The efficiency of this system is 30 to 50 percent for a zinc regenerative fuel cell (ZRFC) system and about 20 to 40 percent for a hydrogen regenerative fuel cell system [7]. The chemical reactions that occur in this cell are shown next:

Anode: $Zn\ (g) + 4OH^-\ (aq) \rightarrow Zn(OH)^{2-}_4 + 2e^-$

Cathode: $\frac{1}{2}O_2 + H_2O + 2e^- \rightarrow 2OH^-$

Overall: $Zn + 2OH^- + \frac{1}{2}O_2 + \frac{1}{2}H_2O \rightarrow Zn(OH)^{2-}_4$

A diagram of the zinc-air fuel cell is shown in Figure 3-7.

The zinc–air fuel cell technology is classified under fuel cells and batteries. The main advantage that zinc–air technology has over other battery technologies is its high specific energy. ZAFCs have been used to power electronic vehicles (EVs), and have proven to deliver longer driving distances between refuels than any other EV batteries of similar weight. The material costs for ZRFCs and zinc–air batteries are low because of the abundance of zinc as an element. Due to this, zinc–air technology can be potentially used for a wide range of applications, such as EVs and consumer electronics to military applications.

3.8 Protonic Ceramic Fuel Cells (PCFCs)

The protonic ceramic fuel cell (PCFC) is a new type of fuel cell based upon a ceramic electrolyte material that exhibits high protonic conductivity at elevated temperatures [5]. This type of fuel cell is fundamentally different than other fuel cells because it relies on the conduction of hydrogen ions (protons) through the electrolyte at much higher temperatures than is possible with other proton-conducting fuel cell types. PCFCs have the same thermal and kinetic advantages as molten carbonate and solid oxide fuel cells because of their high temperature operation (700°C), but this fuel cell also exhibits the benefits of proton conduction like in PEMFCs and PAFCs [5].

To achieve very high electrical fuel efficiency with hydrocarbon fuels, a high operating temperature is necessary. PCFCs can operate at high temperatures and electrochemically oxidize fossil fuels directly to the anode. This eliminates the intermediate step of producing hydrogen through the costly reforming process. The hydrocarbon fuel is absorbed onto the surface of the anode in the presence of water vapor, then the hydrogen atoms are removed and absorbed into the electrolyte. The carbon dioxide is the primary reaction product. The main company that is researching this fuel cell type is CoorsTek.

The hydrogen oxidation reaction that produces the electrical energy occurs at the anode (fuel side), which is the opposite of the other high temperature fuel types. In PCFC, the fuel at the cathode is removed by the flow of air which makes complete fuel utilization possible [6]. When hydrocarbon fuels are used, a variety of reaction paths can take place, but the dilution of the fuel with wastewater makes it difficult to achieve high fuel utilization in SOFCs without expensive post-processing of the fuel stream leaving the stack. Fuel dilution does not occur in PCFCs. Additionally, PCFCs have a solid electrolyte so the membrane cannot dry out as with PEM fuel cells, and liquid cannot leak out as with PAFCs. The reactions with protonic and zirconia electrolytes are

Protonic Electrolyte:
Anode: $CH_4 + 2H_2O \rightarrow CO_2 + 8e^- + 8H^+ + 8H^+$
Cathode: $8H^+ + 8e^- + 2O_2 \rightarrow 4H_2O1p6.25$

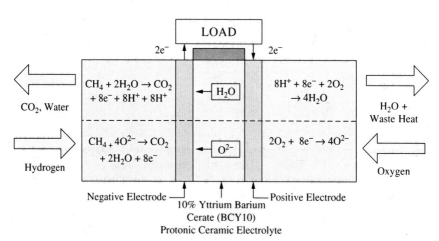

Figure 3-8 A protonic ceramic fuel cell (PCFC).

Zirconia Electrolyte:
Anode: $CH_4 + 4O^{2-} \rightarrow CO_2 + 2H_2O + 8e^-$
Cathode: $2O_2 + 8e^- \rightarrow 4O^{2-}$

A diagram of the PCFC is shown in Figure 3-8.

3.9 Biological Fuel Cells (BFCs)

A biological fuel cell is a device that directly converts biochemical energy into electricity. In biological fuel cells (BFC), there is the redox reaction of a carbohydrate substrate such as glucose and methanol using a microorganism or enzyme as a catalyst. The biological fuel cell works like any other fuel cell, as illustrated in Figure 3-9. The main difference is that the catalyst in the biological fuel cell is a microorganism or enzyme. Therefore, noble metals are not needed for the catalyst and its operating conditions are typically neutral solutions and room temperature. The fuel cell operates in liquid media, and advantages like low temperature and a near neutral environment can be obtained. The possible potential applications of such fuel cells are (1) to develop new practical low power energy sources; (2) to manufacture a specific sensor based upon direct electrode interactions; and (3) to electrochemically synthesize some chemicals.

In the research and development of biological fuel cells, improvements have been made in several areas. The areas include the selection of the microorganisms, the use of different mediators to improve electron transfer, and the investigation of the kinetics in the process. In addition,

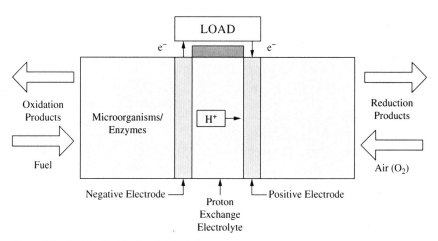

Figure 3-9 A biological fuel cell (BFC).

various types of electrodes have been studied for efficient reactions [13]. A reaction diagram of a biological fuel cell is shown in Figure 3-9.

In bacteria cells, mitochondria serve as the energy storage unit by accumulating or releasing chemical energy in the form of substances like nicotinamide adenine dinucleotide with high-energy hydrogen (NADH) or nicotinamide adenosine dinucleotide phosphate (NADPH). The NADH and NADPH act as electron transfer paths from the substrate to the metabolites. The NADH/NAD ratio increases as oxygen limitation becomes more severe [14].

Most of the substrate is turned into an electroactive substance through proper control of bacteria metabolism [13]. However, the biological reactions mentioned earlier only take place in diluted aqueous media, which is not suitable for different charge-transfer reactions. The electron transfer from these electroactive substances to the electrode is a slow process. Therefore, a suitable redox mediator is needed to improve the electron transfer and the electrode reaction [4].

In a study done by the Helinski University of Toronto, a mediator, 2-hydroxy-1,4-naphtho-quinone (HNQ), was used to improve the electron transfer. When this mediator substance is present in the system, the color in the biofilm reactor changes [13]. The changes in the color occur because of the metabolic activity of the biofilm reactor and the electron transfer in the fuel cell anode. It offers a simple method of monitoring and controlling functions of the fuel cell [14].

Experimental tests of the bacterial fuel cells have resulted in several conclusions. First, the complex reactions make the conversion rate of electrical energy lower (15 to 25 percent) compared to that of chemical fuel cells. Secondly, the current density per anode volume increases when the size of the fuel cell decreases. Finally, the power output increases if the bacteria are immobilized. The next stage of research and development will be with an enzymatic fuel cell. Replacing the bacteria with an enzyme should make the process easier to control because enzymatic reactions are simpler. According to the estimation of the researchers at the Helinski University of Technology, the conversion rate of the Enzymatic Fuel Cell (EFC) is expected to be more than 50 percent [13]. Experiments have shown that the rate is in the range of 40 to 55 percent for the bacterial fuel cell when the substrate is glucose. Similarly, the current density of the EFC is likely to improve significantly compared to that of the bacterial fuel cell.

Chapter Summary

This chapter briefly covered the main fuel cell types: PEMFC, AFC, PAFC, SOFC, MCFC, DMFC, ZAFC, PCFC and BFCs. The PCFC and BFC are newer fuel cell types that are still in the R&D stage. The ZAFC is often noted in the literature as a fuel cell or a battery. These fuel cell types are differentiated from one another based upon electrolyte and fuel type used. The fuel cell electrolyte determines many other parameters in the fuel cell system, such as: operating temperatures, cell materials, and cell and stack designs. The differences lead to important characteristics, advantages, and disadvantages of each fuel cell type. All fuel cell types can run well on pure hydrogen gas, but high temperature fuel cells can also use hydrocarbon-based fuels or carbon monoxide. Out of all the fuel cell types currently being researched, the PEMFC seems to currently be best suited for portable, backup, transportation and small stationary applications, while SOFCs seem to be the best suited for larger-scale stationary applications.

Problems

1. What are the advantages of high-temperature verses low-temperature fuel cells?

2. What fuel cell type(s) would you select if you wanted to use gasoline as the fuel and do not want to use an external reformer?

3. What fuel cell type(s) would you select for an automobile? Generator? Cell phone? Why?

4. What fuel cell types are currently commercially used?

5. Discuss the important differences in reaction chemistry, fuel cell materials, and operating conditions for each fuel cell type.

Bibliography

[1] Malone, E. et al. 2004. "Freeform Fabrication of Zinc-Air Batteries and Electromechanical Assemblies." *Rapid Prototyping Journal*, Vol. 10, No. 1, pp. 58–69.
[2] O'Hayre, Ryan, Suk-Won Cha, Whitney Colella, and Fritz B. Prinz. *Fuel Cell Fundamentals*. 2006. New York: John Wiley & Sons.
[3] Barbir, Frano. *PEM Fuel Cells: Theory and Practice*. 2005. Burlington, MA: Elsevier Academic Press.
[4] Shukla, A. K., P. Suresh, S. Berchmans, and A. Rajendran. "Biological Fuel Cells and Their Applications." *Current Science*, Vol. 87, No. 4. August 25, 2004.
[5] Coors, W.G. "Protonic Ceramic Fuel Cells for High-Efficiency Operation with Methane." *Journal of Power Sources*, Vol. 118, Issue 1–2, 2003, pp. 150–156.
[6] Coors, W.G. "Steam Reforming and Water-Gas Shift by Steam Permeation in a Protonic Ceramic Fuel Cell." *Journal of the Electrochemical Society*, Vol. 151, Issue 7, 2004, pp. A994–A997.
[7] Smedley, S. "A Regenerative Zinc–Air Fuel Cell for Industrial and Specialty Vehicles." Fifteenth Annual Battery Conference on Applications and Advances (Cat. No.00TH8490), 1999, pp. 65–70.
[8] Mikkola, Mikko. "Experimental Studies on Polymer Electrolyte Membrane Fuel Cell Stacks." Helinski University of Technology, Department of Engineering Physics and Mathematics, Masters Thesis, 2001.

[9] Lin, Bruce. "Conceptual Design and Modeling of a Fuel Cell Scooter for Urban Asia." Princeton University, Masters Thesis, 1999.
[10] "Fuel Cell Basics." *Fuel Cells 2000*. Last accessed September 13, 2006. http://www.fuelcells.org/basics/how.html.
[11] Rossi, C et al. "A New Generation of MEMS based Microthrusters for Microspacecraft Applications." *Proc. MicroNanotechnology for Space Applications*. Vol.1. April 1999.
[12] Hirschenhofer, J.H. et al. *The Fuel Cell Handbook*, 4th ed. 1998. Reading, PA: Parsons Corporation.
[13] Halme, Aarne, Xia-Chang Zhang, and Niko Rintala. "Study of Biological Fuel Cells." Automation Technology Laboratory, Helinski University of Technology, 1998. http://www.automation.hut.fi/research/bio/sfc00pos.htm. Last accessed September 19, 2006.
[14] Heller, Adam. "Biological Fuel Cell and Method." TheraSense, Inc. U.S. Patent # 6,294,281. September 25, 2001.

Chapter 4

Fuel Cell Applications

Certain fuel cell types are better suited for particular power applications than others. For instance, certain fuel cell types are more suitable for small portable technologies and backup power, and others are better suited for automobile, and stationary power applications. Fuel cells for each of these categories have already been demonstrated and there are some commercial fuel cells available for some of these applications. Fuel cell design may be slightly different for each application, and depends upon the power output required, the water and heat balance, efficiency, size, weight, and fuel supply.

The specific topics covered in this chapter include the following:

- Portable Power Applications
- Backup Power Applications
- Transportation Applications
- Stationary Power Applications

The basic fuel cell applications will be described, along with a detailed summary of the fuel cell technologies for each application demonstrated.

4.1 Portable Power

Portable fuel cell systems can be divided into two power ranges: (1) battery replacements under 100 watts, and (2) portable power generators up to 1 kW. The greatest benefit of using portable fuel cells is that they are a compact, lightweight, efficient, long-lasting, portable power source that prolongs the amount of time a device can run without the need for recharging. Most secondary (rechargeable) batteries have battery charger systems that consist of AC chargers that must be plugged into an outlet to be charged, or DC chargers that will recharge your batteries from other batteries. These options are not viable for many military and future portable electronic applications because they are heavy, impractical, and do not meet the current power requirements. Some applications for portable fuel cells include the following:

- Laptops
- Power tools
- Cellular phones
- Video cameras
- Military equipment
- Battery chargers
- Computers
- Unattended sensors
- Unmanned aerial and underwater vehicles

One of the most the critical issues for all fuel cell types is the fuel and its storage. Hydrogen sometimes presents a problem for portable fuel cells because of the volume of the storage container. Other fuel types such as metal hydrides, methanol, formic acid, and ethanol provide reasonable options because a small volume can be put into a container for use with the fuel cell. Methanol and ethanol can be directly supplied to the fuel cell as fuel, or a fuel reformer can also be attached to the fuel cell package. Table 4-1 lists a few of the companies currently working on portable fuel cells. As seen in the table, the most common type of fuel cell for small portable applications is the DMFC.

TABLE 4-1 Examples of Portable Fuel Cells

Manufacturer	Fuel type	Fuel cell used	Power	Model
MTI MicroFuel Cells	Methanol	DMFC	Up to 30 W	Mobion cord-free power packs Mobion-30
Medis Technologies	Ethanol or methanol	Direct fuel cell (no proton membrane)	1.3 W	"24/7" power packs
Smart Fuel Cell, GmbH	Methanol	DMFC	40 W	N/A
PolyFuel	Methanol	DMFC	15 to 40 W	N/A
Direct Methanol Fuel Cell Corporation	Methanol	DMFC	1 W to 1.4 kW	N/A

One of the most attractive markets for portable fuel cells during the initial commercialization period is the military. The military accepts higher prices for the trade-off of better performance and weight. A preferred fuel for military applications would be one such as JP-8 because it is readily available. However, there are many difficulties associated with reforming this fuel. Development of small fuel cells for portable power applications has resulted in a myriad of stack configurations. Some stacks are a smaller version of the stacks used in automobiles, while others are MEMS-based (micro-electrical mechanical systems) fuel cells designed using traditional and new photolithography techniques (total volume is less than 1 cm^3). Chapters 11 through 14 will describe popular materials and designs for these portable fuel cells.

4.2 Backup Power

Backup power systems are devices that provide power when the primary power source fails or is disrupted. Fuel cell requirements for backup power vary significantly from other applications. These systems come in many sizes and types, and usually use hydrogen as their fuel. Backup fuel cells can be commercialized relatively quickly compared to fuel cells for other applications because they do not depend upon the implementation of a hydrogen infrastructure. Some of the applications for backup power include:

- Computer systems
- Telecommunications systems
- Manufacturing processes
- Security systems
- Homes
- Utility substations

Fuel cell system efficiency is not as critical in this application as portable and transportation applications, but it does determine the size of the fuel cell. The critical parameter in backup power is the response time or the ability to start immediately upon power outage. The fuel cell can meet this requirement if there is an uninterrupted supply of fuel. Depending upon the design of the fuel cell, a battery may also bridge the gap in time. Table 4-2 shows some examples of commercially available backup power fuel cells.

The PEMFC fuel cell with compressed hydrogen fuel is the most popular fuel cell type used for backup power applications. At power levels between 1 to 500 kW, the backup power market is currently greater than 100,000 units annually [5]. A fuel cell power backup must be equipped

TABLE 4-2 Examples of Commercial Backup Power Fuel Cells

Manufacturer	Fuel type	Fuel cell used	Power	Model
ReliOn (formerly Avista Labs)	Compressed hydrogen	PEMFC	1 to 1.2 kW	T-1000, Independence 1000
Ballard Power Systems	Compressed hydrogen	PEMFC	1 to 5 kW	Nexa Power Module, Mark 1020 ACS
Plug Power Corp.	Compressed hydrogen	PEMFC	0.5 to 500 kW	GenCore Systems
Nuvera	Compressed hydrogen, natural gas, propane	PEMFC	2 to 25 kW	PowerFlow, Avanti

with hydrogen storage sufficient for system operation for the required period. There are many fuel options available: such as hydrogen bottles, a hydrogen generator, propane, natural gas, as well as many other fuel types. The backup fuel cell system should be used where electricity is available, and not necessarily used with other fuels. An electrolyzer system is a good option for backup power applications because it can produce hydrogen on demand. The electrolyzer can be used with electricity generated by solar panels, a wind source, a nuclear source, or electricity generated by the local power company. The electrolyzer must be sized to generate the required amount of hydrogen in a given time period. Section 4.2.1 has some basic electrolyzer design calculations. Chapter 15 includes more details about designing electrolyzers.

4.2.1 Basic electrolyzer calculations

There are many different types of electrolyzers. The designs vary widely, and many types of electrolytes can be used. The most commonly used electrolyte is a proton exchange membrane (the same types that are used for PEM fuel cells). Important parameters for designing an electrolyzer include the power required the efficiency and the hydrogen pressure in the compressed hydrogen bottle. The required electrolyzer power is [3]:

$$P_{EL} = \frac{P_{FC} \cdot \tau_{FC} \cdot CF_{FC}}{\eta_{FC,sys} \cdot \eta_{EL,sys} \cdot \tau_{EL} \cdot CF_{EL}} \quad (4\text{-}1)$$

where $P_{FC,norm}$ is the fuel cell nominal power (kW), τ_{FC} and τ_{EL} is the duration of operation in the fuel cell and electrolyzer modes, respectively (hours), CF_{FC} and CF_{EL} are the capacity factors of fuel cell and electrolyzer defined as a ratio between average and nominal power, and $\eta_{FC,sys}$ and $\eta_{EL,sys}$ is the system efficiency of the fuel cell and the electrolyzer.

The electrolyzer efficiency can be stated as [3]:

$$\eta_{EL} = \frac{1.482}{V_{el\_cell}} \frac{i_{el} - i_{loss}}{i_{el}} \qquad (4\text{-}2)$$

where V_{el_cell} is the electrolyzer cell voltage, i_{el} is the electrolyzer current density (A cm^{-2}) and i_{loss} is the current and hydrogen loss (A cm^{-2}), which is typically negligible at low pressures and high operating current densities. The electrolyzer system efficiency is [12]:

$$\eta_{EL,sys} = \frac{1.482}{V_{el\_cell}} \frac{i_{el} - i_{loss}}{i_{el}} \frac{\eta_{DC}}{1+\zeta} \qquad (4\text{-}3)$$

where η_{DC} is the efficiency of power conversion (AC/DC or DC/DC), and ζ is the ratio between parasitic power and net power consumed by the electrolyzer.

The equations used for fuel cell designs can also be used for electrolyzer designs (as presented in Chapters 5–10, 13, 15, 16). To accurately model and design an electrolyzer, the electrolyzer stack design can be calculated using the equations in Chapter 13. The water and hydrogen flow rates can be calculated by altering the equations in Chapter 16 with water as the input, and hydrogen as the output. If a PEM (polymer) membrane is being used, the membrane water content can be estimated using the equations in Chapters 8 or 10. The power/voltage can be estimated using the equations in Chapter 6. The hydrogen produced by an electrolyzer is stored in a hydrogen bottle, and the following calculation can be used to calculate the hydrogen pressure in a compressed hydrogen bottle:

$$P_b - P_{bi} = z * \frac{n_{H_2} R T_b}{m_{H_2} V_b} \qquad (4\text{-}4)$$

where P_{bi} is the initial H_2 pressure in the bottle, n_{H2} is moles of hydrogen, m_{H2} is the molecular weight of hydrogen, z is the compressibility factor of hydrogen (usually 1 when the pressure is below 2000 psi and room temperature), T_b is the bottle temperature and V_b is the bottle volume. The process of filling the bottle with hydrogen is slow, therefore, the temperature is usually assumed to be constant [13].

4.3 Transportation Applications

Fuel cells could be used for many transportation applications. The applications that will be covered in this section include the following:

- Automobiles
- Buses

- Utility vehicles
- Scooters and bicycles

Many fuel cell demonstration vehicles have been created for each of these vehicle types. The fuel cells used for each vehicle type will be summarized in Sections 4.3.1–4.3.4.

4.3.1 Automobiles

Most automobile manufacturers have been developing fuel cell vehicles for at least a decade, and have demonstrated at least one prototype vehicle. The major reasons for developing automotive fuel cell technology are their efficiency, low or zero emissions, and fuel that could be reproduced from local sources rather than imported. Some automobile manufacturers are working on their own fuel cell technology. Some examples include General Motors, Toyota, and Honda. Other automobile manufacturers buy their fuel cells from fuel cell companies such as Ballard, DeNora, and UTC fuel cells. These automobile manufacturers include Ford, Mazda, DaimlerChrysler, Mazda, Hyundai, Fiat, and Volkswagen.

Table 4-3 shows a summary of fuel cell vehicles demonstrated by the major automobile manufacturers. Notice in the table that many auto companies demonstrated their first fuel cell vehicles in the late 1990s. The fuel type most often used was compressed hydrogen, although many manufacturers also demonstrated a fuel cell vehicle with an alternative fuel type such as methanol. The types of fuel cells commonly used are the PEMFC and the DMFC. About 79 percent of the fuel cell vehicles demonstrated used fuel cells developed from a fuel cell company instead of manufacturing their own fuel cell.

Automotive fuel cells can have one or all of the following characteristics:

- A fuel cell is sized to provide all of the power to a vehicle. A battery may be present for startup.
- A fuel cell typically supplies a constant amount of power, so for vehicle acceleration and other power spikes, additional devices are typically switched on such as batteries, ultra or supercapacitors, and so on.
- Sometimes a fuel cell is used as the secondary power source. A system is set up where batteries power the vehicle, and the fuel cell just recharges the batteries when needed.
- A fuel cell can run part or all of the vehicle's electrical system. Sometimes another engine is used for propulsion.

TABLE 4-3 Fuel Cell Vehicles Produced by Major Manufacturers (Compiled from *Fuel Cells 2000* [6])

Manufacturer	Year(s) demonstrated	Fuel type	Fuel cell used	Power	Fuel cell supplier	Name of vehicle(s)
Ford	1999 to 2002	Compressed hydrogen, methanol	PEMFC, DMFC	75 to 85 kW	Ballard	P2000 HFC, Focus HCV, THINK HC5, Advanced Focus FCV
General Motors	1997 to 2005	Compressed hydrogen, liquid hydrogen, methanol, metal hydrides, gasoline	PEMFC, DMFC	50 to 100 kW	GM	Sintra, Zafira, Precept FCEV, Hydrogen 1, 3, Chevy S-10, Autonomy, HV-wire, Diesel hybrid military truck, Sequel, Phoenix
Toyota	1996 to 2003	Compressed hydrogen, methanol metal hydrides, gasoline	Hybrid PEMFC, DMFC	20 to 90 kW	Toyota	RAV 4 FC-EV, FC-HV3-5, FINE-S
Honda	1999 to 2003	Compressed hydrogen, metal hydrides, methanol	PEMFC/PMFC with batteries/ PEMFC with ultracapacitors, DMFC	60 to 85 kW	Ballard/Honda	FCX, FCX-V1 – FCX- V4, Kiwami
Nissan	1999 to 2003	Compressed hydrogen, methanol	DMFC, PEMFC	10 to 85 kW	Ballard/UTC	R'nessa, Effis

(*Continued*)

TABLE 4-3 Fuel Cell Vehicles Produced by Major Manufacturers (Compiled from *Fuel Cells 2000* [6]) *(Continued)*

Manufacturer	Year(s) demonstrated	Fuel type	Fuel cell used	Power	Fuel cell supplier	Name of vehicle(s)
Mazda	1997 to 2001	Metal hydride/ methanol	DMFC, PEMFC	20 kW	Ballard	Demio, Premacy FC-EV
Daimler Chrysler	1994 to 2003	Compressed hydrogen, liquid methanol, liquid hydrogen, chemical hydride	DMFC, PEMFC	50 to 85 kW	Ballard	NECAR 1-5.2, Jeep Commander 2, Jeep Teo
Volkswagen	1999 to 2002	Compressed hydrogen, methanol, liquid hydrogen	PEMFC with battery, PEMFC with capacitors, DMFC	15 to 75 kW	Ballard/Paul Scherer Institute	EU Capri, HyMotion, HyPower
Hyundai	2000 to 2004	Compressed hydrogen	PEMFC	75 to 80 kW	UTC	Santa Fe SUV, Tucson
Mitsubishi	2001 to 2003	Methanol / compressed hydrogen	PEMFC with Battery, DMFC	40 to 68 kW	Ballard	Spaceliner, Grandis FCV

Suzuki	2001 to 2003	Compressed hydrogen	PEMFC	Unknown	GM	Covie, Mobile Terrance
BMW	2000	Gasoline/liquid hydrogen	PEMFC	5 kW	UTC	Series 7 (745H)
Audi	2004	Gaseous hydrogen	PEMFC	66 kW	Ballard	A2
Kia	2004	Compressed hydrogen	PEMFC with battery	80 kW	UTC	Sportage
Renault	1997	Liquid hydrogen	PEMFC with battery	30 kW	Nuvera	EU FEVER
ESORO	2001	Compressed hydrogen	PEMFC with battery	6.4 kW	Nuvera	Hycar
Daihatsu	1999 to 2001	Methanol/compressed hydrogen	PEMFC with battery	16, 30 kW	Toyota	MOVE EV-FC, MOVE FCV-K2
Fiat	2001 to 2003	Compressed hydrogen	PEMFC with battery	7 kW	Nuvera	Seicento Electtra H2 Fuel Cell
PSA Peugeot Citron	2001 to 2002	Compressed hydrogen	PEMFC with battery	30 to 55 kW	Nuvera/H-Power	Peugeot HydroGen, Fuel Cell Cab

The main components of a fuel cell system are shown in Figure 4-1. The operating temperature of the fuel cell stack for an automobile ranges from 60 to 80°C. Operating temperatures above 100°C would improve the heat transfer and simplify stack cooling, but most automotive fuel cells use PEMFCs or DMFCs, which have a polymer membrane, which limits the operation to temperatures below 100°C.

Issues to address. Although considerable efforts have been spent on fuel cells during the last decade, major challenges still need to be addressed before fuel cell vehicles become a viable option. Some of these challenges include [1]:

- Current manufacturing methods for the mass production of fuel cells are cost prohibitive. New techniques, mass fabrication methods, and materials need to be created to lower the cost of producing fuel cells.
- If fuels other than hydrogen are used, CO poisoning of the catalyst may become an issue. The catalyst may need to be replaced or refreshed over time.
- The size and weight of the fuel tanks.
- A robust, efficient fuel cell system that can withstand long-term frequent use.

A hydrogen fuel cell does not generate any pollution. The only by-product is pure water. The amount of water produced by a fuel cell system is comparable to the amount produced by the internal combustion engine. If another fuel type is used with the fuel cell, the emissions will still be much lower than the emissions produced by the internal combustion engine.

An automobile fuel cell system can use a variety of fuels, and the selection of a fuel depends on factors such as the fuel supply infrastructure,

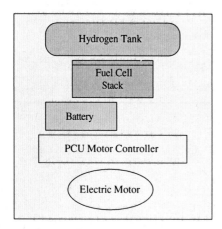

Figure 4-1 Illustration of the fuel cell system in a fuel cell vehicle.

the cost of fuel, the complexity and cost of storage, safety, environmental implications, and the national energy policy.

The biggest obstacle in the introduction of fuel cell vehicles is the lack of hydrogen infrastructure. Establishing a new fuel infrastructure would be extremely costly (but not any more costly than establishing a methanol or ethanol infrastructure) [16]. There are, however, already hundreds of hydrogen refueling stations in the United States, Japan, and Germany (see Table 2-2 in Chapter 2). As discussed in Chapter 2, hydrogen that is produced from natural gas can be cheaper than gasoline. Hydrogen produced from water and electricity via hydrolysis is more expensive than gasoline using conventional methods, unless low-cost off-peak electricity is used or solar panels are employed.

The proper storage of hydrogen is critical in commercializing fuel cell vehicles. Hydrogen can be stored in many forms, such as compressed gas, liquid, or as metal hydrides. Tanks that hold compressed gases are bulky. The average fuel efficiency of new cars is between 20 and 30 mpg; European and Japanese cars average even less. Current vehicles hold 10 to 16 gallons of gasoline, or 30 to 45 liters of space. Since hydrogen has twice the efficiency of gasoline vehicles, they would store between 5 to 8 kg of hydrogen, which is equivalent to between 200 and 400 L—a sizable reduction in the space needed for fuel. Liquid hydrogen tanks are also less bulky, but they must be stored at extremely low temperatures [3].

The problems associated with the storage of hydrogen have made certain automobile manufacturers resort to using other fuels for fuel cells, but most of the manufacturers still choose hydrogen as the best option. Onboard reforming of various fuels can be conducted, as well as directly feeding various fuel types into the fuel cell. Onboard reforming may eliminate the problem of onboard storage and the nonexistence of the hydrogen infrastructure. However, bringing fuel cell vehicles to market with onboard processors presents many problems [3]:

- The vehicles do not have zero emissions.
- Reformed hydrogen is not pure, and therefore decreases the fuel cell's efficiency.
- Onboard reforming increases the complexity, size, weight, and cost of the entire propulsion system.
- The long-term effects of fuel impurities. The next section displays some basic calculations that need to be considered when designing fuel cells for automotive systems.

Basic automotive fuel cell calculations. The power required for an automotive engine depends upon many factors, such as vehicle mass, frontal cross-sectional area, drag coefficient, rolling resistance coefficient, and the efficiency of the drive train. The power required also changes with the vehicle speed, acceleration, and the road slope. The equation for the efficiency of an automobile engine as specific fuel consumption ($f_{sp\_fuel\_cons}$) (gkWh^{-1}) is:

$$f_{sp\_fuel\_cons} = \frac{3.6 \times 10^6}{\eta_{vehicle\_sys} LHV_{fuel}} \tag{4-5}$$

where LHV_{fuel} is the lower heating value of fuel (kJ/kg), and $\eta_{vehicle\_sys}$ is the vehicle efficiency, which is a product of the fuel cell system efficiency, traction efficiency, and electric drive efficiency.

The specific fuel consumption for a gasoline internal combustion engine is about 240 gkWh^{-1}. One gram of hydrogen has the same amount of energy as 2.73 g of gasoline based upon the lower heating value. Although hydrogen seems superior to gasoline based upon the heating values, the efficiency of fuel cells versus internal combustion engines should not be compared because the two technologies are very different and have different power characteristics [3].

When designing a fuel cell for an automobile (or any other vehicle type), there are many forces that affect a vehicle at an incline angle, and the fuel system needs to provide enough power to overcome these forces to move the vehicle on an incline. These forces include: air resistance, the rolling resistance of the wheels, gravity, and the normal force of the ground acting upon the vehicle. This concept can be demonstrated by a free body diagram of a vehicle, as shown in Figure 4-2.

These forces must have a sum of zero if the vehicle is to maintain a constant velocity. If the vehicle needs to accelerate, the net forward acceleration times the mass can provide a quick estimate. The various power

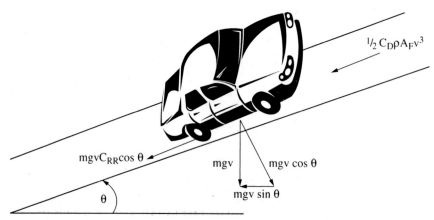

Figure 4-2 Free-body diagram of an automobile on an incline.

demands can be summarized to a total mechanical power ($P_{req\_vehicle}$) demanded by the motion of the vehicle:

$$P_{req\_vehicle} = m_{veh} a_{veh} v_{veh} + m_{veh} g v_{veh} \sin\theta + m_{veh} g v_{veh} C_{RR} \cos\theta$$
$$+ \frac{1}{2} \rho_{air} C_D A_F v_{veh}^3 \qquad (4\text{-}6)$$

where m_{veh} is the total mass of the vehicle, θ is the angle of the slope, a_{veh} is the acceleration of the vehicle, v_{veh} is the velocity of the vehicle, C_{RR} is the coefficient of the tire rolling resistance, ρ_{air} is the density of the air, C_D is the drag coefficient, and A_F is the frontal area. The drag coefficient (C_D) is a dimensionless constant that summarizes an object's resistance to flow. C_D varies from 0.20 for automobiles to 1.2 for bicycles.

The vehicle fuel economy (B_e) (in gm^{-1}) is:

$$B_e = \frac{\int f_{sp\_fuel\_cons} v_{speed} \times P_{req\_vehicle}}{3.6 \times 10^6 \int v_{speed} dt} \qquad (4\text{-}7)$$

$f_{sp\_fuel\_cons}$ is the specific fuel consumption in $gkWh^{-1}$ from equation 4–5, and v_{speed} is the vehicle speed (ms^{-1}).

The amount of power output can be calculated by applying the inefficiencies in the system:

$$P_{vehicle\_output} = \frac{P_{req\_vehicle}}{\eta_{drivetrain}} + P_{auxiliary} + P_{parasitics} \qquad (4\text{-}8)$$

where $P_{auxiliary}$ is the power needed by auxiliary systems, such as headlights, dashboard and other internal systems, $\eta_{drivetrain}$ is the efficiency of the electric motor and controller subsystem, and $P_{parasitics}$ is the parasitic power needed by the fuel cell system, such as blowers, fans etc.

An automotive fuel cell system replacing an internal combustion engine must be of similar size and weight. The size of the stack depends on the nominal cell voltage and the stack voltage efficiency. The stack specific volume (m^3 per KW) is [3]:

$$V_s = 0.1 \frac{n_{cells} d_{cell} + 2 d_{ep}}{\alpha_{act} n_{cells} V_{cell} i} \qquad (4\text{-}9)$$

where n_{cells} is the number of cells in a stack, d_{cell} is the individual cell thickness, including the cooling arrangements (m), d_{ep} is the thickness of the end plates, α_{act} is the ratio of cell active area and bipolar plate area,

V_{cell} is the cell potential at nominal power (V), and i is the current density at nominal power (A cm^{-2}).

The cell potential, V_{cell}, and the current density, i, are connected by the cell polarization curve. Chapter 13 goes into the details of the fuel cell stack design equations for automobiles and all other vehicle types.

4.3.2 Buses

Buses are considered the best candidate for the early introduction of fuel cells into the commercial vehicle market. The difference between buses and automobiles are the power requirements, space availability, operating regimen, and refueling sites. Buses obviously require more power than automobiles and get more wear due to constant stops and starts. Despite this, the average fuel economy of a bus is about 15 percent better than a diesel engine [10]. Buses can be refueled in a central facility, which makes refueling with hydrogen much easier. Large quantities of hydrogen can also be stored onboard easily because of the large area of a bus. Hydrogen is usually stored in a composite compressed gas cylinder located on the roof. This is a safe place to store the tank since hydrogen is lighter than air, and it is not near critical engine components.

These buses have a major advantage over transitional diesel buses because they have zero emissions. This is critical in heavily populated and polluted cities. The Clean Urban Transport for Europe (CUTE) program's fuel cell buses are currently running in Amsterdam, Barcelona, Hamburg, London, Luxembourg, Madrid, Porto, Reykjavik, Stockholm, and Stuttgart. The Sunline transit authority in California has been running fuel cell buses for several years [11].

Table 4-4 shows a summary of fuel cell buses demonstrated by major bus manufacturers. As shown in the table, many bus manufacturers began demonstrating their first fuel cell buses in the early 1990s. Like the fuel cell automobiles, the fuel type most often used is compressed hydrogen, although methanol and zinc were also demonstrated. The most common type of fuel cell used is the PEMFC, but DMFCs, PAFCs, and ZAFCs were also used. All of the fuel cell buses demonstrated used fuel cells developed from a fuel cell company (they were not developed in-house).

The main obstacles for commercialization of fuel cell buses are fuel cell cost and lifetime. Fewer fuel cell buses are being manufactured; therefore, the cost is higher per bus than the traditional combustion engine buses.

4.3.3 Utility vehicles

Many utility vehicles may be able to adapt fuel cell technology earlier than automobiles because the competing technology for these vehicles

TABLE 4-4 Fuel Cell Buses Demonstrated to Date (Compiled from *Fuel Cells 2000* [7])

Manufacturer	Year(s) demonstrated	Fuel type	Fuel cell used	Power	Fuel cell supplier	Model
Bus Manufacturing U.S.A., Inc.	1994 to 1995	Methanol	PAFC	50 kW	Fuji Electric	30-foot transit bus
NovaBus Corporation (a subsidiary of Volvo)	1998 to 2001	Methanol compressed hydrogen, zinc	PEMFC, DMFC, ZAFC	100 kW	UTC/Ballard/Arotech	40-foot heavy-duty transit buses, 15.3 meterlong double-decker
New Flyer Industries Ltd.	1993 to 2006	Compressed hydrogen	PEMFC	90 to 205 kW	Ballard/Hydrogenics	Full-sized, 40-foot, H40LF models
Enova Systems	2004	Compressed hydrogen	PEMFC	20 kW	Hydrogenics	N/A
EvoBus: a Daimler Chrysler company	1997 to 2003	Compressed hydrogen	PEMFC	205 kW	Ballard	Nebus, Zebus, Mercedes Benz Citaro
Gillig Corporation	2004	Compressed hydrogen	PEMFC	205 kW	Ballard	N/A
Irisbus: a Renault V.I.and Iveco Co.	2001	Compressed hydrogen	PEMFC	60 kW	UTC fuel cells	40 foot

(Continued)

TABLE 4-4 Fuel Cell Buses Demonstrated to Date (Compiled from *Fuel Cells 2000* [7]) (*Continued*)

Manufacturer	Year(s) demonstrated	Fuel type	Fuel cell used	Power	Fuel cell supplier	Model
MAN	2000 to 2006	Compressed hydrogen, liquid hydrogen	PEMFC	30 to 68 kW	Ballard/Nuvera	Bavria I", 40-foot MAN low floor
Neoplan	1999 to 2000	Compressed hydrogen	PEMFC	40 to 80 kW	Nuvera/ Proton Motor Fuel Cell GmbH	N8012 – 33-seat bus
Thor Industries (ThuderPower LLC)	2001	Compressed hydrogen	PEMFC	75 kW	UTC	30 ft. Low Floor El Dorado National E-Z Rider
Van Hool	1995 to 2006	Compressed hydrogen, liquid hydrogen	PEMFC, PAFC	78 to 120 kW	UTC/Elenco	18 meter City Bus, 40 foot
Macchi-Ansaldo (EC project EQHHPP)	1997	Liquid hydrogen	PEMFC	45 kW	Nuvera	Full size regular floor city bus
Hino Motors Ltd. (Toyota subsidiary)	2001 to 2005	Compressed hydrogen	PEMFC	160 to 180 kW	Toyota, Nuvera	FCHVBUS1, FCHVBUS2

is usually lead-acid batteries that often require charging and have maintenance issues. Demonstrations of fuel cell utility vehicles show that they offer lower operating cost, reduced maintenance, lower downtime, and extended range. Fuel cell–powered utility vehicles can also be operated indoors because there are no emissions. Examples of utility vehicles that can be powered by fuel cells are

- Golf carts
- Lawn maintenance vehicles
- Forklifts
- Airport movers
- Wheelchairs
- Unmanned vehicles
- Boats
- Small planes
- Submarines
- Small military vehicles

Table 4-5 shows a summary of demonstrated fuel cell utility vehicles. As noted in the table, many manufacturers began demonstrating their first fuel cell utility vehicles in the early 2000s, which is later than many of the fuel cell automobile and bus demonstrations. Like the fuel cell automobiles, the fuel type most often used is compressed hydrogen, although methanol, metal hydrides, and sodium borohydride were also demonstrated. The most common type of fuel cell used is the PEMFC, but DMFCs and AFCs were also used.

4.3.4 Scooters and bicycles

In countries with very large populations, scooters and bicycles are popular forms of transportation. Fuel cells may provide a solution in these countries, and have already been positively demonstrated. Power requirements are much less, and prototypes have been demonstrated with compressed hydrogen and methanol. Hydrogen storage is still an issue for these vehicles; therefore, metal hydrides and electrolyzers are also being considered.

Table 4-6 shows a summary of demonstrated fuel cell scooters and bicycles. As presented in the table, many manufacturers began demonstrating their first fuel cell scooters and bicycles in the early 2000s, which is later than many of the fuel cell automobile and bus demonstrations. Like the fuel cell automobiles, the fuel type most often used is compressed hydrogen, although methanol, metal hydrides, and zinc

TABLE 4-5 Utility Vehicles Demonstrated to Date (Compiled from *Fuel Cells 2000* [8])

Manufacturer	Year(s) demonstrated	Fuel type	Fuel cell used	Power	Model
Aerovironment/ NASA Dryden Flight Research Center	2003	Compressed hydrogen	PEMFC	10 to 25 kW	Helios–unmanned aerial vehicle
Anuvu, Inc./ Millennium Cell, Duffy Electric Boats, Seaworthy Systems	2003	Hydrogen from sodium borohydride	PEMFC	41.5 kW	Fuel cell boat
Astris Energi	2001	Compressed hydrogen	AFC	1 kW	Freedom golf cart
Besel S.A./ MEYRA, ONCE	2003	Metal hydride onboard hydrogen storage	PEMFC	N/A	Wheelchair
Cellex Power Products/BOC/ Wal-Mart	2004	Compressed hydrogen	PEMFC	N/A	Forklift
Deere & Company/ Hydrogenics,/ Dynatek	2003 to 2005	Compressed hydrogen	PEMFC	20 kW	Commercial Work Vehicle (CWV), ProGator^{TM}
EIVD/MW-Line/Paul Scherrer Institute (PSI)	1999 to 2003	Compressed hydrogen	PEMFC	300 W to 3 kW	Hydroxy 3000 – Fuel Cell Boat, Hydroxy 300
FASTec/ ATP/UQM Technologies, NASA, American Ghiles Aircraft, Giner Electrochemical Systems, Satcon Technology Corp., Diamond Aircraft, Analytic Energy Systems, Lockwood Aviation, Lynntech	2001 to 2005	Compressed hydrogen	PEMFC	2 to 75 kW	Fuel cell–powered electric plane
H2 Logic ApS/ A. Flensborg - Herning A/S	2005	Low pressure hydride storage	PEMFC	N/A	H2 Truck

(*Continued*)

TABLE 4-5 Utility Vehicles Demonstrated to Date (Compiled from *Fuel Cells 2000* [8]) (*Continued*)

Manufacturer	Year(s) demonstrated	Fuel type	Fuel cell used	Power	Model
HaveBlue LLC, Texaco Ovonic Hydrogen Systems, Hydrogenics, Catalina Yachts	2004	Metal hydride onboard hydrogen storage	PEMFC	10 kW	X/V-1 Sailboat
Howaldtswerke-Deutsche WerftAG (HDW)	2002 to 2006	N/A	N/A	300 kW	U 33 Class 212 A submarine
Hydrogenics	2004	Compressed hydrogen	PEMFC	10 to 65 kW	Airport tow tractor, forklift
Kurimoto/ APFCT	2003	Compressed hydrogen	PEMFC	250 W	Fuel cell wheelchair
Los Alamos National Laboratory (LANL)/ Ergenics	2003	Metal hydride onboard hydrogen storage	PEMFC	140 W	Personal mobility vehicle
Manhattan Scientifics/ U.S. Army ERDC/ CERL,FC*Tec*, Concurrent Technologies Corporation	2003	Compressed hydrogen	PEMFC	700 W	Fuel cell Segway
MTU Friedrichshafen, Ballard Power Systems	2003	Compressed hydrogen	PEMFC	4.8 kW	"No. 1" Yacht
Proton Motor Fuel Cell, Linde and STILL	2004	Compressed hydrogen	PEMFC	N/A	Fuel cell lift truck
Quantum Fuel Systems Technologies, U.S. Army TARDEC	2004	Compressed hydrogen	PEMFC	10 kW	Quantum AMV "Aggressor" military vehicle
Research Centre Jülich, Ministry for Science and Research of the Federal State of North Rhine-Westphalia	2004	Methanol	DMFC	1.3 kW	JuMOVe fuel cell electric vehicle

(*Continued*)

TABLE 4-5 Utility Vehicles Demonstrated to Date (Compiled from *Fuel Cells 2000* [8]) (*Continued*)

Manufacturer	Year(s) demonstrated	Fuel type	Fuel cell used	Power	Model
Siemens/KWU	1999	Compressed hydrogen	PEMFC	10 kW	Forklift
Toyota Industries Corp.	2005	Compressed hydrogen	PEMFC	30 kW	Forklift
Vehicle Projects, LLC, DoT Volpe Center, DoD, TTCI, Nuvera, NAC, Aero-Vironment, WSRC, HERA Hydrogen Systems	2002 to 2004	Metal hydride	PEMFC	17 kW to 1.2 MW	Army locomotive, mine locomotive, mine loader

were also demonstrated. The most common type of fuel cell used is the PEMFC, but DMFCs and ZAFCs were also used.

4.4 Stationary Power Applications

Fuel cells for stationary applications present greater commercial potential than fuel cells for automobiles. The main differences in these fuel cell systems are the choice of fuel cell and fuel, and the heating and cooling of the stacks. A variety of stationary power systems are being developed. Stationary fuel cells may be used for many applications:

- As the primary power source, competing with the grid, or being used in places where the grid cannot reach
- To provide supplemental power
- In hybrid power systems with photovoltaics, batteries, capacitors, or wind turbines, providing primary or secondary power
- As a backup or energy power generator providing power when the grid is down

A stand-alone system may require another power source for peak periods. These can be batteries and supercapacitors, or a combination of both. The fuel cell is designed and built to provide a certain power output.

Table 4-7 shows a summary of demonstrated fuel cell stationary power systems. Many manufacturers began demonstrating their stationary power stations in the 1990s. Unlike other fuel cell applications, the fuel type most often used is natural gas. Other common fuel types employed

TABLE 4-6 Scooters and Bicycles Demonstrated to Date (Compiled from *Fuel Cells 2000* [8])

Manufacturer	Year(s) demonstrated	Fuel type	Fuel cell used	Power	Model
Asia Pacific Fuel Cell Technologies (APFCT)	2002 to 2003	Metal hydride onboard hydrogen storage	PEMFC	35 to 58 kW	ZES IV, ZES III
Besel S.A./ Derbi	2003 to 2004	Metal hydride onboard hydrogen storage	PEMFC	N/A	Scooter
Bombardier Recreational Products	N/A	N/A	N/A	N/A	Urban transport vehicle – Embrio
ECN/Piaggio & C SpA, Selin Sistemi SpA and Commissiariat a l' Energie Atomique	2004	Compressed hydrogen	PEMFC	12 kW	FRESCO scooter
FAAM/Beijing Fuyuan	2003	Metal hydride onboard hydrogen storage	PEMFC	400 W	Camaleo hydrogen bicycle
Honda	2004	Compressed hydrogen	PEMFC	N/A	Fuel cell scooter
Hydrogenics	TBD	Compressed hydrogen	PEMFC	5 kW	GEM hybrid mini-car
Intelligent Energy/ Seymourpowell	2005	Compressed hydrogen	PEMFC	1 kW	ENV – fuel cell motorbike
Manhattan Scientifics/ Aprilia s.P.a.	2000 to 2002	Compressed hydrogen	PEMFC	600 W to 3 kW	Mojito FC scooter, Enjoy FC bicycle, Hydrocycle
Masterflex AG/ Veloform	2004	Metal hydride stored hydrogen	PEMFC	250 W	Fuel cell bike
MES-DEA, Aprilia	2004	Compressed hydrogen	PEMFC	2 to 3 kW	Aprilia Atlantic Zero Emission fuel cell scooter

(*Continued*)

TABLE 4-6 Scooters and Bicycles Demonstrated to Date (Compiled from *Fuel Cells 2000* [8]) (*Continued*)

Manufacturer	Year(s) demonstrated	Fuel type	Fuel cell used	Power	Model
Palcan Fuel Cells Ltd., MOU with Celco Profil S.R.L.	2002 to 2003	Metal hydride onboard hydrogen storage	PEMFC	500 W to 2 kW	Scooter, E-bike
PEM Technologies, Inc.	2003	Compressed hydrogen	PEMFC	1 kW	PemPower-04 3-wheel Motorcycle, PemPower-03 2-wheel motorcycle
Peugeot, MES-DEA	2004	Compressed hydrogen	PEMFC	1.5 kW	Quark concept vehicle
Powerzinc Electric	2002 to 2003	Zinc	ZAFC	N/A	Motorcycle, fuel cell/ electric bicycle, scooter
University of Tasmania, Technical University of Nurenburg, Germany	2004	Metal hydride onboard hydrogen storage	PEMFC	N/A	Fuel cell scooter
Vectrix, Parker Hannifin, Protonex, and NGen	2003 to 2004	Methanol	DMFC	500 to 800 W	Scooter, fuel cell/ electric hybrid scooter
Yahama Motor Company, Yuasa Corporation	2003	Methanol	DMFC	500 W	Fuel cell motorcycle

with power stations are propane, compressed hydrogen, biogas, methanol, oil-based fuels, town gas, synthesis gas, digester gas, and land fill gas. The most common type of fuel cell used is the PEMFC, but SOFCs, MCFCs, AFCs, and PAFCs are also utilized. The countries with the highest number of stationary fuel cell power stations are the U.S., Germany, and Japan, which is consistent with the amount of fuel cell funding for these countries. The power range for stationary fuel cells in Table 4-7 are from 500 W to 5 MW.

Basic stationary fuel cell calculations. In order to evaluate and compare stationary fuel cell systems, the following efficiencies can be calculated

TABLE 4-7 Stationary Power Systems Demonstrated to Date (Compiled from *Fuel Cells 2000* [9])

Manufacturer	No. of stations	Year(s) started	Fuel type	Fuel cell used	Power	Location
Accumetrics Corporation	8	2002 to 2005	Natural gas/propane	SOFC	2 to 10 kW	U.S.: 7; Japan: 1
American Fuel Cell Corp.	16	N/A	Natural gas	N/A	3 kW	Various locations
Ansaldo Fuel Cells Spa	6	1998 to 2006	Biogas/Diesel/Natural Gas	MCfC	100 to 500 kW	Italy: 3; Spain: 2; Turkey: 1
Apollo Energy Systems Inc.	1	2002	N/A	AFC	10 kW	U.S.: 1
Astris Energi Inc.	1	N/A	N/A	AFC	N/A	Italy: 1
Ballard	25	1999 to 2005	Hydrogen, oil, natural gas, methane, town gas	PEMFC	1 to 250 kW	U.S.: 2; Canada: 3; Japan: 14; Germany: 4; Switzerland: 1; Belgium: 1
Bharat Heavy Electricals Ltd. (BHEL)	1	2000	Hydrogen	PAFC	50 kW	India: 1
Brennstoffzellen-technik GmbH (ZBT)	1	2005	Natural gas	PEM	4 kW	Germany: 1
Ceramic Fuel Cells Ltd. (CFCL)	6	2005 to 2006	Natural gas	SOFC	1 kW	Australia: 2; Germany: 2; New Zealand: 2
Dais Analytic	5	1999 to 2000	Natural gas	PEMFC	3 kW	Germany: 5
DCH Technologies	2	2002	Natural gas or hydrogen	PEMFC	1.5 to 5 kW	U.S.: 2
European Fuel Cell GmbH	2	2005	Natural gas	PEMFC	1.5 kW	Germany: 1; U.K.:
Fuel Cell Technologies (FCT)	19	2002 to 2005	Biogas, natural gas, methanol	SOFC	5 kW	U.S.: 9; Canada: 5; Sweden: 1; Belgium: 1; Brazil: 1; Japan: 1; Germany: 1

(Continued)

TABLE 4-7 Stationary Power Systems Demonstrated to Date (Compiled from *Fuel Cells 2000* [9]) *(Continued)*

Manufacturer	No. of stations	Year(s) started	Fuel type	Fuel cell used	Power	Location
FuelCell Energy	65	1996 to 2006	Natural gas, methanol, synthesis gas, methane, digester gas	MCFC	3 kW to 2 MW	U.S.: 39; South Korea: 2; Japan: 9; Germany: 15
Fuji Electric	35	1991 to 2004	Town gas, methane, digester gas	PEMFC, PAFC	1 kW to 5 MW	Japan: 30; Sweden: 2; Italy: 1; Spain: 1; Netherlands: 1
GenCell	1	2005	Natural gas	MCFC	40 kW	U.S.: 1
General Electric	1	2002	Natural gas	PEMFC	5 kW	U.S.: 1
Global Thermoelectric	4	2001 to 2003	Natural gas, methane, propane	SOFC	2 to 5 kW	U.S.: 3; Canada: 1
GM	3	2001 to 2004	Hydrogen, gas, methane, natural	PEMFC	78 kW	U.S.: 3
H-Power Corp.	25	2000 to 2003	Natural gas, hydrogen, propane	PEMFC	500 W to 4.5 kW	U.S.: 13; Canada: 1; France: 7; Japan: 2; Sweden: 1; Finland: 1
Hydrogenics Corp.	9	2002 to 2005	Hydrogen	PEMFC	5 to 25 kW	U.S.: 5; Japan: 2; Canada: 2
IdaTech	21	1998 to 2005	Natural gas, propane, methanol	PEMFC	1.2 to 6 kW	U.S.:16; France: 2; Germany: 2; Europe: 1
Industrial Research laboratory (IRL)	1	2002	Hydrogen	AFC	6 kW	Australia: 1

Company	#	Year	Fuel	Type	Power	Location
Intelligent Energy	1	2003	Hydrogen	PEMFC	2 kW	U.K.: 1
Ishikawajima-Harima Heavy Industries (IHI)	11	1996 to 2005	Waste gas, LP gas natural gas	PEMFC, MCFC 1 MW	1 kW to	Japan: 11
Ishikawajima-Shibaura Machinery Co. (ISm)	1	2005	Natural gas	PEMFC	1 kW	Japan: 1
M-C Power Corporation	3	1995 to 2002	Natural gas	MCFC, SOFC	75 to 250 kW	U.S.: 3
Minaton (Russian Ministry of Atomic Energy)	1	2003	Natural Gas	SOFC	1 kW	Russia: 1
Matsushita Electrical Industrial Co.	4	2002 to 2004	Natural gas	PEMFC	1 to 1.3 kW	Japan: 4
Mitsubishi Electric Corp.	2	2002	Natural gas	PEMFC	1 kW	Japan: 2
Mitsubishi Heavy Industries (MHI)	10	1993 to 2006	LP gas, natural gas, city gas	PEMFC, SOFC	10 to 200 kW	Japan: 10
Mosaic Energy	2	2000 to 2001	Natural gas, naptha	PEMFC	3 to 5 kW	Japan: 1; U.S.: 1
MTU CFC Solutions	5	2002 to 2004	Biogas, landfill gas	MCFC	300 kW	Austria: 1; Slovak Republic: 1; Spain: 2; Germany: 1
Nuvera	10	2001 to 2006	Biogas, natural gas, ethanol	PEMFC	3.3 to 15 kW	U.S.: 7; Japan: 1; Germany: 2
Plug Power, Inc.	150	1998 to 2005	Hydrogen, propane, natural gas, petroleum gas	PEMFC	500 W to 7 kW	U.S.: 84; Japan: 10; Germany: 23; France: 6; Austria: 5; Luxemberg: 2; Puerto Rico: 1; Iceland: 1; U.K.: 4; Netherlands: 10

(Continued)

TABLE 4-7 Stationary Power Systems Demonstrated to Date (Compiled from *Fuel Cells 2000* [9]) (*Continued*)

Manufacturer	No. of stations	Year(s) started	Fuel type	Fuel cell used	Power	Location
Proton Energy Systems	1	2004	Hydrogen	PEMFC	1 kW	U.S.: 1
ReliOn	27	2002 to 2005	Hydrogen, methanol, natural gas, propane	PEMFC	200 W to 4 kW	U.S.: 24; New Zealand: 1; Bahamas: 1; Italy: 1
Sanyo Electric Company	6	2001 to 2005	Town gas	PEMFC, SOFC	750 W to 100 kW	Japan: 6
Schatz Energy Research Center	1	1999	Hydrogen	PEMFC	100 W	U.S.: 1
Siemens Power Generation, Inc	22	1986 to 2005	H2 + CO, natural gas, jet fuel, diesel fuel	SOFC	3 kW to 1 MW	U.S.: 8; Italy: 2; Japan: 4; Germany: 3; Canada: 3; Norway: 1; Austria: 1
Smart Fuel Cell	1	2005	Methanol	DMFC	120 W	Germany: 1
Sulzer Hexis	28	1997 to 2003	Hydrogen, natural gas, city gas, biogas	SOFC	1 kW	Germany: 15; France: 1; Switzerland: 6; Austria: 1; Netherlands: 2; Japan: 1; Spain: 1

Company	Count	Years	Fuel	Type	Power	Countries
Teledyne Energy Systems Inc.	3	2002 to 2005	Hydrogen, natural gas	PEMFC	5 to 12 kW	U.S.: 3
Tokyo Gas	1	1998	City gas	SOFC	N/A	Japan: 1
Toshiba	10	2000 to 2005	Petroleum gas, biogas, town gas	PEMFC	1 to 700 kW	Japan: 10
Toyota	1	2001	N/A	PEMFC	1 kW	Japan: 1
UTC Power	175	1983 to 2006	Natural gas, dimethyl esther, digester gas, petroleum gas, hydrogen, methane, landfill gas	PAFC	40 kW to 4.5 MW	U.S.: 116; Japan: 15; Germany: 20; Canada: 2; France: 2; Austria: 1; U.K.: 1; Russia: 1; Brazil: 4; China: 1; Italy: 3
Zentrum für Sonnenenergie- und Wasserstoff- Forschung (ZSW) (Center for Solar Energy and Hydrogen Research)	1	2003	N/A	PEMFC	500 W	France: 1
ZTEK Corp.	4	1994 to 2005	Natural gas, hydrogen	SOFC	1 to 25 kW	U.S.: 2; U.K.:1; Japan: 1

for all of the components and aspects of the fuel cell system. The total efficiency of the fuel cell system is defined as:

Total efficiency = (electric power output + thermal output) / fuel consumption

or
$$\eta_{total} = \frac{P_{net} + Q_{net}}{HHV_{fuel} * n_{fuel}} \qquad (4\text{-}10)$$

where P_{net}, and Q_{net} is the usable power and heat amounts respectively, and n_{fuel} is the amount of fuel input into the fuel cell system. The total efficiency of the fuel cell system can also be calculated by multiplying the efficiencies of the individual components, and the ratio of parasitic power and fuel cell gross power output (ξ_p). An example of this equation is as follows:

$$\eta_{total} = \sum \eta - \xi_p \qquad (4\text{-}11)$$

The electrical efficiency of the stationary fuel system is:

$$\eta_{electrical} = \frac{P_{net}}{HHV_{fuel} * n_{fuel}} \qquad (4\text{-}12)$$

where
$$P_{net} = P_{AC} - P_{aux\_equipment}$$

and
$$P_{aux\_equipment} = P_{compressor} + P_{pump} + P_{control} \qquad (4\text{-}13)$$

where P_{AC} is the usable AC power generated, $P_{aux\_equipment}$ is the power required by auxiliary equipment, $P_{compressor}$ is the power required by compressor, P_{pump} is the power required by the pump, and $P_{control}$ is the power required by the control system.

The thermal efficiency of the stationary fuel cell system is:

$$\eta_{thermal} = \frac{Q_{net}}{HHV_{fuel} * n_{fuel}} \qquad (4\text{-}14)$$

If the stationary fuel cell system uses a fuel processor, the efficiency of the fuel processor is:

$$\eta_{fuel\_processor} = \frac{HHV_{H_2} * n_{H_2}}{HHV_{fuel} * n_{fuel}} \qquad (4\text{-}15)$$

The DC/AC efficiency of the stationary fuel cell system is:

$$\eta_{DC/AC} = \frac{P_{AC}}{P_{stack}} \qquad (4\text{-}16)$$

For any auxiliary equipment used, the efficiency is:

$$\eta_{aux\_equipment} = \frac{P_{stack} - P_{aux\_equipment}}{P_{stack}} \quad (4\text{-}17)$$

The efficiency of the fuel cell stack is:

$$\eta_{stack} = \frac{P_{net}}{HHV_{H_2} * n_{H_2}} \quad (4\text{-}18)$$

Economics of stationary fuel cell systems. In order for stationary fuel cell systems to be cost-effective, they must compete with the cost of electricity from utilities. The purchase price of the fuel cell system must save money for the owner over the lifetime of the system. A simple calculation that can be used to determine the cost-effectiveness of the fuel cell system is the payback time. This is simply the ratio between the purchase price and the annual savings in electricity. This calculation does not take into account the interest lost or spent during that period, inflation, and changes in electric and gas prices. The equation for simple payback time is

SPT = purchase price of fuel cell system / annual savings on electricity

or

$$SPT = \frac{P_{fc,norm} * C_{fc}}{AEP * COE - \frac{C_{ng}}{\eta_{fc}}} \quad (4\text{-}19)$$

where $P_{fc,norm}$ is the fuel cell power system nominal power (kW), C_{fc} is the specific cost of fuel cell power system per kW of nominal power ($ kW^{-1}), AEP is the annual electricity produced by the fuel cell power system per kW of nominal power (kWhyr^{-1}), η_{fc} is the average annual efficiency of the fuel cell system, COE is the cost of electricity ($ kWh^{-1}), and C_{ng} is the cost of natural gas ($ kWh^{-1}) using the lower heating value of natural gas.

The annual electricity produced (AEP) and the average annual efficiency are functions of the load profile. Load profiles vary depending upon application, day of the week, and the season. A fuel cell power system can be sized to generate any type of power with a minimum and maximum range. The amount of electricity produced (kW) annually (AEP) by a fuel cell system is [15]:

$$AEP = P_{fc\_time} * h_{year\_operating} \quad (4\text{-}20)$$

where P_{fc_time} is the fuel cell power at a given time and $h_{year\text{-}operating}$ is the yearly operating hours of fuel cell (in hours). This equation can only be used when the load profile is known. A capacity factor, CF, can make estimation easier because it is a ratio between the electricity actually produced in a time period and the electricity that could have been produced at nominal power for the entire period. The amount of electricity produced annually (in kW) is

$$AEP = h_{year\_operating} * CF * P_{fc,norm} \qquad (4\text{-}21)$$

where CF is the capacity factor.

A load itself has a certain capacity factor. A fuel cell that is sized to provide all of the power required by the load would therefore operate with a 25-percent capacity factor. The average annual efficiency ($\overline{\eta}_{fc}$) of the fuel cell system (in percent) is:

$$\overline{\eta}_{fc} = \frac{P_{fc} * h_{year\_operating}}{\left(\dfrac{P_{fc} * h_{year\_operating}}{\eta_{fc}}\right)} \qquad (4\text{-}22)$$

where P_{fc} and η_{fc} are the fuel cell system power output and system efficiency.

This equation can only be used if the annual load profile is known. The annual fuel costs (AFC) and the cost of the fuel cell (C_{FC}) can be calculated using the following equations:

$$AFC = \frac{(C_f * AEP)}{\overline{\eta}_{fc}} \qquad (4\text{-}23)$$

$$C_{FC} = C_{fix} + (C_{cell} * N_{cell})$$

where

$$N_{cell} = \frac{P_{FC}}{V_{cell} i A_{cell}} \qquad (4\text{-}24)$$

where C_f is the fuel cost (U.S. dollars per kilowatt hour), C_{fix} is the costs associated with the fuel cell installation excluding the costs of individual cells, but including the balance of plant and installation costs (in U.S. dollars), C_{cell} is the cost of each individual cell (U.S. dollars per year), N_{cell} is the number of individual cells, P_{FC} is the power output of the fuel cell in kilowatts, V_{cell} is the cell nominal voltage, i is the current density (A/cm^2) and A_{cell} is the cell active area (cm^2). The capital recovery factor (CRF) (yr^{-1}) is:

$$CRF = \frac{i_r(1 + i_r)^L}{(1 + i_r)^L - 1} \qquad (4\text{-}25)$$

where i_r is the annual interest rate (in percent) and L is the lifetime of the individual cells (in years).

The capital recovery factor is used to calculate the capital recovery cost (CRC):

$$CRC = C_{rep}CRF + (C_{FC} - C_{rep})i_r \qquad (4\text{-}26)$$

where C_{rep} is the replacement cost of the fuel cell at end of lifetime ($C_{cell} * N_{cell}$).

The cost of electricity produced by a fuel cell can be calculated using the following equation:

$$COE = \frac{CRC + AFC + AMC - S_{env}}{AEP} \qquad (4\text{-}27)$$

This calculation of the COE includes the annual maintenance cost (AMC) and a term that includes the environmental cost associated with using fossil fuels (S_{env}) (both in U.S. dollars per year). S_{env} is only used with fuel cell types that use hydrogen from non-fossil fuel based sources as the fuel, and can be defined as follows [15]:

$$S_{env} = AEP * C_p(\delta - \varepsilon) \qquad (4\text{-}28)$$

where C_p is the environmental damage cost (in U.S. dollars per kilowatt-hour), δ is the ratio of hydrogen utilization efficiency to that of fossil fuels, and ε is the ratio of environmental impact of hydrogen utilization to that of fossil fuels. The values used for these factors by Nelson et al [15] are $C_p = 0.0216$ dollars per kWh, $\delta = 1.358$, and $\varepsilon = 0.04$.

The economics of a fuel cell may be improved by either exporting electricity back to the grid, or utilizing the heat produced by the fuel cell. If electricity is exported back to the grid, the simple payback time would be

$$SPT = \frac{P_{fc,norm} * C_{fc}}{AEP_{int} * \left(COE - \dfrac{C_{ng}}{\eta_{fc}}\right) + AEP_{exp}\left(COE_{exp} - \dfrac{C_{ng}}{\eta_{fc}}\right)} \qquad (4\text{-}29)$$

where AEP_{int} is the amount of electricity consumed internally (kWhyr^{-1}), AEP_{exp} is the amount of electricity exported back to the grid (kWhyr^{-1}), and COE_{exp} is the price of electricity exported back to the grid ($ kWh^{-1}).

Chapter Summary

Certain fuel cell types are better suited for small portable technologies, backup power, automobiles, or stationary power applications. Fuel cells for each of these applications have been demonstrated, and there are some fuel cells commercially available in these cataegories. DMFCs and PEMFCs are most commonly used for portable applications (~1 to 1 kW). For very small applications, such as cellular phones, the DMFC is preferred because of the favorable storage characteristics of a liquid fuel compared with hydrogen gas. Backup power applications usually use PEM fuel cells for the power range of 1 to 20 kW. This mid-sized fuel cell system works well with compressed hydrogen storage. Automobile applications have the potential to use several fuel cell types, but the ones most commonly used are the PEMFCs and DMFCs, which range from 5 to 100 kW. Buses that are powered by fuel cells use several fuel cell types, such as PAFCs, PEMFCs, DMFCs, and ZAFCs. The power range is typically from 20 to 205 kW. Other vehicle types such as utility vehicles, scooters, and bicycles primarily use PEMFCs with a power range of 140 W to 1.2 MW. Other fuel types are also sometimes used, such as ZAFC, AFC, and DMFCs. Stationary power applications use SOFCs, MCFCs, AFCs, and PEMFCs in the power range of 1 kW to 5 MW. The fuel cell design parameters such as power output, heat balance, efficiency, size, weight, and fuel supply may be slightly different for each application, and must be customized to suit the required load.

Problems

1. Design a fuel cell electrolyzer for a backup power fuel cell system that has a fuel cell nominal power of 1 kW.

2. Calculate the fuel cell stack size for an automobile fuel cell system that has bipolar and end plate dimensions of 12" × 12" × 0.25". There are 25 cells in the stack, and the cooling is built into the bipolar plates.

3. Calculate the payback time of a stationary fuel cell system that has a nominal power of 50 kW, and a cost of $2000 per kW for the area you live in.

4. Which type of fuel cell would you select for an automobile? For stationary power applications? Why?

5. What countries seem to be embracing fuel cells for stationary and automobile applications? Why?

Bibliography

[1] Haraldson, Kristina. 2005. *On Direct Hydrogen Fuel Cell Vehicles Modeling and Demonstration*. KTH Royal Institute of Technology, Department of Chemical Engineering and Technology, PhD thesis, Stockholm, Sweden.
[2] O'Hayre, Ryan, Suk-Won Cha, Whitney Colella, and Fritz B. Prinz. *Fuel Cell Fundamentals*. 2006. New York: John Wiley & Sons.
[3] Barbir, Frano. *PEM Fuel Cells: Theory and Practice*. 2005. Burlington, MA: Elsevier Academic Press.

[4] Barbir, F. "System Design for Stationary Power Generation." W. Vielstich, A. Lamm, and H. Gasteiger, eds. Handbook of Fuel Cell Technology: Fundamentals, Technology, and Applications. Vol. 4. New York: Wiley. 2003. pp. 683–692.
[5] Barbir, F., T. Maloney, T. Molter, and F. Tombaugh. "Fuel Cell Stack System Development: Matching Market to Technology Status." Proc. 2002 Fuel Cell Seminar. Palm Springs, CA. November 18 to 21, 2002. pp. 948–951.
[6] "Fuel Cell Vehicles." *Fuel Cells 2000*. http://www.fuelcells.org/info/charts/carchart.pdf Last accessed September 18, 2006.
[7] "Fuel Cell Buses." *Fuel Cells 2000*. http://www.fuelcells.org/info/charts/buses.pdf Last accessed September 18, 2006.
[8] "Fuel Cell Specialty Vehicles." *Fuel Cells 2000*. http://www.fuelcells.org/info/charts/specialty.pdf. Last accessed September 18, 2006.
[9] "World Wide Fuel Cell Installations." *Fuel Cells 2000*. http://www.fuelcells.org/info/charts/FCInstallationChart.pdf. Last accessed September 18, 2006.
[10] Hoogers, G. "Automotive Applications." Fuel Cell Technology Handbook. 2003, Boca Raton, FL: CRC Press.
[11] Adamson, Kerry-Anne. Fuel Cell Market Survey: Buses. *Fuel Cell Today*. November 2004. http://www.fuelcelltoday.com/FuelCellToday/FCTFiles/FCTArticleFiles/Article_916_BusSurvey%20Final%20Version.pdf. Last accessed September 15, 2006.
[12] Barbir, F., and T. Molter. Regenerative Fuel Cells for Energy Storage: Efficiency and Weight Tradeoffs. IEEE A&E Systems Magazine, March 2005.
[13] Gorgun, Haluk. Dynamic Modeling of a Proton Exchange membrane (PEM) Electrolyzer.International Journal of Hydrogen Energy, 31 (2006) 29–38.
[14] Kerfoot, Katie, Amanda Christiana, and Thomas Horan. Final Report: MiSo Solar Hydrogen Production. APD 2004-01 Dec.13, 2004.
[15] Nelson, Don B., Hashem Nehrir, and Victor Gerez. Economic Evaluation of Grid-Connected Fuel Cell Systems. IEEE Transactions on Energy Conversion, Vol. 20, No. 2, June 2005.
[16] Thomas, C.E., Brian D. James, Frank D. Lomax, Jr., and Ira F. Kuhn. "Fuel Options for the Fuel Cell Vehicle: Hydrogen, Methanol, or Gasoline?" *International Journal of Hydrogen Energy*. Vol. 25, 2000, pp. 551–567.
[17] Wallmark, Cecilia. Design and Evaluation of Stationary Polymer Electrolyte Fuel Cell Systems. Doctoral Thesis 2004. Royal Institute of Technology, Stockholm, Sweden.

Chapter 5

Basic Fuel Cell Thermodynamics

Thermodynamics is the study of energy transforming from one state to another. The concepts presented in this chapter are important for understanding and predicting fuel cell performance since fuel cells transform chemical energy into electrical energy. Thermodynamic calculations give the theoretical predictions for electrical potential, temperature, and pressure variations within a fuel cell. The specific topics to be covered are

- Basic Thermodynamic Concepts
- Theoretical Fuel Cell Potential
- Fuel Cell Efficiency
- Pressure Effects on the Fuel Cell
- Temperature Effects on the Fuel Cell

The thermodynamic theory in this section will be further developed in subsequent chapters after additional concepts have been introduced. The additional concepts will help predict and design the fuel cell potential as well as many of the internal and external fuel cell properties and operating conditions.

5.1 Basic Thermodynamic Concepts

A few basic thermodynamic concepts that are helpful in analyzing fuel cell reactions are the absolute enthalpy, specific heat, and higher and lower heating values. The absolute enthalpy can be defined as the enthalpy that includes chemical and sensible thermal energy [9]. The enthalpy of formation (h_f) is associated with the energy of the chemical bonds, and the sensible thermal energy (Δh_s) is the enthalpy difference between the given and reference state.

Chapter Five

The enthalpy of formation of a substance is the amount of heat absorbed or released when one mole of the substance is formed from its elemental substances at the reference state. The enthalpy of substances in their naturally occurring state is defined as zero at the reference state (reference state is typically referred to as $T_{ref} = 25°C$ and $P_{ref} = 1$ atm). For example, hydrogen and oxygen at reference state are diatomic molecules (H_2 and O_2) and therefore, the enthalpy of formation for H_2 and O_2 at $T_{ref} = 25°C$ and $P_{ref} = 1$ atm is equal to zero. The enthalpy of formation is typically determined by laboratory measurements, and can be found in various thermodynamic tables. Appendix B lists some values for the most common fuel cell substances.

The sensible energy change (Δh_s) is a function of temperature, and can be calculated using the specific heat at a constant temperature as shown in equation 5-1:

$$\Delta h_s(T, P) = \int_{T_{ref}}^{T} c_p(T) dT \qquad (5\text{-}1)$$

where $c_p(T)$ is the specific heat at a constant pressure. The specific heat is available in many thermodynamic property tables, and Appendix C-E shows the specific heats for some of the common fuel cell substances. The average specific heat can be approximated as a linear function of temperature:

$$\overline{c_p} = c_p\left(\frac{T + T_{ref}}{2}\right) \qquad (5\text{-}2)$$

where $\overline{c_p}$ is the average specific heat at a constant pressure, T is the given temperature and T_{ref} 5 25°C.

Example 5-1 Calculate the average specific heat at a constant pressure for H_2 at temperatures between T = 55°C and T = 100°C at a pressure of 1 atm. Assume that H_2 is an ideal gas, and use (a) the average temperature, and (b) the average of the specific heats.

(a) The specific heat at a constant pressure, $c_p(t)$, for H_2 is given as a function of temperature. At the average temperature:

$$T = \frac{T + T_{ref}}{2} = \frac{328 + 373}{2} = 350.5 K$$

From Appendix D, the average specific heat at a constant pressure per mol with $c_p = 14.42 \frac{kJ}{kgK}$ and the molecular weight of H_2 is $2.016 \frac{kg}{mol}$:

$$c_p = 14.427 \frac{kJ}{kgK} \times 2.016 \frac{kg}{mol} = 29.084 \frac{kJ}{kmolK}$$

(b) From Appendix B and interpolating, $c_p(328\ K) = 14.374 \frac{kJ}{kgK}$ and $c_p(373\ K) = 14.450 \frac{kJ}{kgK}$

Therefore, the average specific heat at a constant pressure, c_p is

$$c_p = \frac{1}{2}\Big(c_p(328K) + c_p(373K)\Big)$$

$$= \frac{1}{2}\left(14.374\frac{kJ}{kgK} + 14.450\frac{kJ}{kgK}\right) = 14.412\frac{kJ}{kmolK}$$

$$c_p = 14.412 \frac{kJ}{kmolK} * 2.016 \frac{kg}{mol} = 29.054 \frac{kJ}{kmolK}$$

Example 5-2 Determine the absolute enthalpy of H_2, O_2, and water (H_2O) at the pressure of 1 atm at a temperature of (a) 25°C, (b) 75°C, and (c) 800°C. Calculate the absolute enthalpy for the vapor and liquid form if applicable.

(a) For T = 25°C = 298 K, from Appendix B, the enthalpy of formation at 25°C and 1 atm is

$$h_{f,H_2} = 0,\ h_{f,O_2} = 0,\ h_{f,H_2O(l)} = -285{,}826\ (J/mol),\ h_{f,H_2O(g)} = -241{,}826\ (J/mol)$$

(b) For T = 75°C = 348 K, the average temperature is (298 + 348)/2 = 323 K. From Appendix D, after interpolation and converting to a per mol basis:

$$c_{p,H_2} = 14.362 \frac{kJ}{kgK} \times 2.016 \frac{kg}{kmol} = 28.954 \frac{kJ}{kmolK}$$

$$c_{p,O_2} = 0.923 \frac{kJ}{kgK} \times 31.999 \frac{kg}{kmol} = 29.535 \frac{kJ}{kmolK}$$

$$c_{p,H_2O(g)} = 33.845 \frac{kJ}{kmolK}$$

For liquid water from the Appendix E:

$$c_{p,H_2O(l)} = 4.181 \frac{kJ}{kgK} \times 18.015 \frac{kg}{kmol} = 75.321 \frac{kJ}{kmolK}$$

The absolute enthalpy is determined as follows:

$$h_{H_2} = h_{f,H_2} + c_{p,H_2}(T - T_{ref}) = 0 + 28.954 \times (348 - 298) = 1447.7 \frac{J}{mol}$$

$$h_{O_2} = h_{f,O_2} + c_{p,O_2}(T - T_{ref}) = 0 + 29.535 \times (348 - 298) = 1476.8 \frac{J}{mol}$$

$$h_{H_2O(l)} = h_{f,H_2O(l)} + c_{p,H_2O(l)}(T - T_{ref})$$

$$= -285,826 + 75.321 \times (348 - 298) = -282,059.95 \frac{J}{mol}$$

$$h_{H_2O(g)} = h_{f,H_2O(g)} + c_{p,H_2O(g)}(T - T_{ref})$$

$$= -241,826 + 33.845 \times (348 - 298) = -240,133.75 \frac{J}{mol}$$

(c) For T = 800°C = 1073 K, the average temperature is (298 + 1073)/2 = 685.5 K. From Appendix D, after interpolation and converting to a per mol basis:

$$c_{p,H_2} = 14.594 \frac{kJ}{kgK} \times 2.016 \frac{kg}{kmol} = 29.422 \frac{kJ}{kmolK}$$

$$c_{p,O_2} = 1.027 \frac{kJ}{kgK} \times 31.999 \frac{kg}{kmol} = 32.863 \frac{kJ}{kmolK}$$

$$c_{p,H_2O(g)} = 37.360 \frac{kJ}{kmolK}$$

The absolute enthalpy is determined as follows:

$$h_{H_2} = h_{f,H_2} + c_{p,H_2}(T - T_{ref}) = 0 + 29.422 \times (1073 - 298) = 22,802.05 \frac{J}{mol}$$

$$h_{O_2} = h_{f,O_2} + c_{p,O_2}(T - T_{ref}) = 0 + 32.863 \times (1073 - 298) = 25,468.82 \frac{J}{mol}$$

$$h_{H_2O(g)} = h_{f,H_2O(g)} + c_{p,H_2O(g)}(T - T_{ref})$$

$$= -241,826 + 37.360 \times (1073 - 298) = -212,872.00 \frac{J}{mol}$$

This problem shows that as the temperature increases, the specific heat at constant temperature increases very slowly. Specific heat is a very weak function of temperature. The increase in specific heat is the smallest for H_2, which is the smallest molecule in the periodic table.

Another useful thermodynamic calculation is the higher and lower heating values of fuels. The higher heating value (HHV) measures the amount of heat released (heat of formation) by a constant quantity of fuel between identical initial and final temperatures (25°C). The fuel is initially at 25°C, it is ignited and combusted, then returned to 25°C. The lower heating value (LHV) is determined by stopping the cooling at 150°C (or some other arbitrary temperature). The reaction heat will only be partially recovered. The difference between the higher and lower heating value is equal to the enthalpy of vaporization (or enthalpy of condensation), and depends upon the chemical composition of the fuel.

Example 5-3 Determine the higher and lower heating values for methane and propane reacting with O_2. The reaction occurs at 25°C and 1 atm.

(a) For 1 mol of methane, the reaction equation can be written as:

$$CH_4\ (g) + 2O_2 \rightarrow 2H_2O + CO_2$$

The enthalpy of the reaction can be written as:

$$\Delta h_{rx} = 2h_{H_2O} + h_{CO_2} - (h_{CH_4} + 2h_{O_2})$$

From Appendix B, for water in the liquid and vapor state:

$$\Delta h_{rx} = *(-285{,}826) + (-393{,}522) - (-74{,}850 + 2*0)$$

$$= -1{,}040{,}024\ \frac{J}{mol}\ CH_4$$

$$\Delta h_{rx} = 2*(-241{,}826) + (-393{,}522) - (-74{,}850 + 2*0)$$

$$= -802{,}324\ \frac{J}{mol}\ CH_4$$

The heating values are therefore:

$$HHV = -\Delta h_{rx} = 1{,}040{,}024\ \frac{J}{mol}\ CH_4$$

$$LHV = -\Delta h_{rx} = 802{,}324\ \frac{J}{mol}\ CH_4$$

(b) For 1 mol of propane, the reaction equation is

$$C_3H_8\ (g) + 5O_2 \rightarrow 4H_2O + 3CO_2$$

The enthalpy of the reaction can be written as:

$$\Delta h_{rx} = 4h_{H_2O} + 3h_{CO_2} - (h_{C_3H_8} + 5h_{O_2})$$

From Appendix B, for water in the liquid and vapor state:

$$\Delta h_{rx} = 4*(-285{,}826) + 3*(-393{,}522) - (-130{,}850 + 5*0)$$

$$= -2{,}193{,}020 \ \frac{J}{mol} \ C_3H_8$$

$$\Delta h_{rx} = 4*(-241{,}826) + 3*(-393{,}522) - (-130{,}850 + 5*0)$$

$$= -2{,}017{,}020 \ \frac{J}{mol} \ C_3H_8$$

The heating values are therefore:

$$HHV = -\Delta h_{rx} = 2{,}193{,}020 \ \frac{J}{mol} \ C_3H_8$$

$$LHV = -\Delta h_{rx} = 2{,}017{,}020 \ \frac{J}{mol} \ C_3H_8$$

As the carbon content increases, the difference between the high and low heating value decreases.

5.2 Fuel Cell Reversible and Net Output Voltage

The maximum electrical energy output, and the potential difference between the cathode and anode is achieved when the fuel cell is operated under the thermodynamically reversible condition. This maximum possible cell potential is the reversible cell potential. The net output voltage of a fuel cell at a certain current density is the reversible cell potential minus the irreversible potential which is discussed in this section, and can be written as [1]:

$$V(i) = V_{rev} - V_{irrev} \qquad (5\text{-}3)$$

where $V_{rev} = E_r$ is the maximum (reversible) voltage of the fuel cell, and V_{irrev} is the irreversible voltage loss (overpotential) occurring at the cell.

The maximum electrical work (W_{elec}) a system can perform at a constant temperature and pressure process is given by the negative change in Gibbs free energy change (ΔG) for the process. This equation in molar quantities is:

$$W_{elec} = -\Delta G \qquad (5\text{-}4)$$

The Gibbs free energy represents the net energy cost for a system created at a constant temperature with a negligible volume, minus the energy from the environment due to heat transfer. This equation is valid at any constant temperature and pressure for most fuel cell systems. From the second law of thermodynamics, the maximum useful work (change in free energy) can be obtained when a "perfect" fuel cell operating irreversibly is dependent upon temperature. Thus, W_{elec}, the electrical power output is:

$$W_{elec} = \Delta G = \Delta H - T\Delta S \qquad (5\text{-}5)$$

where G is the Gibbs free energy, H is the heat content (enthalpy of formation), T is the absolute temperature, and S is entropy. Both reaction enthalpy and entropy are also dependent upon the temperature. The absolute enthalpy can be determined by the system temperature and pressure and is usually defined as combining both chemical and thermal bond energy. The change in the enthalpy of formation for the chemical process can be expressed from the heat and mass balance:

$$\Delta H = \sum_i m_i h_i - \sum_j m_j h_j \qquad (5\text{-}6)$$

where $\Sigma_i m_i h_i$ the summation of the mass times the enthalpy of each substance leaving the system, and $\Sigma_j m_j h_j$ is the summation of the mass times the enthalpy of each substance entering the system. A simple diagram of the heat and mass balance is shown in Figure 5-1. Detailed discussions of mass and heat (energy) balances are described in Chapters 16 and 9 respectively.

The potential of a system to perform electrical work by a charge, Q (coulombs) through an electrical potential difference, E in volts is [2]:

$$W_{elec} = EQ \qquad (5\text{-}7)$$

If the charge is assumed to be carried out by electrons:

$$Q = nF \qquad (5\text{-}8)$$

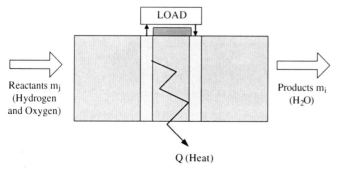

Figure 5-1 Fuel cell heat and mass balance.

where n is the number of moles of electrons transferred and F is the Faraday's constant (96,485 coulombs per mole of electrons). Combining the last three equations, the maximum reversible voltage provided by the cell can be calculated:

$$\Delta G = -nFE_r \qquad (5\text{-}9)$$

where n is the number of moles of electrons transferred per mol of fuel consumed, F is Faraday's constant, and E_r is the standard reversible potential.

The relationship between voltage and temperature is derived by taking the free energy, linearizing about the standard conditions of 25°C, and assuming that the change in enthalpy (ΔH) does not change with temperature:

$$E_r = -\frac{\Delta G_{rxn}}{nF} = -\frac{\Delta H - T\Delta S}{nF}$$

$$\Delta E_r = \left(\frac{dE}{dT}\right)(T - 25) = \frac{\Delta S}{nF}(T - 25)$$

(5-10)

where E_r is the standard-state reversible voltage, and ΔG_{rxn} is the standard free-energy change for the reaction. The change in entropy is negative; therefore, the open circuit voltage output decreases with increasing temperature. The fuel cell is theoretically more efficient at low temperatures. However, mass transport and ionic conduction is faster at higher temperatures and this more than offsets the drop in open-circuit voltage.

In the case of a hydrogen–oxygen fuel cell under standard-state conditions:

$$H_2\,(g) + {}^1\!/_2\,O_2\,(g) \rightarrow H_2O\,(l)$$

$$(\Delta H = -285.8\ KJ/mol;\ \Delta G = -237.3\ KJ/mol)$$

$$E_{H_2/O_2} = -\frac{-237.3\ KJ/mol}{2\ mol * 96{,}485\ C/mol} = 1.229V$$

At standard temperature and pressure, this is the highest voltage obtainable from a hydrogen–oxygen fuel cell. Most fuel cell reactions have theoretical voltages in the 0.8 to 1.5 V range. To obtain higher voltages, several cells have to be connected together in series.

For nonstandard conditions, the reversible voltage of the fuel cell may be calculated from the energy balance between the reactants and the products [7]. The theoretical potential E_t for an electrochemical reaction is expressed by the Nernst equation:

$$E_t = E_r - \frac{RT}{nF} \ln \left[\prod_i a_i^{v_i} \right] \tag{5-11}$$

where R is the ideal gas constant, T is the temperature, a_i is the activity of species i, v_i is the stoichiometric coefficient of species i, and E_r is the standard-state reversible voltage, which is a function of temperature and pressure.

The hydrogen–oxygen fuel cell reaction is written as follows using the Nernst equation:

$$E = E_r - \frac{RT}{2F} \ln \frac{a_{H_2O}}{a_{H_2} a_{O_2}^{1/2}} \tag{5-12}$$

where E is the actual cell voltage, E_r is the standard-state reversible voltage, R is the universal gas constant, T is the absolute temperature, N is the number of electrons consumed in the reaction, and F is Faraday's constant. If the fuel cell is operating under 100°C, the activity of water can be set to 1 because liquid water is assumed. At a pressure of 1.00 atmospheres absolute (as it is at sea level on a normal day), and if the acid electrolyte has an effective concentration of 1.00 moles of H^+ per liter, the ratio of $1.00^{1/2}/1.00 = 1$, and ln 1 = 0. Therefore, $E = E_r$. The standard electrode potential is that which is realized when the products and reactants are in their standard states.

At standard temperature and pressure, the theoretical potential of a hydrogen–air fuel cell can be calculated as follows:

$$E = 1.229 - \frac{8.314(J/(mol*K))*298.15}{2*9,6485(C/mol)} \ln \frac{1}{1*0.21^{1/2}} = 1.219 V$$

The potential between the oxygen cathode where the reduction occurs and the hydrogen anode at which the oxidation occurs will be 1.229 volts at standard conditions with no current flowing. When a load connects the two electrodes, the current will flow as long as there is hydrogen and oxygen gas to react. If the current is small, the efficiency of the cell (measured in voltages) could be greater than 0.9 V, with an efficiency greater than 90 percent. This efficiency is much higher than the most complex heat engines such as steam engines or internal combustion engines, which can only reach a maximum 60 percent thermal efficiency.

By assuming the gases are ideal (the activities of the gases are equal to their partial pressures, and the activity of the water phase is equal to unity), equation 5-11 can be written as:

$$E_t = E_r - \frac{RT}{nF} \ln \left[\prod_i \left(\frac{p_i}{p_0} \right)^{v_i} \right] \tag{5-13}$$

where p_i is the partial pressure of species i, and p_0 is the reference pressure. For ideal gases or an estimate for a nonideal gas, partial pressure of species A, p_A* can be expressed as a product of total pressure P_A and molar fraction χ_A of the species:

$$p_A* = x_A P_A \qquad (5\text{-}14)$$

If the molar fraction for the fuel is unknown, it can be estimated by taking the average of the inlet and outlet conditions [11]:

$$x_A = \frac{1 - x_{C,Anode}}{1 + \left(\dfrac{x_{Anode}}{2}\right)\left(1 + \left(\dfrac{\zeta_A}{(\zeta_{A-1})}\right)\right)} \qquad (5\text{-}15)$$

where ζ_A is the stoichiometric flow rate, x_{Anode} is the molar ratio of species 2 to 1 in dry gas, and $x_{C,Anode}$ is:

$$x_{C,Anode} = \frac{P_{sat}}{P_A} \qquad (5\text{-}16)$$

The molar fractions are simply ratios of the saturation pressure (P_{sat}) at a certain fuel cell temperature to the anode and cathode pressures.

The water saturation temperature is a function of cell operating temperature. For a PEM hydrogen–oxygen fuel cell, P_{sat} can be estimated using [10]:

$$\log_{10} P_{sat} = -2.1794 + 0.02953*T - 9.1837 \\ \times 10\text{--}5*T^2 + 1.4454 \times 10^{-7}*T^3 \qquad (5\text{-}17)$$

where T is the cell operating temperature in °C.

If the current is large, the cell voltage falls fairly rapidly due to various nonequilibrium effects. The simplest of these effects is the voltage drop due to the internal resistance of the cell itself. According to Ohm's law, the voltage drop is equal to the resistance times the current flowing. At maximum current density of 1 amp/cm^2, the cell can drop 0.5 volts.

If the total energy based upon the higher heating value could be converted into electrical energy, then a theoretical potential of 1.48 V per cell could be obtained. The theoretical potential based upon the lower heating value is also shown. Because of the TΔS limitation, the maximum theoretical potential of the cell is 1.229 V. This is the voltage that could be obtained if the free energy could be converted entirely to electrical energy without any losses. The actual work in the fuel cell is less than the maximum useful work because of other irreversibilities in the process. These irreversibilities (irreversible voltage losses) are the activation over potential (v_{act}), ohmic overpotential (v_{ohmic}), and concentration overpotential (v_{conc}). This is shown by the following equation:

$$V_{irrev} = v_{act} + v_{ohmic} + v_{conc} \quad (5\text{-}18)$$

V_{irrev} in equation 5-18 is substituted back into equation 5-3 to account for the irreversible voltage losses to obtain an accurate fuel cell net output voltage. The variables in equation 5-18 will be discussed in more detail in Chapters 6–8. Chapter 6 covers fuel cell electrochemistry and discusses activation potential, Chapter 7 covers fuel cell charge transport and discusses ohmic overpotential, and Chapter 8 covers fuel cell mass transport and concentration overpotential. Figure 5-2 illustrates the fuel cell voltage losses that need to be considered when designing fuel cells.

Example 5-4 Determine the reversible cell potential as a function of temperature at 650°C and 1 atm. Assume that T_{ref} = 25°C. The reaction is

$$H_2 + \tfrac{1}{2} O_2 \rightarrow H_2O$$

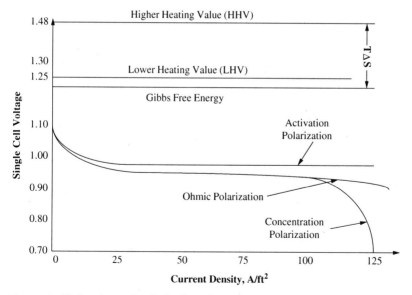

Figure 5-2 Hydrogen–oxygen fuel cell performance curve at equilibrium.

The entropy for the fuel cell reaction at the standard reference temperature and pressure, and the reversible cell potential is:

$$\Delta G = \Delta H - T\Delta S$$

$$\Delta G = \left(h_{H_2O(g)} - \left(h_{H_2} + \frac{1}{2}h_{O_2}\right)\right) - T*\left(s_{H_2O(g)} - \left(s_{H_2} + \frac{1}{2}s_{O_2}\right)\right)$$

$$\Delta G = \left(-241,826\,\frac{J}{mol\_H_2O} - \left(0 + \frac{1}{2}*0\right)\right) - T*\left(188.83\,\frac{J}{mol\_H_2O}\right.$$

$$\left. - \left(130.68\,\frac{J}{mol\_H_2O} + \frac{1}{2}*205.14\,\frac{J}{mol\_O_2}\right)\right)$$

$$\Delta G = -241,826\,\frac{J}{molK} - (298K)*44.42\,\frac{J}{molK} = -228,588.84\,\frac{J}{molK}$$

$$E_r = \frac{\Delta G}{nF} = \frac{-228,588.84\,\frac{J}{molK}}{\left(\frac{2mol\_e^-}{mol\_fuel}*96,487\,\frac{C}{mol\_e^-}\right)} = 1.185\,\frac{J}{C} = 1.185\,V$$

For gaseous water vapor at T = 25°C and 1 atm:

$$\frac{\Delta s(T_{ref}, P)}{nF} = \frac{-44.42\,J/(mol\_fuel*K)}{2mol\_e^-/mol\_fuel \times 96,487\,C/mol\_e^-} = -0.2302 \times 10^{-3}\,V/K$$

Therefore, the desired expression is:

$$E_r(T, P) = 1.185\,V - 0.2302 \times 10^{-3}\,\frac{V}{K}*(T - T_{ref})$$

For every degree of temperature increase, reversible cell potential is reduced by 0.2302 mV. At a temperature of 650°C:

$$E_r(650,1) = 1.185\,V - 0.2302 \times 10^{-3}\,\frac{V}{K}*(650 - 25)K = 1.041\,V$$

The reversible potential is reduced from 1.185 to 1.041 V when the temperature increases from 25°C to 650°C.

Example 5-5 Determine the reversible cell potential for the following reaction:

$$H_2(g) + \tfrac{1}{2}O_2\,(g) + H_2O\,(l)$$

The molar fraction of H_2 in the fuel stream is 0.5 and the molar fraction of O_2 in the oxidant stream is 0.21. The remaining species are chemically inert.

Since the fuel cell operates at the standard temperature and pressure, T = 25°C and P is 1 atm. Since the reactant streams are not pure, the reversible cell potential for the reaction is:

$$E_r(T, P_i) = E_r(T, P) - \frac{RT}{nF} \ln K$$

The molar fraction of the reactants are $X_{H_2} = 0.5$ and $X_{O_2} = 0.21$, and K can be calculated as follows:

$$K = \prod_{i=1}^{N} (X_i)^{(v_i'' - v_i')/v_F'} = X_{H_2}^{(0-1)/1} X_{O_2}^{(0-1)/2/1} = X_{H_2}^{-1} X_{O_2}^{-1/2}$$

The reversible cell potential at standard pressure and temperature when pure H_2 and O_2 are used as reactants is 1.229 V.

$$E_r(T, P_i) = 1.229\,V - \frac{8.314\,J/mol\,K \times 298\,K}{2\,mol\_e^-/mol fuel \times 96{,}487\,C/mol\_e^-}$$

$$\ln 0.5^{-1} \times 0.21^{-1/2} = 1.210\,V$$

The cell potential is decreased as a result of dilute reactant products, but not as drastically decreased as one would expect.

5.3 Theoretical Fuel Cell Efficiency

The efficiency of a chemical process must be evaluated differently than the conventional heat engine. Efficiency can be defined in two ways:

$$\eta_{\Delta G} \equiv \text{(actual useful work)/(maximum useful work)}$$
$$= \text{(power} \times \text{time)}/\Delta G$$

$$\eta_{\Delta H} \equiv \text{(actual useful work)/(maximum useful work)}$$
$$= \text{(power} \times \text{time)}/\Delta H$$

Since $\Delta G = \Delta H - T\Delta S$, $\eta_{\Delta H} < \eta_{\Delta G}$ for the same power output.

Efficiency of an ideal fuel cell is based upon heat content ΔH and is obtained by dividing the maximum work out by the enthalpy input, so the fuel cell efficiency is

$$\eta_{fuel\_cell} = \Delta G / \Delta H \qquad (5\text{-}19)$$

Using the standard free energy and enthalpy given previously ($\Delta G = -237.2$ KJ/mol, $\Delta H = -285.8$ KJ/mol), equation 5-19 shows the maximum thermodynamic efficiency under standard conditions is 83 percent.

The fuel cell directly converts chemical energy into electrical energy. The maximum theoretical efficiency can be calculated using the following equation:

$$\eta_{max} = 1 - T * \Delta S/\Delta H \qquad (5\text{-}20)$$

Therefore, even an ideal fuel cell operating reversibly and isothermally will have an efficiency ranging from 60 to 90 percent. The heat quantity $T\Delta S$ is exchanged with the surroundings.

The efficiency is not a major function of device size. Energy consumed is measured in terms of the higher heating value of the fuel used. In other words for hydrogen:

$$\eta = \frac{power_{out}}{power_{in}} = \frac{n_{electrons}FV_{output}}{n_{hydrogen}\Delta H_{HHV}} = \frac{2FV_{output}}{\Delta H_{HHV}} \qquad (5\text{-}21)$$

where $n_{electrons}$ and $n_{hydrogen}$ is the flow rates in moles per second, F is Faraday's constant, V_{output} is the voltage of the cell output, and ΔH_{HHV} is -285.8 KJ/mol. The higher heating value enthalpy can be converted to an equivalent voltage of 1.481 V [4] so that

$$\eta = V_{output}/1.481 V \qquad (5\text{-}22)$$

This equivalent voltage concept is very useful in calculating efficiency and waste heat. The waste heat generated is simply

$$Q = n\Delta H_{HHV}(1 - \eta) \qquad (5\text{-}23)$$

And the maximum efficiency is a thermodynamically limited 83 percent. (If the assumption that water stays in the liquid form is incorrect, the waste heat that must be rejected decreases because the vaporization of water cools the stack.) Chapter 9 discusses the heat generated by a fuel cell in more detail.

5.3.1 Energy efficiency

Fuel consumption rates can be calculated simply as a function of current density and Faraday's constant:

$$n_{A,reacted} = n_{B,reacted} = \frac{i}{2F} \qquad (5\text{-}24)$$

The mass continuity of two reactants:

$$n_{A,in} = n_{A,reacted} + n_{A,out} \qquad (5\text{-}25)$$

where $n_{A,in}$ is the molar flow rate to the fuel cell, $n_{A,out}$ is the molar flow rate from the fuel cell. The energy efficiency of the fuel cell is

$$\eta_{Energy} = \frac{W_{FC}}{(n_{A,reacted} + n_{A,out}) \times HHV_A} \qquad (5\text{-}26)$$

The fuel consumption rates are explained in more detail in Chapter 16, which includes calculations for all fuel cell operating conditions.

5.4 Fuel Cell Temperature

The theoretical potential of the fuel cell changes with temperature as shown in the following equation:

$$E = \frac{\Delta H}{nF} - \frac{T \Delta S}{nF} \qquad (5\text{-}27)$$

As seen from equation 5-27, an increase in cell temperature results in a lower fuel cell potential. Many fuel cell types, such as solid oxide, run at high temperatures (600–1000°C). Both ΔH and ΔS are functions of temperature:

$$\Delta H_T = h_{298.15} + \int_{298.15}^{T} \Delta c_p dT \qquad (5\text{-}28)$$

$$\Delta S_T = s_{298.15} + \int_{298.15}^{T} \Delta c_p dT \qquad (5\text{-}29)$$

The specific heat (c_p) is a function of temperature as explained in section 5.1. A commonly used relation is

$$c_p = a + bT + cT^2 \qquad (5\text{-}30)$$

where a, b, and c are coefficients specified for a particular gas. If these equations are substituted into each other, this yields for a hydrogen–oxygen fuel cell operating at 25°C:

$$\Delta H_T = \Delta H_{298.15} + \Delta a(T-298.15) + \Delta b \frac{(T-298.15)^2}{2} + \Delta c \frac{(T-298.15)^3}{3}$$

$$\Delta S_T = \Delta S_{298.15} + \Delta a \ln\left(\frac{T}{298.15}\right) + \Delta b(T-298.15) + \Delta c \frac{(T-298.15)^2}{2}$$

where Δa, Δb, and Δc are the differences between coefficients a, b, and c for the products and reactants:

$$\Delta a = a_{H_2O} + a_{H_2} - \frac{1}{2}a_{O_2} \qquad (5\text{-}31)$$

$$\Delta b = b_{H_2O} + b_{H_2} - \frac{1}{2}b_{O_2} \qquad (5\text{-}32)$$

$$\Delta c = c_{H_2O} + c_{H_2} - \frac{1}{2}c_{O_2} \qquad (5\text{-}33)$$

The changes in C_p, ΔH, and ΔS are very small, but at higher temperatures these equations cannot be neglected.

5.5 Fuel Cell Pressure

Fuel cells can operate at any pressure, and often it is advantageous to operate the fuel cell at pressures above atmospheric. The typical pressure range for fuel cells is atmospheric pressure to 6 to 7 bar. The change in Gibbs free energy as related to pressure can be written as:

$$dG = V_m dP \qquad (5\text{-}34)$$

where V_m is the molar volume (m³/mol), and P is the pressure in Pascals. For an ideal gas:

$$PV_m = RT \qquad (5\text{-}35)$$

Therefore:
$$dG = RT\frac{dP}{P} \qquad (5\text{-}36)$$

After integration:
$$G = G_0 + RT\ln\left(\frac{P}{P_0}\right) \qquad (5\text{-}37)$$

where G_0 is the Gibbs free energy at standard pressure and temperature (25°C and 1 atm), and P_0 is the standard pressure (1 atm). This is another form of the Nernst equation. For any chemical reaction:

$$jA + kB \rightarrow mC + nD$$

The change in Gibbs free energy between the products and reactants is:

$$\Delta G = mG_c + nG_D - jG_A - kG_B$$

If this is substituted into the equation:

$$G = G_0 + RT \ln \frac{\left(\frac{P_C}{P_0}\right)^m \left(\frac{P_D}{P_0}\right)^n}{\left(\frac{P_A}{P_0}\right)^j \left(\frac{P_B}{P_0}\right)^k} \quad (5\text{-}38)$$

where P is the partial pressure of the reactant or product species, and P_0 is the reference pressure.

For the hydrogen-oxygen fuel cell reaction, the Nernst equation becomes

$$G = G_0 + RT \ln \left(\frac{P_{H_2} P_{O_2}^{0.5}}{P_{H_2O}} \right)$$

Therefore, the cell potential as a function of temperature and pressure is:

$$E_{T,P} = \left(\frac{\Delta H}{nF} - \frac{T\Delta S}{nF} \right) + RT \ln \left(\frac{P_{H_2} P_{O_2}^{0.5}}{P_{H_2O}} \right)$$

Example 5-6 Determine the inlet and outlet Nernst potential for the following reaction:

$$H_2 + \tfrac{1}{2}O_2 \rightarrow H_2O \text{ (g)}$$

at a temperature of 25°C and 750°C and a pressure of 1 atm. Assume the fuel is pure H_2 and O_2 from the air supplied to the cell. The H_2 utilization is zero, the oxygen utilization is 0.5, and the reactant product water is formed on the oxidant side.

The inlet and outlet Nernst potential can be calculated based upon the reactant composition at the cell inlet and outlet. The reversible cell potential at the cell inlet is:

$$E_r(T, P_i) = E_r(T, P) - \frac{RT}{nF} \ln X_{O_2,in}^{-1/2}$$

At the cell inlet: $X_{H_2,in} = 1, X_{O_2,in} = 0.21, X_{N_2,in} = 0.79, P_{O_2,in} + P_{N_2,in} = 1$ atm

From Example 5-5, $E_r(25°C, 1 \text{ atm}) = 1.185$ V

From Appendix F, at 1023 K and 1 atm it is

$$E_r = \frac{\Delta G}{nF} = \frac{-191{,}375.04 \, \frac{J}{molK}}{\left(\frac{2mol\_e^-}{mol\_fuel} * 96{,}487 \, \frac{C}{mol\_e^-} \right)} = 0.9917 \, \frac{J}{C} = 0.9917 V$$

The inlet Nernst potential is

$$E_r(25,1) = 1.185V - \frac{8.314\ J/molK \times 298K}{2\ mol\_e^-\ mol/fuel \times 96{,}487C/mol\_e^-} \ln 0.21^{-1/2}$$

$$= 1.175\ V$$

$$E_r(750,1) = 0.9917V - \frac{8.314\ J/molK \times 1023K}{2\ mol\_e^-\ mol/fuel \times 96{,}487C/mol\_e^-} \ln 0.21^{-1/2}$$

$$= 0.9573\ V$$

At the cell outlet, $X_{H_2,out} = 1$, $X_{O_2,out} = 0.095$, $P_{O_2,out} + P_{H_2,out} + P_{N_2,out}$
$= P_{out} = P_{in} = 1$ atm

The Nernst potential becomes

$$E_r(T, P_{O_2,out}) = E_r(T, P) - \frac{RT}{nF} \ln X_{O_2,out}^{-1/2}$$

$$E_r(25,1) = 1.185 - 0.01284 \times \ln 0.095^{1/2} = 1.170V$$

$$E_r(750,1) = 0.9917 - 0.04407 \times \ln 0.095^{1/2} = 0.9398V$$

Chapter Summary

Thermodynamics provides the theoretical limits and many of the necessary equations for predicting fuel cell performance. Some of these theoretical limits explored in this chapter include the theoretical fuel cell potential, fuel cell efficiency, and net output voltage. It also provides the basis for evaluating the properties of fuel cell systems, especially the pressure and temperature effects of fuel cells.

Problems

1. Calculate the theoretical cell potential for a hydrogen–oxygen fuel cell operating at 75°C with the reactant gases at 1 atm and 20°C with liquid water as a product.

2. Calculate and compare the differences in theoretical cell potential between three hydrogen–oxygen fuel cells: (1) operating at 20°C and 1 atm, (2) operating at 50°C and 2 atm, and (3) operating at 80°C and 3 atm.

3. Determine the inlet and outlet Nernst potential, as well as the associated Nernst loss for the following reaction:

$$H_2 + \tfrac{1}{2}O_2 \rightarrow H_2O\ (l)$$

at a temperature of 60°C and a pressure of 3 atm. Assume pure O_2 and H_2, the reaction product water is formed at the oxidant side.

4. For the reaction in Problem 3, what is the reversible cell efficiency and the cell current efficiency if the cell operates at a cell voltage of 0.65 V and a current of 0.8 A with a fuel flow of 5 mL/min.

5. For the reaction in Problem 3, what is the amount of entropy generation, the amount of cell potential loss, and the amount of waste heat at 25°C and 1 atm, if the cell operates at a cell voltage of 0.7 V.

Bibliography

[1] Hussain, M.M., J.J. Baschuk, X. Li, and I. Dincer. "Thermodynamic Analysis of a PEM Fuel Cell Power System." *International Journal of Thermal Sciences*. 2005. Vol. 44, pp. 903–911.
[2] O'Hayre, Ryan, Suk-Won Cha, Whitney Colella, and Fritz B. Prinz. *Fuel Cell Fundamentals*. 2006. New York: John Wiley & Sons.
[3] Barbir, Frano. *PEM Fuel Cells: Theory and Practice*. 2005. Burlington, MA: Elsevier Academic Press.
[4] Lin, Bruce. "Conceptual Design and Modeling of a Fuel Cell Scooter for Urban Asia." Princeton University, Masters Thesis, 1999.
[5] Mench, M.M., Z.H. Wang, K. Bhatia, and C.Y. Wang. "Design of a Micro-direct Methanol Fuel Cell," Electrochemical Engine Center, Department of Mechanical and Nuclear Engineering, Pennsylvania State University. 2001.
[6] Mench, Matthew M., Cao-Yang Wang, and Stephan T. Tynell. "An Introduction to Fuel Cells and Related Transport Phenomena." Department of Mechanical and Nuclear Engineering, Pennsylvania State University. Draft. *http://mtrl1.mne.psu.edu/Document/jtpoverview.pdf* Last Accessed March 4, 2007.
[7] Ay, M., A. Midilli, and I. Dincer. "Thermodynamic Modeling of a Proton Exchange Membrane Fuel Cell." *Int. J. Exergy*. 2006. Vol. 3, No. 1, pp. 16–44.
[8] Simon, William Emile. *Transient Thermodynamic Analysis of a Fuel-Cell System*. NASA Technical Note. NASA TN D-4-4601. Manned Spacecraft Center, Houston, Texas. June, 1968.
[9] Li, Xianguo. *Principles of Fuel Cells*. 2006. New York: Taylor & Francis Group.
[10] Springer et al. "Polymer Electrolyte Fuel Cell Model." *J. Electrochem. Soc.* 1991. Vol. 138, No. 8, pp. 2334–2342.
[11] Rowe, A. and X. Li. Mathematical Modeling of Proton Exchange Membrane Fuel Cells." *Journal of Power Sources*. 2001. Vol. 102, pp. 82–96.

Chapter 6

Fuel Cell Electrochemistry

This chapter will cover the basic electrochemistry needed in order to predict or model electrode kinetics, activation overpotential, currents, and potentials in a fuel cell. It will specifically cover how fast a reaction occurs, how reactants create products, and how much energy loss occurs during the electrochemical reaction. The specific topics to be covered are:

- Electrode Kinetics
- Voltage Losses
- Fuel Cell Potential and Current
- Polarization Curves

The theory introduced in this section will be further developed in subsequent chapters in order to accurately predict the fuel cell net output voltage and operating characteristics.

All electrochemical processes involve the transfer of electrons between an electrode and a chemical species with a change in Gibbs free energy. The electrochemical reaction occurs at the interface between the electrode (catalyst & backing layer or porous electrode layer) and the electrolyte, as shown in Figure 6-1.

Each charge must overcome an activation energy barrier in order to move through the electrolyte, electrode, or bipolar plate. The speed of the electrochemical reaction is dependent upon the rate that electrons are created or consumed. Therefore, the current is a direct measure of

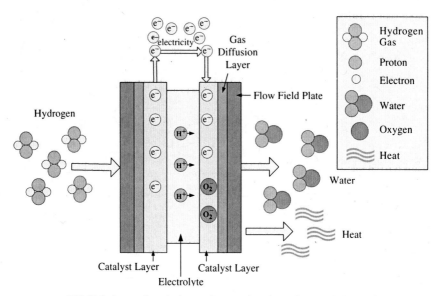

Figure 6-1 PEMFC electrochemical reactions at the electrolyte and electrode.

the rate of the electrochemical reactions. From Faraday's law, the rate of charge transfer is:

$$i = \frac{dQ}{dt} \qquad (6\text{-}1)$$

where Q is the charge and t is the time. If each electrochemical reaction results in the transfer of n electrons per unit of time, then

$$i = nF\frac{dN}{dt} \qquad (6\text{-}2)$$

where dN/dt is the rate of the electrochemical reaction (mol/s) and F is Faraday's constant (96,400 C/mol). Integrating this equation gives

$$\int_0^t i\, dt = Q = nFN \qquad (6\text{-}3)$$

This equation states that the total amount of electricity produced is proportional to the number of moles of material times the number of electrons times Faraday's constant.

Example 6-1 A hydrogen/oxygen fuel cell with the reaction

$$H_2 + O_2 \rightarrow H_2O$$

is operating at 75°C and 1 atm. The fuel cell runs for 72 hours. How much current does the fuel cell produce at a flow rate of 10 sccm? How many moles of H_2 are consumed?

H_2 can be treated as an ideal gas; the molar flow rate is related to the volumetric flow rate via the ideal gas law:

$$\frac{dN}{dt} = \frac{P(dV/dt)}{RT}$$

$$\frac{dN}{dt} = \frac{1 \; atm \times (0.01 \; L/min)}{(0.082 \; L/atm/(molK)) \times (348.15 \; K)} = 3.50 \times 10^{-4} \frac{molH_2}{min}$$

Since two moles of electrons are transferred for every mol of H_2 gas reacted, n = 2.

$$i = nF\frac{dN}{dt} = 2 \times (96,400 \; C) * \left(3.50 \times 10^{-4} \frac{molH_2}{min}\right) * 1\frac{min}{60 \; s} = 1.126 \; A$$

The total amount of electricity produced is calculated by integrating the current load over the operation time.

$$Q_{tot} = i_1 t_1 = (1.126 A) * \left(72 \; hours * \frac{3600 \; sec}{1 \; hr}\right) = 291,859 \; C$$

Since 2 moles of electrons are needed for every mol of H_2 reacted, n = 2. The total number of moles of H_2 processed by the fuel cell is

$$N_{H_2} = \frac{Q_{tot}}{nF} = \frac{291,859 \; C}{2 \times 96,400 \; C/mol} = 1.514 \; mol\_H_2$$

The molar mass of H_2 is 2.016 g/mol, which equals 3.052 g of H_2.

An electrochemical reaction occurring at the electrode between oxidized and reduced forms of a chemical species takes the following form:

$$Ox + e^- \leftrightarrow Re \tag{6-4}$$

If the potential (voltage) of the electrode is made more negative than the equilibrium potential, the reaction will form more Re. If the potential of the electrode is more positive than the equilibrium potential, it will create more Ox [2]. The forward and backward reactions take place simultaneously. The consumption of the reactant is proportional to the surface concentration [3]. For the forward reaction, the flux is:

$$j_f = k_f C_{Ox} \tag{6-5}$$

where k_f is the forward reaction rate coefficient, and C_{Ox} is the surface concentration of the reactant species.

The backward reaction of the flux is described by:

$$j_B = k_b C_{Rd} \tag{6-6}$$

where k_b is the backward reaction rate coefficient, and C_{Rd} is the surface concentration of the reactant species.

These reactions either release or consume electrons. The difference between the electrons released and consumed equals the net current generated [3]:

$$i = nF(k_f C_{Ox} - k_b C_{Rd}) \qquad (6\text{-}7)$$

At equilibrium, the net current should equal zero because the reaction will proceed in both directions simultaneously at the same rate [3]. This reaction rate at equilibrium is called the current exchange density.

6.1 Electrode Kinetics

According to the Transition State Theory, an energy barrier needs to be overcome for the reaction to proceed [1]. An illustration of the Gibbs free energy function with the distance from the interface is shown in Figure 6-2. The magnitude of the energy barrier to be overcome is equal to the Gibbs free energy change between the reactant and product [9].

The heterogeneous rate coefficient, k, can be calculated for an electrochemical reaction that is a function of the Gibbs free energy:

$$k = \frac{k_B T}{h} \exp\left(\frac{-\Delta G}{RT}\right) \qquad (6\text{-}8)$$

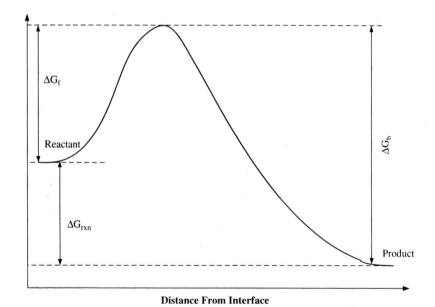

Figure 6-2 Gibbs free energy change compared with distance from the interface.

where k_B is the Boltzmann's constant (1.38049×10^{-23} J/K), h is Plank's constant (6.621×10^{-34} Js), and ΔG is the Gibbs energy of activation (kJ/mol).

The Gibbs free energy can be considered to consist of both chemical and electrical terms because it occurs in the presence of an electrical field [1, 2]. For a reduction reaction:

$$\Delta G = \Delta G_{AC} + \alpha_{Rd} FE \quad (6\text{-}9)$$

For an oxidation reaction:

$$\Delta G = \Delta G_{AC} - \alpha_{Ox} FE \quad (6\text{-}10)$$

where ΔG_{AC} is the activated complex of the Gibbs free energy, α is the transfer coefficient, F is Faraday's constant, and E is the potential.

Some confusion exists in the literature between the transfer coefficient (α) and the symmetry factor (β). The symmetry factor is typically used for single-step reactions, but the typical process is multistep; therefore, an experimental parameter is used called the transfer coefficient [3]. The value of α depends upon the activation barrier, and is typically between 0 and 1, but is specifically between 0.2 to 0.5 for most electrochemical reactions [2]. The relationship between α_{Rd} and α_{ox} is:

$$\alpha_{Rd} - \alpha_{Ox} = \frac{n}{v_{times}} \quad (6\text{-}11)$$

where n is the number of electrons transferred, and v_{times} is the number of times the stoichiometric step must take place for the reaction to occur [3].

The forward and backward oxidation reaction rate coefficients are:

$$k_f = k_{0,f} \exp\left[\frac{-\alpha_{Rd} FE}{RT}\right] \quad (6\text{-}12)$$

$$k_b = k_{0,b} \exp\left[\frac{-\alpha_{Ox} FE}{RT}\right] \quad (6\text{-}13)$$

If these equations are introduced into equation 6-7, the net current is:

$$i = nF\left\{k_{0,f} C_{Ox} \exp\left[\frac{-\alpha_{Rd} FE}{RT}\right] - k_{0,b} C_{Rd} \exp\left[\frac{\alpha_{Ox} FE}{RT}\right]\right\} \quad (6\text{-}14)$$

Since the reaction proceeds in both directions simultaneously, the net current at equilibrium is equal to zero [3]. As mentioned previously, the exchange current density is the rate at which these reactions proceed at equilibrium:

$$i_0 = nF k_{0,f} C_{Ox} \exp\left[\frac{-\alpha_{Rd} FE_r}{RT}\right] = nF k_{0,b} C_{Rd} \exp\left[\frac{-\alpha_{Ox} FE_r}{RT}\right] \quad (6\text{-}15)$$

The Butler-Volmer equation is the relationship between the current density and the potential:

$$i = i_0 \left\{ \exp\left[\frac{-\alpha_{Rd}F(E - E_r)}{RT}\right] - \exp\left[\frac{\alpha_{Ox}F(E - E_x)}{RT}\right] \right\} \quad (6\text{-}16)$$

The Butler-Volmer equation is valid for both anode and cathode reaction in a fuel cell. It states that the current produced by an electrochemical reaction increases exponentially with activation overpotential [2]. It also says that if more current is needed from a fuel cell, voltage will be lost. The Butler-Volmer equation applies to all single-step reactions, and some modifications to the equation must be made in order to use it for multistep approximations.

The exchange current density is the rate constant for electrochemical reactions, and is a function of temperature, catalyst loading, and catalyst specific surface area. The effective exchange current density at any temperature and pressure is given by the following equation:

$$i_0 = i_0^{ref} a_c L_c \left(\frac{P_r}{P_r^{ref}}\right)^{\gamma} \exp\left[-\frac{E_c}{RT}\left(1 - \frac{T}{T_{ref}}\right)\right] \quad (6\text{-}17)$$

where i_0^{ref} is the reference exchange current density per unit catalyst surface area (A/cm^2), a_c is the catalyst specific area, L_c is the catalyst loading, P_r is the reactant partial pressure (kPa), P_r^{ref} is the reference pressure (kPa), γ is the pressure coefficient (0.5 to 1.0), E_c is the activation energy (66 kJ/mol for O_2 reduction on Pt), R is the gas constant (8.314 J/(mol*K)), T is the temperature, K, and T_{ref} is the reference temperature (298.15 K).

The exchange current density measures the readiness of the electrode to proceed with the chemical reaction. The higher the exchange current density, the lower the barrier is for the electrons to overcome, and the more active the surface of the electrode.

6.2 Voltage Losses

The typical voltage losses seen in a fuel cell were first described in Chapter 5, and are illustrated in Figure 6-3. The single fuel cell provides a voltage dependent on operating conditions such as temperature, applied load, and fuel/oxidant flow rates. The standard measure of performance for fuel cell systems is the polarization curve, which represents the cell voltage behavior against the operating current density.

Electrical energy is obtained from a fuel cell only when current is drawn, and the cell voltage drops due to several irreversible loss mechanisms. The loss is defined as the deviation of the cell potential (V_{irrev}) from the theoretical potential (V_{rev}) as first mentioned in Chapter 5:

$$V(i) = V_{rev} - V_{irrev} \quad (6\text{-}18)$$

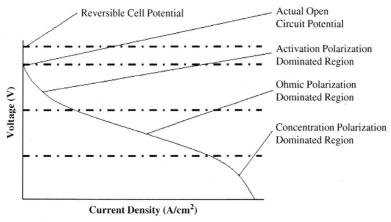

Figure 6-3 Generalized polarization curve for a fuel cell.

The actual open circuit voltage of a fuel cell is lower than the theoretical model due to species crossover from one electrode through the electrolyte and internal currents. The three major classifications of losses that result in the drop from open circuit voltage is (1) activation polarization, (2) ohmic polarization, and (3) concentration polarization [6]. Therefore, equation 6-18 can be expanded by including these voltage losses:

$$V(i) = V_{rev} - v_{act\_anode} - v_{act\_cath} - v_{ohmic} - v_{conc\_anode} - v_{conc\_cath} \quad (6\text{-}19)$$

where v_{act}, v_{ohmic}, v_{conc} represent activation, ohmic (resistive), and mass concentration polarization. As seen in equation 6-19, activation and concentration polarization occurs at both the anode and cathode, while the resistive polarization represents ohmic losses throughout the fuel cell.

Substituting the equations for activation, ohmic and concentration polarization into equation 6-19, the relationship between the fuel cell potential and current density (fuel cell polarization curve) as illustrated in Figure 6-3 and 6-4 can be shown as:

$$E = E_r - \frac{RT}{\alpha_c F}\ln\left(\frac{i}{i_{0,c}}\right) - \frac{RT}{\alpha_a F}\ln\left(\frac{i}{i_{0,a}}\right) - \frac{RT}{nF}\ln\left(\frac{i_{L,c}}{i_{L,c} - i}\right)$$

$$- \frac{RT}{nF}\ln\left(\frac{i_{L,a}}{i_{L,a} - i}\right) - iR_i \quad (6\text{-}20)$$

The shorter version of the equation is:

$$E = E_r - \frac{RT}{\alpha F}\ln\left(\frac{i_{ext} + i_{loss}}{i_0}\right) - \frac{RT}{nF}\ln\left(\frac{i_L}{i_L - i}\right) - iR_i \quad (6\text{-}21)$$

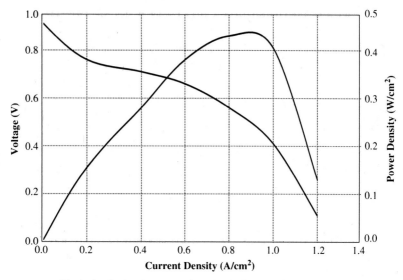

Figure 6-4 Typical polarization curve for a hydrogen PEM fuel cell.

Activation polarization is the voltage overpotential required to overcome the activation energy of the electrochemical reaction on the catalytic surface [6]. This type of polarization dominates losses at low current density, and measures the catalyst effectiveness at a given temperature. This is a complex three-phase interface problem, since gaseous fuel, the solid metal catalyst, and electrolyte must all make contact. The catalyst reduces the height of the activation barrier, but a loss in voltage remains due to the slow oxygen reaction. The total activation polarization overpotential is 0.1 to 0.2 V, which reduces the maximum potential to less than 1.0 V even under open-circuit conditions [4]. Activation overpotential expressions can be derived from the Butler-Volmer equation. The activation overpotential increases with current density and can be expressed as:

$$\Delta V_{act} = E_r - E = \frac{RT}{\alpha F}\ln\left(\frac{i}{i_0}\right) \quad (6\text{-}22)$$

where i is the current density, and i_0, is the reaction exchange current density. Exchange current density represents the reaction rate.

The activation losses can be expressed simply as the Tafel equation:

$$\Delta V_{act} = a + b\ln(i) \quad (6\text{-}23)$$

where $a = -\frac{RT}{\alpha F}\ln(i_0)$ and $b = -\frac{RT}{\alpha F}$

The equation for the anode plus cathode activation overpotential can be represented by:

$$v_{act\_anode} + v_{act\_cath} = \frac{RT}{nF\alpha}\ln\left(\frac{i}{i_o}\right)\bigg|_{anode} + \frac{RT}{nF\alpha}\ln\left(\frac{i}{i_o}\right)\bigg|_{cath} \quad (6\text{-}24)$$

where n is the number of exchange protons per mole of reactant, F is Faraday's constant, and α is the charge transfer coefficient used to describe the amount of electrical energy applied to change the rate of the electrochemical reaction [6]. The exchange current density, i, is the electrode activity for a particular reaction at equilibrium. In PEMFC, the anode i_o for hydrogen oxidation is very high compared to the cathode i_o for oxygen reduction, therefore, the cathode contribution to this polarization is often neglected. DMFCs have approximately equal i_o for the anode and cathode. In high temperature fuel cells, the operating temperatures are so high that the activation losses are very low. It seems that the activation polarization should increase linearly with temperature based upon equation 6-24, but the purpose of increasing temperature is to decrease activation polarization.

The ohmic and concentration polarization regions are discussed in detail in Chapter 7 (Fuel Cell Charge Transport) and Chapter 8, (Fuel Cell Mass Transport) respectively.

Example 6-2 A 50 cm^2 active area hydrogen–oxygen fuel cell operates at a temperature of 75°C and 1 atm. For the polarization curve, the transfer coefficient, α, is 1, the reference exchange current density, i_0, is 0.001 mA/cm^2, and the internal resistance, R, is 0.1.8 Ωcm^2. (a) Calculate the nominal power output, (b) Calculate the new power output with an increased operating temperature of 90°C and a cell voltage of 0.75 V.

The power output is W_{el} = E × I × A. The current density needs to be determined from the polarization curve. The fuel cell polarization curve can be approximated from the following equation since no hydrogen crossover, internal current losses, and no limiting current are given:

$$E = E_r - \frac{RT}{\alpha F}\ln\left(\frac{i}{i_0}\right) - iR_i$$

$$E_r = 1.482 - 0.000845T + 0.0000431T\ln(P_{H_2}P_{O_2}^{0.5})$$

$$E_r = 1.482 - 0.000845*348.15 + 0.0000431*348.15*\ln(0.21)^{0.5} = 1.176\,V$$

$$0.75V = 1.176V - \frac{(8.314)*(348.15\,K)}{(1)*(96,485\,C/mol)}\ln\left(\frac{i}{0.001\,mA/cm^2}\right) - i*0.18\,\Omega cm^2$$

The current density must be calculated iteratively, by plotting the polarization curve or by linear approximation. With a voltage of 0.75 V, the fuel cell power output is

$$W_{el} = E \times i \times A = 0.75 * 1.2 * 50 = 45 \; W$$

(b) For the increase in temperature, a new iteration, linear approximation, or polarization curve needs to be created. After doing this, the new fuel cell power output is

$$E_r = 1.482 - 0.000845T + 0.0000431 T \ln(P_{H_2} P_{O_2}^{0.5})$$

$$E_r = 1.482 - 0.000845 * 363.15 + 0.0000431 * 363.15 * \ln(0.21)^{0.5} = 1.163 \; V$$

$$0.75 \; V = 1.163 \; V - \frac{(8.314)*(363.15 \; K)}{(1)*(96{,}485 \; C/mol)} \ln\left(\frac{i}{0.001 \; mA/cm^2}\right) - i*0.18 \; \Omega cm^2$$

$$W_{el} = E \times i \times A = 0.75 * 1.1 * 50 = 41.25 \; W$$

Example 6-3 Use the Tafel equation to calculate the activation losses for a fuel cell operating at a current density of 0.7 A/cm^2 at standard pressure and temperature (25°C and 1 atm).

Using the simplified version of the Tafel equation:

$$\Delta V_{act} = a + b\ln(i), \text{ where } a = -\frac{RT}{\alpha F} \ln(i_0) \text{ and } b = -\frac{RT}{\alpha F}$$

The Tafel slope is

$$\Delta V_{actA} = 0.14 + 0.005\ln(i)$$

$$\Delta V_{actC} = 0.2 + 0.007\ln(i)$$

$$\Delta V_{actA} = 0.14 + 0.005\ln(700) = 0.173 \; V$$

$$\Delta V_{actC} = 0.2 + 0.007\ln(700) = 0.246 \; V$$

6.3 Internal Currents and Crossover Currents

Despite the fact that the electrolyte is not electrically conductive and is typically impermeable to gases, some hydrogen and electrons diffuse through the electrolyte. Each hydrogen molecule that diffuses through the electrolyte results in fewer electrons that travel to the external circuit. These losses are usually minor during fuel cell operation, but can sometimes be significant when the fuel cell operates at low current densities, or when it is at open circuit voltage. If the total electrical current is the sum of the current that can be used and the current that is lost, then:

$$i = i_{ext} + i_{loss} \qquad (6\text{-}25)$$

The current density (A/cm²) in the fuel cell is i = i/A. If the total current density is used in equation 6-22, then:

$$E = E_r - \frac{RT}{\alpha F}\ln\left(\frac{i_{ext} + i_{loss}}{i_0}\right) \qquad (6\text{-}26)$$

The activation potential is described by the second half of equation 6-26, and was also shown in equations 6-20 and 6-21. Hydrogen crossover and internal currents have different effects in the fuel cell. Hydrogen that diffuses through the electrolyte typically will form water and reduce the cell potential. Hydrogen crossover is a function of the electrolyte properties, such as permeability, thickness, and partial pressure [3]. A very low open-circuit potential (< 0.9 V) typically indicates a hydrogen leak or an electrical short [3].

6.4 Improving Kinetic Performance

There are several methods that can be used to increase kinetic performance:

- Increase the temperature
- Increase the reactant concentration
- Decrease the activation barrier
- Increase the number of reaction sites

Increasing the temperature of the reaction will increase the fuel cell performance kinetics. An increased temperature will cause the molecules in the system to vibrate and reach the activated state more easily, which increases the rate of reaction. However, at high overvoltage levels, increasing the temperature can decrease the current density.

The reactants can be increased in concentration by using a purer reactant and/or by increasing the pressure/flow rate of the reactant. For example, most PEM fuel cells use air instead of pure oxygen at the cathode. Reactants tend to decrease at the electrodes quickly under high current-density operation due to mass transfer limitations (Chapter 8 explains this phenomenon in more detail). This causes a temporary shortage of the reactant at the cathode, which leads to further kinetic losses.

Another way of improving fuel cell kinetic performance is to decrease the activation barrier of the fuel cell. This can be accomplished by using a highly catalytic electrode. The type of catalyst used should have an adequate bond between the electrode surface, and should be neither too strong nor too weak. A bond that is too weak will not adequately bond with hydrogen, and a bond that is too strong will bond extremely well to the surface, and the H^+ ions will not be able to be easily separated.

Increasing the number of reaction sites will also improve the kinetic performance of the fuel cell. The reaction sites can be increased by creating a greater catalyst surface area for the reaction to proceed. A common method of achieving this is to roughen the electrode surface.

Chapter Summary

This chapter covered the basic electrochemistry needed to predict electrode kinetics, activation losses, currents, and potentials in a fuel cell. The catalyst activation barrier has to be overcome in order to convert products into reactants. In order to lower this activation barrier, a portion of the voltage is lost, and this is called activation overvoltage. The Butler-Volmer equation describes the relationship between the current density and activation overvoltage. The equations in this chapter help to predict how fast reactants respond in order to produce electric current and power, how the reactants form products and produce current in the cell, and how much energy loss occurs during the actual electrochemical reaction. In order to calculate the actual fuel cell voltage, the concepts in this chapter will be combined with the fuel cell charge and mass transport concepts in Chapters 7 and 8.

Problems

1. If a portable electronic device draws 1 A of current at a voltage of 4 V, what is the power requirement for the device? (b) You would like to design a device to have an operating lifetime of 80 hours. Assuming 100-percent fuel utilization, what is the minimum amount of H_2 fuel (in grams) required?

2. A hydrogen–oxygen fuel cell has the following polarization curve parameters: $i_0 = 0.003$, $\alpha = 0.5$, and $R_i = 0.15$ Ohm·cm^2. The fuel cell operates at 65°C and 1 bar. (a) Calculate the cell voltage at 0.8 A/cm^2. (b) Calculate the voltage gain if the cell is going to be operated at 3 bar.

3. Calculate the expected current density at 0.75 V if an MEA is prepared with a catalyst specific area of 640 cm^2/mg and with Pt loading of 0.4 mg/cm^2, where the cell operates at 70°C and 200 kPa. The cathode exchange current density is 1×10^{-10} A/cm^2 of platinum surface. What potential gain may be expected at the same current density if the Pt loading on the cathode is increased to 2 mg/cm^2?

4. A hydrogen/oxygen fuel cell operates at 65°C and 2 atm. The exchange current density at these conditions is 0.002 mA/cm^2 of the electrode area. Pt loading is 0.3 mg/cm^2 and the charge transfer coefficient is 0.5. The electrode area is 50 cm^2. (a) Calculate the theoretical fuel cell potential at these conditions. (b) The open circuit voltage for this fuel cell is 1.1 V. Calculate the current density loss due to hydrogen crossover or internal currents.

5. What is the limiting current density of a 20 cm^2 fuel cell with a hydrogen flow rate of 0.1 g/s?

Bibliography

[1] Atkins, P.W. *Physical Chemistry*, 6th ed. 1998. Oxford: Oxford University Press.
[2] O'Hayre, Ryan, Suk-Won Cha, Whitney Colella, and Fritz B. Prinz. *Fuel Cell Fundamentals*. 2006. New York: John Wiley & Sons.
[3] Barbir, Frano. *PEM Fuel Cells: Theory and Practice*. 2005. Burlington, MA: Elsevier Academic Press.
[4] Lin, Bruce. "Conceptual Design and Modeling of a Fuel Cell Scooter for Urban Asia." Princeton University, Masters Thesis. 1999.
[5] Mench, M.M., Z.H. Wang, K. Bhatia, and C.Y. Wang. "Design of a Micro-direct Methanol Fuel Cell," Electrochemical Engine Center, Department of Mechanical and Nuclear Engineering, Pennsylvania State University. 2001.
[6] Mench, Matthew M., Cao-Yang Wang, and Stephan T. Tynell. "An Introduction to Fuel Cells and Related Transport Phenomena." Department of Mechanical and Nuclear Engineering, Pennsylvania State University. Draft. http://mtrl1.mne.psu.edu/Document/jtpoverview.pdf Last Accessed March 4, 2007.
[7] Sousa, Ruy Jr., and Ernesto Gonzalez. "Mathematical Modeling of Polymer Electrolyte Fuel Cells." *Journal of Power Sources*. Vol. 147, 2005, pp. 32–45.
[8] You, Lixin, and Hongtan Liu. "A Two-Phase Flow and Transport Model for PEM Fuel Cells." *Journal of Power Sources*. Vol. 155, 2006, pp. 219–230.
[9] Li, Xianguo. *Principles of Fuel Cells*. 2006. New York: Taylor & Francis Group.
[10] Springer et al. "Polymer Electrolyte Fuel Cell Model." *J. Electrochem. Soc.* Vol. 138, No. 8, 1991, pp. 2334–2342.
[11] Rowe, A. and X. Li. "Mathematical Modeling of Proton Exchange Membrane Fuel Cells." *Journal of Power Sources*. Vol. 102, 2001, pp. 82–96.
[12] Appleby, A and F. Foulkes. *Fuel Cell Handbook*. 1989. New York: Van Nostrand Reinhold.
[13] Bard, A.J., and L.R. Faulkner. *Electrochemical Methods*. 1980. New York: John Wiley & Sons.
[14] Chen, E., Thermodynamics and Electrochemical Kinetics, in G. Hoogers (editor), *Fuel Cell Technology Handbook*. Boca Raton, FL: CRC Press, 2003.

Chapter 7

Fuel Cell Charge Transport

This chapter will cover the fuel cell electronic and ionic charge transport and voltage losses due to transport resistance. The specific topics to be covered are

- Voltage Loss Due to Charge Transport
- Microscopic Conductivity of Metals
- Ionic Conductivity of Aqueous Electrolytes
- Ionic Conductivity of Polymer Electrolytes
- Ionic Conductivity of Ceramic Electrolytes

Electronic charge transport describes the movement of charges from the electrode where they are produced, to the load where they are consumed. The two major types of charged particles are electrons and ions. Ionic transport is far more difficult to predict and model than fuel cell electron transport. The transfer of ions occurs when O^{2-}, OH^-, CO_3^{2-} or H^+ ions travel through the electrolyte. Resistance to charge transport results in a voltage loss for fuel cells (ohmic loss). Ohmic losses can be minimized by making electrolytes as thin as possible, and employing high conductivity materials which are connected well to each other.

7.1 Voltage Loss Due to Charge Transport

Conductors have an intrinsic resistance to charge flow, which results in a loss in cell voltage. This phenomenon is called "ohmic polarization," and it occurs because of the electrical resistance in the cell components. The cell components that contribute to the electrical resistance are the electrolyte, the catalyst layer, the gas diffusion layer, bipolar plates, interface contacts and terminal connections. The reduction in voltage is dominated by internal ohmic losses through the fuel cell. This voltage loss is called "ohmic loss," and includes the electronic (R_{elec}) and ionic (R_{ionic}) contributions to fuel cell resistance. Ohmic polarization can be described by Ohm's law:

$$v_{ohmic} = IR_{ohmic} = I(R_{elec} + R_{ionic}) \qquad (7\text{-}1)$$

R_{ionic} dominates the reaction in equation 7-1 since ionic transport is more difficult than electronic charge transport. R_{ionic} is the ionic resistance of the electrolyte, and R_{elec} includes the electrical resistance of bipolar plates, cell interconnects, contacts, and other cell components through which electrons flow.

The conductivity of a material is the material's ability to support the flow of charge through the material. The electrical resistance of the fuel cell components is often expressed in the literature as conductance (σ), which is the reciprocal of resistance:

$$\sigma = \frac{1}{\rho} \qquad (7\text{-}2)$$

where ρ is the resistivity. Resistance is characteristic of the size, shape and properties of the material, as expressed by equation 7-3:

$$R = \frac{L_{cond}}{\sigma A_{cond}} = \frac{L_{cond}\rho}{A_{cond}} \qquad (7\text{-}3)$$

where L_{cond} is the length (cm) of the conductor, A_{cond} is the cross-sectional area (cm^2) of the conductor, and σ is the electrical conductivity (ohm^{-1}cm^{-1}). The current density, i, (A/cm^2) can be defined as:

$$i = \frac{I}{A_{cell}} \qquad (7\text{-}4)$$

or

$$i = n_{carriers} q v_{drift} = \sigma \xi \qquad (7\text{-}5)$$

where A_{cell} is the active area of the fuel cell, $n_{carriers}$ is the number of charge carriers (carriers/cm^3), q is the charge on each carrier (1.602×10^{-19} C), v_{drift} is the average drift velocity (cm/s) where the charge carriers move and ξ is the electric field. The conductivity for most materials is:

$$\sigma = nq\frac{v}{\xi} \qquad (7\text{-}6)$$

The term $\frac{v}{\xi}$ can be defined as the mobility, u_i. A more specific equation for material conductivity can be characterized by two major factors: the number of carriers available, and the mobility of those carriers in the material. This can be expressed as:

$$\sigma_i = F\sum |z_i| c_i u_i \qquad (7\text{-}7)$$

where c_i is the number of moles of charge carriers per unit volume, u_i is the mobility of the charge carriers within the material (cm^2/Vs), z_i is the charge number (valence electrons) for the carrier, and F is Faraday's constant.

If the fuel cell resistance is decreased, fuel cell performance will improve [2]. The fuel cell resistance changes with area. When studying ohmic losses, it is helpful to compare resistances on a per-unit basis using current density. Ohmic losses can be calculated from current density using

$$v_{ohmic} = i(ASR_{ohmic}) = i(A_{cell} R_{ohmic}) \qquad (7\text{-}8)$$

where ASR_{ohmic} is area specific resistance of the fuel cell. This parameter allows fuel cells of different sizes to be compared. There are differences in the conduction mechanisms for electronic versus ionic conduction. In a metallic conductor, valence electrons associated with the atoms of the metal become detached, and are free to move around the metal. In a typical ionic conductor, the ions move into the interstitials or vacancies in the crystallographic lattice. The ions move from site to site, hopping through defects in the material. The number of charge carriers in an electronic conductor is much higher than a ionic conductor. Electron and ionic transport is shown in Figures 7-1 and 7-2.

Table 7-1 shows a summary and comparision of electronic and ionic conductors, and the fuel cell components that are classified under each type.

When the cell is designed properly, the ohmic polarization is typically dominated by electrolyte conductivity, which is primarily a function temperature and, in the case of PEM fuel cells, water content. Ohmic

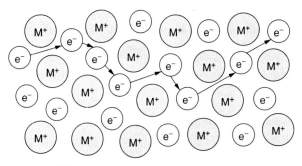

Figure 7-1 Electron transport in a metal.

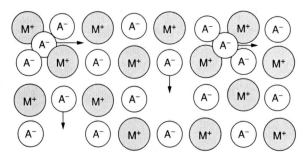

Figure 7-2 Ionic transport in a crystalline ionic conductor.

losses can be reduced through advanced conductive materials, thinner electrolytes, or an optimal temperature/water balance.

One of the most effective methods of reducing ohmic loss is to use a better ionic conductor for the electrolyte layer, or a thinner electrolyte layer since the electrolyte component of a fuel cell dominates the ohmic losses. In the case of PEM fuel cells, thinner membranes are also advantageous because they keep the anode electrode saturated through "back" diffusion of water from the cathode. At very high current densities (fast fluid flows), mass transport causes a rapid drop-off in the voltage, because oxygen and hydrogen simply cannot diffuse through the electrode and ionize quickly enough, and products cannot be moved out at the necessary speed [4].

Since the ohmic overpotential for the fuel cell is mainly due to ionic resistance in the electrolyte, this can be expressed as:

$$v_{ohmic} = IR_{ohmic} = iA_{cell}\left(\frac{\delta_{thick}}{\sigma A_{fuelcell}}\right) = \frac{i\delta_{thick}}{\sigma} \qquad (7\text{-}9)$$

TABLE 7-1 Comparison of Electronic and Ionic Conduction for Fuel Cell Components

Materials	Conductivity	Fuel cell components
Electronic Conductors		
Metals	$10^3 - 10^7$	Bipolar plates, gas diffusion layer, contacts, interconnects, end plates
Semiconductors	$10^{-3} - 10^4$	Bipolar plates, end plates
Ionic Conductors		
Ionic Crystals	$10^{-16} - 10^{-2}$	SOFC YSZ electrolyte
Solid/polymer Electrolytes	$10^{-1} - 10^3$	PEMFC Nafion electrolyte, DMFC Nafion electrolyte,
Liquid Electrolytes	$10^{-1} - 10^3$	AFC KOH electrolyte, PAFC $H_3PO_4/H_4P_2O_7$ electrolyte, MCFC $Li_2CO_3 + K_2CO_3$ electrolyte

where A_{cell} is the active area of the fuel cell, δ_{thick} is the thickness of the electrolyte layer and σ is the conductivity. As mentioned previously, equation 7-9 shows that the ohmic potential can be reduced by using a thinner electrolyte layer and using a higher ionic conductivity electrolyte.

Example 7-1 Determine the ohmic voltage loss for a 25 cm² PEMFC that has an electrolyte membrane with a conductivity of 0.20 Ω^{-1} cm^{-1} and a thickness of 75 microns. The current density is 0.8 A/cm² and R_{elec} for the fuel is 0.005 Ω.

First, calculate R_{ionic} based upon electrolyte dimensions to calculate v_{ohmic}. The current of the fuel cell is

$$I = iA = 0.8 A/cm^2 \times 25\ cm^2 = 20\ A$$

$$R = \frac{L}{\sigma A} = \frac{0.0075\ cm}{(0.10\ \Omega^{-1} cm^{-1}) * (25\ cm^2)} = 0.003\ \Omega$$

$$v_{ohmic} = I(R_{elec} + R_{ionic}) = 20\ A * (0.005\ \Omega + 0.003\ \Omega) = 0.16\ V$$

If this equation is calculated for thinner and thicker membranes, one will notice that the ohmic loss is reduced with thinner membranes.

Example 7-2 Calculate the ohmic voltage losses for two fuel cell sizes at a current density of 0.7 A/cm²: (a) $A_1 = 5\ cm^2$, $R_1 = 0.05\ \Omega$; (b) $A_1 = 25\ cm^2$, $R_1 = 0.02\ \Omega$.

(a)

$$ASR_1 = R_1 A_1 = (0.05\ \Omega)(5\ cm^2) = 0.25\ \Omega cm^2$$

The ohmic loss can be calculated as follows:

$$v_{ohmic1} = i(ASR_1) = \left(0.7\ \frac{A}{cm^2}\right) 0.25\ \Omega cm^2 = 0.175\ V$$

Convert the current densities into fuel cells with currents:

$$I_1 = iA_1 = 0.7\ \frac{A}{cm^2} * 5\ cm^2 = 3.5\ A$$

The ohmic voltage losses are

$$v_{ohmic1} = I_1(R_1) = 3.5\ A * 0.05\ \Omega = 0.175\ V$$

(b)

$$ASR_2 = R_2 A_2 = (0.02\ \Omega)(25\ cm^2) = 0.50\ \Omega cm^2$$

The ohmic loss can be calculated as follows:

$$v_{ohmic2} = i(ASR_2) = \left(0.7\ \frac{A}{cm^2}\right) 0.50\ \Omega cm^2 = 0.350\ V$$

Convert the current densities into fuel cells with currents:

$$I_1 = iA_2 = 0.7 \frac{A}{cm^2} * 25 \ cm^2 = 17.5 \ A$$

The ohmic voltage losses are

$$v_{ohmic2} = I_2(R_2) = 17.5 \ A * 0.02 \ \Omega = 0.35 \ V$$

7.2 Microscopic Conductivity in Metals

The electron conductivity of metals used in a fuel cell is an important property to consider because it affects the charge transfer of electrons. Typical components made of metals are bipolar plates and current collectors. The conductivity, σ, of a material can be related to microscopic parameters that describe the motion of electrons. Recall the expression for conductivity (equation 7-7):

$$\sigma_i = F \sum |z_i| c_i u_i$$

where u_i is the mobility of free electrons in a metal conductor limited by impurities, defects, lattice vibrations and photon scattering; and can be expressed by:

$$u = \frac{q\tau}{m_e} \tag{7-10}$$

where τ is the relaxation time (time between collisions), m_e is the effective mass of the electron (m = 9.11×10^{-31} kg) and q is the elementary electron charge in coulombs (q = 1.602×10^{-19} C). Inserting equation 7-10 into the equation for conductivity (7-7) gives:

$$\sigma = \frac{|z_e| c_e q \tau}{m_e} \tag{7-11}$$

Equation 7-11 shows that the carrier concentration in a metal can be calculated from the density of free electrons [2].

7.3 Ionic Conductivity in Aqueous Electrolytes

There are several fuel cell types that use aqueous electrolytes. MCFCs use molten $(K/Li)_2CO_3$, PAFCs use aqueous or concentrated H_3PO_4, and AFCs use an aqueous KOH electrolyte. The aqueous electrolyte is trapped in a matrix material for support. The matrix material not only mechanically holds the electrolyte, but it also prevents crossover of reactant gases, and creates a short distance between electrodes (typically 0.1–1 mm).

As an ion moves through the liquid, it encounters two forces: an electrical force, F_ξ, and a frictional force, F_D. The two forces act in opposite directions, and the ions quickly reach a terminal speed. The electric field force is given by equation 7-12:

$$F_\xi = |z_i| q \frac{dV}{dx} \qquad (7\text{-}12)$$

where $|z_i|$ is the charge number of the ion, and q is the fundamental electron charge (1.602×10^{-19} C). The frictional force is:

$$F_D = 6\pi \mu r_{ion} v_{ion} \qquad (7\text{-}13)$$

where r_{ion} is the radius of the ion, μ is the viscosity of the liquid and v_{ion} is the velocity of the ion. If these forces are made equal to each other, the mobility can be predicted:

$$u_i = v_{ion}\frac{dV}{dx} = \frac{|z_i| q}{6\pi \mu r_{ion}} \qquad (7\text{-}14)$$

Equation 7-14 shows that mobility is a function of ion size and liquid viscosity, and gives an estimate for dilute aqueous solutions. If the ions are too large, or the liquid is too thick, the ion mobility is reduced. For more concentrated ionic solutions, the conductivity is more difficult to calculate because of the multitude of electrical interactions that occurs.

The conductivity can be written in terms of the mobilities of all ions present:

$$\sigma_i = F \sum |z_i| c_i u_i = F \sum |z_i| c_i \left(\frac{|z_i| q}{6\pi \mu r_{ion}} \right) \qquad (7\text{-}15)$$

Equation 7-15 shows that the conductivity reflects the charge and mobility of all ions present in the solution, and their concentrations [10,13].

7.4 Ionic Conductivity of Polymer Electrolytes

In order for a polymer to be a good conductor, it should have a fixed number of charge sites and open space. The charged sites have the opposite charge of the moving ions, and provide a temporary resting place for the ion. Increasing the number of charged sites raises the ionic conductivity, but an excessive number of charged side chains may degrade

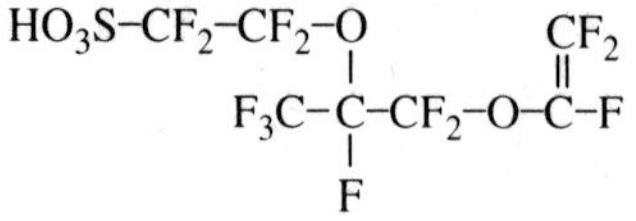

HO$_3$S-CF$_2$-CF$_2$-O
                |
F$_3$C-C-CF$_2$-O-C-F
     |
     F

Figure 7-3 Chemical structure of Nafion.

the stability of the polymer. Polymers usually have a certain amount of free volume, and increasing the free volume allows more space for the ions to move across the polymer. Ions are also transported through the polymer membrane by hitching onto water molecules that move through the membrane. A persulfonated polytetrafluoroethylene (PTFE)-based polymer, known as Nafion, has high conductivity, and is currently the most popular membrane used for PEMFCs and DMFCs. Nafion has a similar structure to Teflon, but includes sulfonic acid groups ($SO_3^-H^+$) that provide sites for proton transport. Figure 7-3 shows the chemical structure of Nafion.

Nafion has to be fully hydrated with water in order to have high conductivity. Hydration can be achieved by humidifying the gases, or through fuel cell design to allow product water to hydrate the membrane. In the presence of water, the protons form hydronium complexes (H_3O^+), which transport the protons in the aqueous phase. When the Nafion is fully hydrated, the conductivity is similar to liquid electrolytes.

Nafion can hold significant amounts of water—its volume will increase by about 22 percent when fully hydrated [2]! Since the conductivity and the hydration of the membrane are correlated, determining the water content correlates with determining the conductivity of the membrane. The membrane water content at the electrode/membrane interface is determined by the activity of water vapor assuming equilibrium. The water vapor activity, a_{water_vap}, can be defined by [8]:

$$a_{water\_vap} = \frac{p_w}{p_{sat}} \qquad (7\text{-}16)$$

where p_w represents the partial pressure of water vapor in the system, and p_{sat} represents saturation water vapor pressure for the system at the temperature of operation [2,8].

The conductivity of a membrane is highly dependent upon the structure and water content of the membrane. The amount of water uptake in the membrane also depends upon the membrane properties and pretreatment. For example, at high temperatures, the water uptake by the Nafion membrane is much less due to changes in the polymer at high temperatures. The relationship between water activity on the faces of the membrane and water content can be described by [8]:

$$\lambda = 0.043 + 17.18 a_{water\_vap} - 39.85 (a_{water\_vap})^2 + 36 (a_{water\_vap})^3 \qquad (7\text{-}17)$$

where $0 < a_{water\_vap} \leq 1$. One of the main reasons water content varies in Nafion is because the protons usually have one or more water molecules associated with them. This phenomenon is called the electroosmotic drag (n_{drag}), which is the number of water molecules accompanying the movement of each proton [2,8]:

$$n_{drag} = n_{drag}^{Sat} \frac{\lambda}{22} \tag{7-18}$$

where n_{drag}^{Sat} is the electroosmotic drag and λ is the water content. The electroosmotic drag, n_{drag}^{Sat}, is assumed to be 2.5 in [8], but was experimentally found to range from 2.3 to 2.7 for a fully hydrated membrane in equilibrium with liquid water at 30° to 50°C [8]. The water content, λ, ranges from 0 to 22 water molecules per sulfonate group, and $\lambda = 22$ when Nafion is fully hydrated [2]. The water drag flux from the anode to the cathode with a net current i is [2]:

$$J_{H_2O,drag} = 2n_{drag} \frac{i}{2F} \tag{7-19}$$

where $J_{H_2O,drag}$ is the molar flux of water due to the electroosmotic drag (mol/scm^2), and j is the current density of the fuel cell (A/cm^2).

The electroosmotic drag moves water from the anode to the cathode, and when the water builds up at the cathode, some water travels back through the membrane. This is known as back diffusion, and it usually happens because the amount of water at the cathode is many times greater than at the anode. The water back-diffusion flux can be determined by:

$$J_{H_2O,backdiffusion} = \frac{\rho_{dry}}{M_m} D_\lambda \frac{d\lambda}{dz} \tag{7-20}$$

where ρ_{dry} is the dry density (kg/m^3) of Nafion, M_n is the Nafion equivalent weight (kg/mol), D_λ is the water diffusivity and z is the direction through the membrane thickness.

The net water flux in the membrane is a combination of the electroosmotic drag and back diffusion, and can be calculated by using equation 7-21 [2,8]:

$$J_{H_2O,backdiffusion} = 2n_{drag}^{Sat} \frac{i}{2F} \frac{\lambda}{22} - \frac{\rho_{dry}}{M_m} D_\lambda(\lambda) \frac{d\lambda}{dz} \tag{7-21}$$

Water uptake results in the membrane swelling and changes its dimensions and conductivity, which is a significant factor for fuel cell design. When the water profile in the membrane has been determined, the membrane conductivity, resistance and potential drop across it can be calculated. The ionic conductivity, σ (S/cm), can be correlated with the water content and temperature using the following relation [8]:

$$\sigma = (0.005139\lambda - 0.00326)exp\left[1268\left(\frac{1}{303} - \frac{1}{T}\right)\right] \quad (7\text{-}22)$$

Since the conductivity of Nafion can change depending upon water content, the resistance of the membrane also changes with water saturation. The total resistance of a membrane (R_m) is found by integrating the local resistance over the membrane thickness:

$$R_m = \int_0^{t_m} \frac{dz}{\sigma[\lambda(z)]} \quad (7\text{-}23)$$

where t_m is the membrane thickness, λ is the water content of membrane, σ is the conductivity (S/cm) of the membrane.

Example 7-3 A hydrogen fuel cell operates at 60°C at 1 atm. It has a Nafion 112 membrane of 50 microns, and the following equation can be used for the water content across the membrane: $\lambda(z) = 5 + 2 \exp(100z)$. This fuel cell has a current density of 0.8 A/cm². Estimate the ohmic overvoltage loss across the membrane.

The conductivity profile of the membrane is

$$\sigma(z) = 0.005193(5 + 2\exp(100z) - 0.00326) \times \exp 1268\left(\frac{1}{303} - \frac{1}{333}\right)$$

$$\sigma(z) = 0.04107 + 0.01878\exp(100z)$$

The resistance of the membrane is

$$R_m = \int_0^{t_m} \frac{dz}{\sigma(\lambda(z))} = \int_0^{0.0050} \frac{dz}{0.04107 + 0.01878 \exp(100z)} = 0.15\ \Omega cm^2$$

The ohmic overvoltage due to the membrane resistance in this fuel cell is

$$V_{ohm} = I \times R_m = 0.8\ A/cm^2 \times 0.15\ \Omega cm^2 = 0.12\ V$$

7.5 Ionic Conduction in Ceramic Electrolytes

Most medium to high temperature fuel cell types use a ceramic electrolyte for the ion transport. The most popular high temperature electrolyte material is yttria stabilized zirconia (YSZ), which is used in solid oxide fuel cells (SOFCs). Adding yttria to zirconia introduces oxygen vacancies, which enables YSZ to have high ionic conductivity.

Since vacancies are required for ions to move within the ceramic, these are considered charge carriers. Increasing the yttria content results in increased oxygen vacancies, but there is an upper limit to doping. Excessive doping will cause defects to interact with each other, and decrease conductivity. The maximum amount of doping is about 8 percent molar yttria concentration for YSZ [2]. Since mobility is related to the movement of ionic defects, it is directly related to the diffusion coefficient (D) by Einstein's equation [10,13]:

$$u_i = \frac{z_i D}{kT} \qquad (7\text{-}24)$$

where k is the Boltzmann constant ($8.616 * 10^{-5}$ eV/K). Substituting equation 7-24 into equation 7-7:

$$\sigma = \frac{|z_i|^2 c_i D}{kT} \qquad (7\text{-}25)$$

For thermally-generated ionic defects, the concentration and diffusion are activated processes. These can be described by:

$$c \propto \exp\left(\frac{-q_1}{kT}\right) \qquad D \propto \exp\left(\frac{-q_2}{kT}\right) \qquad (7\text{-}26)$$

Conductivity also has the same form:

$$\sigma = \sigma_0 \exp\left(\frac{-E}{kT}\right) \qquad (7\text{-}27)$$

where $E = q_1 + q_2$ is the activation energy for conduction.

Chapter Summary

Charge transport is predominantly driven by conduction. The ohmic resistance losses includes the resistance from the electrodes, electrolyte, and interconnects; however, the total resistance is dominated by the electrolyte resistance. Since the resistance scales with conductivity, developing high conductivity electrolyte and electrode materials is critical. Electrolytes must not only be conductive, but also stable in a chemical environment and able to withstand the required fuel cell temperatures. The electrolyte types commonly used in fuel cells are liquid, polymer, and ceramic. The equations for conductivity for each electrolyte type very slightly depending upon the method of ion transport through that particular medium.

Problems

1. A 10 cm^2 fuel cell has R_{elec} = 0.01 Ω and $\sigma_{electrolyte}$ = 0.10 Ω^{-1}cm^{-1}. If the electrolyte is 100 microns thick, predict the ohmic voltage losses for the fuel cell at j = 500 mA/cm^2.

2. Estimate the ohmic overpotential for a fuel cell operating at 70°C. The external load is 1 A/cm^2, and it uses a 50-micron thick membrane. The humidity levels, $a_{w,anode}$ and $a_{w,cathode}$ are 1.0 and 0.5, respectively.

3. A fuel cell is operating at 0.8 A/cm^2 and 60°C. Hydrogen gas at 30°C and 50 percent relative humidity is provided to the fuel cell at a rate of 2 A. The fuel cell area is 10 cm^2, and the drag ratio of water molecules to hydrogen is 0.7. The hydrogen exhaust exits the fuel cell at 60°C and p = 1 atm.

4. In a PEMFC, the water activities on the anode and cathode sides of a Nafion 115 membrane are 0.7 and 0.9 respectively. The fuel cell is operating at a temperature of 60°C and 1atm with a current density of 0.8 A/cm^2. Estimate the ohmic overvoltage loss across the membrane.

Bibliography

[1] Sousa, Ruy Jr., and Ernesto Gonzalez. "Mathematical Modeling of Polymer Electrolyte Fuel Cells." *Journal of Power Sources*. Vol. 147, 2005, pp. 32–45.
[2] O'Hayre, Ryan, Suk-Won Cha, Whitney Colella, and Fritz B. Prinz. *Fuel Cell Fundamentals*. 2006. New York: John Wiley & Sons.
[3] Barbir, Frano. *PEM Fuel Cells: Theory and Practice*. 2005. Burlington, MA: Elsevier Academic Press.
[4] Lin, Bruce. "Conceptual Design and Modeling of a Fuel Cell Scooter for Urban Asia." Princeton University, Master's Thesis. 1999.
[5] Mench, M.M., Z.H. Wang, K. Bhatia, and C.Y. Wang. "Design of a Micro-direct Methanol Fuel Cell." Electrochemical Engine Center, Department of Mechanical and Nuclear Engineering, Pennsylvania State University. 2001.
[6] Mench, Matthew M., Cao-Yang Wang, and Stephan T. Tynell. "An Introduction to Fuel Cells and Related Transport Phenomena." Department of Mechanical and Nuclear Engineering, Pennsylvania State University. Draft. http://mtrl1.mne.psu.edu/Document/jtpoverview.pdf Last Accessed March 4, 2007.
[7] You, Lixin, and Hongtan Liu. "A Two-Phase Flow and Transport Model for PEM Fuel Cells." *Journal of Power Sources*. Vol. 155, 2006, pp. 219–230.
[8] Springer et al. "Polymer Electrolyte Fuel Cell Model." *J. Electrochem. Soc.* Vol. 138, No. 8, 1991, pp. 2334–2342.
[9] Li, Xianguo. *Principles of Fuel Cells*. 2006. New York: Taylor & Francis Group.
[10] Bard, A.J. and L.R. Faulkner. Electrochemical Methods: Fundamentals and Applications. New York: John Wiley & Sons, 1980.
[11] Rowe, A. and X. Li. "Mathematical Modeling of Proton Exchange Membrane Fuel Cells." *Journal of Power Sources*, Vol. 102, 2001, pp. 82–96.
[12] Kreuer, K.D. On the Development of Proton Conducting Materials for Technological Applications. Solid State Ionics. 97, 1997. pp. 1–15.
[13] Bagotsky, V.S. Fundamentals of Electrochemisty. 2nd ed. Hoboken, NJ: John Wiley & Sons, 2006.

Chapter 8

Fuel Cell Mass Transport

A fuel cell must continuously be supplied with fuel and oxidant in order to produce electricity. Products must also be continuously removed in order to have maximum fuel cell efficiency. The study of mass transfer of uncharged species is important because it can lead to significant fuel cell performance losses if not understood properly. The reactant and product concentrations within the catalyst layer determine the fuel cell performance. Concentration loss can be minimized by optimizing the mass transport in the fuel cell electrodes and flow structures. This chapter covers the following aspects of mass transfer:

- Mass transport in fuel cell electrodes
- Mass transport in fuel cell flow structures

The mass transport in the fuel cell electrodes/fuel structures is dominated by convection and the laws of fluid dynamics since the flow channels are macroscale (usually in millimeters or centimeters). The mass transport of the fuel cell electrodes occur on a microscale and are dominated by diffusion.

The convective forces that dominate mass transfer in the flow channels are primarily imposed by the fuel, while the oxidant flow rates are imposed by the user. High flow rates can ensure a good distribution of reactants, but may cause other problems in the fuel cell stack, such as high pressures, and fuel cell membrane rupture.

The diffusive forces that occur in the electrode/catalyst layer are shielded from the convective forces in the flow channels. The velocity of the reactants tends to slow down near the gas diffusion/catalyst layers where the diffusion regime of the reactants begins. Figure 8-1 illustrates convective flow in the reactant flow channel and diffusive flow through the gas diffusion and catalyst layers.

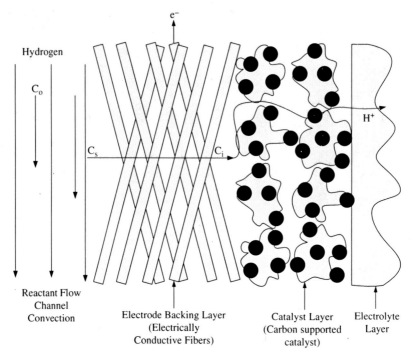

Figure 8-1 Example fuel cell layers (PEMFC and DMFC) that have convective and diffusive mass transport.

8.1 Convective Mass Transport from Flow Channels to Electrode

As shown in Figure 8-1, the reactant is supplied to the flow channel at a concentration C_0, and is transported from the flow channel to the concentration at the electrode surface C_s through convection. The rate of mass transfer is then [9]:

$$\dot{m} = A_{elec} h_m (C_0 - C_s) \quad (8\text{-}1)$$

where A_{elec} is the electrode surface area, and h_m is the mass transfer coefficient.

The value of h_m is dependent upon the channel geometry, the physical properties of species i and j, and the wall conditions. H_m can be found from the Sherwood number:

$$h_m = Sh \frac{D_{i,j}}{D_h} \quad (8\text{-}2)$$

Sh is the Sherwood number, D_h is the hydraulic diameter, and D_{ij} is the binary diffusion coefficient for species i and j. The Sherwood number depends upon channel geometry, and can be expressed as:

$$Sh \equiv \frac{h_H D_h}{k} \qquad (8\text{-}3)$$

where Sh = 5.39 for uniform surface mass flux ($\dot{m}$ = constant)., and Sh = 4.86 for uniform surface concentration (C_s = constant).

The binary diffusion coefficients for many fuel cell fuels are shown in Appendix G. If the binary diffusion coefficient needs to be calculated at a different temperature than what is shown in Appendix G, the following relation can be used [3]:

$$D_{i,j}(T) = D_{i,j}(T_{ref}) * \left(\frac{T}{T_{ref}}\right)^{3/2} \qquad (8\text{-}4)$$

where T_{ref} is the temperature at which the binary diffusion coefficient is given, and T is the temperature of the fuel used in the fuel cell.

8.2 Diffusive Mass Transport in Fuel Cell Electrodes

As shown in Figure 8-1, the diffusive flow occurs at the electrode backing and catalyst layer. The electrochemical reaction in the catalyst layer can lead to reactant depletion, which can affect fuel cell performance as predicted by the Nernst equation (see Chapter 6). To determine the size of the concentration loss, the amount the catalyst layer reactant and product concentrations differ from the bulk values needs to be found.

The rate of mass transfer by diffusion of the reactants to the catalyst layer ($\dot{m}$) can be calculated as shown in equation 8-5:

$$\dot{m} = -D\frac{dC}{dx} \qquad (8\text{-}5)$$

where D is the bulk diffusion coefficient and C is the concentration of reactants.

The diffusional transport through the electrode backing layer at steady-state is:

$$\dot{m} = A_{elec} D^{eff} \frac{C_s - C_i}{\delta} \qquad (8\text{-}6)$$

where C_i is the reactant concentration at the backing layer/catalyst interface, and δ is the electrode-backing layer thickness, and D^{eff} is the effective diffusion coefficient for the porous electrode backing layer, (which is dependent upon the bulk diffusion coefficient D and the pore structure). Assuming uniform pore size, the backing layer is free from flooding of water or liquid electrolyte, and D^{eff} can be defined as:

$$D^{eff} = D\phi^{3/2} \tag{8-7}$$

where ϕ is the electrode porosity. The total resistance to the transport of the reactant to the reaction sites can be expressed by combining equations 8-1 and 8-3 [9]:

$$\dot{m} = \frac{C_0 - C_i}{\left(\dfrac{1}{h_m A_{elec}} + \dfrac{\delta}{D^{eff} A_{elec}}\right)} \tag{8-8}$$

where $\dfrac{1}{h_m A_{elec}}$ is the resistance to the convective mass transfer, and $\dfrac{L}{D^{eff} A_{elec}}$ is the resistance to the diffusional mass transfer through the electrode backing layer.

When the fuel cell is turned on, it begins producing electricity at a fixed current density, i. The reactant and product concentrations in the fuel cell are constant. As soon as the fuel cell begins producing current, the electrochemical reaction leads to the depletion of reactants at the catalyst layer. The flux of reactants and products will match the consumption/depletion rate of reactants and products at the catalyst layer which can be described by the following equation:

$$i = \frac{nF\dot{m}}{A_{elec}} \tag{8-9}$$

where i is the fuel cell's operating current density, F is the Faraday constant, n is the number of electrons transferred per mol of reactant consumed, and $\dot{m}$ is the rate of mass transfer by diffusion of reactants to the catalyst layer. Substituting equation 8-8 into 8-9 yields [9]:

$$i = -nF\frac{C_0 - C_i}{\left(\dfrac{1}{h_m} + \dfrac{\delta}{D^{eff}}\right)} \tag{8-10}$$

The reactant concentration in the backing layer/catalyst interface is less than the reactant concentration supplied to the flow channels, which depends upon i, δ, and D^{eff}. The higher the current density, the worse the concentration losses will be. These concentration losses can be improved if the diffusion layer thickness is reduced, or the effective diffusivity is increased.

The limiting current density of the fuel cell is the point where the current density becomes so large the reactant concentration falls to zero. The limiting current density (i_L) of the fuel cell can be calculated if the minimum concentration at the backing catalyst layer interface is $C_i = 0$ as follows [9]:

$$i_L = -nF \frac{C_0}{\left(\frac{1}{h_m} + \frac{\delta}{D^{eff}}\right)} \quad (8\text{-}11)$$

When designing a fuel cell, the limiting current density can be increased by ensuring that C_0 is high, which is accomplished by designing good flow structures to evenly distribute the reactants, and ensuring that D^{eff} is large and δ is small by optimizing fuel cell operating conditions such as temperature, pressure, electrode structure flooding, and diffusion layer thickness.

The typical limiting current density is 1 to 10 A/cm². The fuel cell will not be able to produce a higher current density than its limiting current density. However, other types of losses may limit the fuel cell voltage to zero before the limiting current density does.

The fuel cell performance is affected by the reactant concentration through the Nernst equation. This is because the thermodynamic voltage of the fuel cell is determined by the reactant and product concentrations at the catalyst sites. Recall the Nernst equation from Chapter 5:

$$E = E_r - \frac{RT}{nF} \ln \frac{\prod a_{products}^{v_i}}{\prod a_{reactants}^{v_i}} \quad (8\text{-}12)$$

In order to calculate the incremental voltage loss due to reactant depletion in the catalyst layer, the changes in Nernst potential using c_R^* values instead of c_R^0 values are represented by the following [2]:

$$v_{conc} = E_{r,Nernst} - E_{Nernst} \quad (8\text{-}13)$$

$$v_{conc} = \left(E_r - \frac{RT}{nF} \ln \frac{1}{C_0}\right) - \left(E_r - \frac{RT}{nF} \ln \frac{1}{C_i}\right) \quad (8\text{-}14)$$

$$v_{conc} = \frac{RT}{nF} \ln \frac{C_0}{C_i} \quad (8\text{-}15)$$

where $E_{r,\text{Nernst}}$ is the Nernst voltage using C_0 values and E_{Nernst} is the Nernst voltage using C_i values. Combining equations 8-7 and 8-8:

$$\frac{i}{i_L} = 1 - \frac{C_i}{C_0} \tag{8-16}$$

Therefore, the ratio C_0/C_i (the concentration at the backing/catalyst layer interface can be written as:

$$\frac{C_0}{C_i} = \frac{i_L}{i_L - i} \tag{8-17}$$

Substituting equation 8-14 into 8-12 yields [2,9]:

$$v_{conc} = \frac{RT}{nF} \ln\left(\frac{i_L}{i_L - i}\right) \tag{8-18}$$

This is the basic expression for concentration polarization loss, and is only valid for $i < i_L$.

Concentration also affects fuel cell performance through reaction kinetics. The reaction kinetics are dependent upon the reactant and product concentrations at the reaction sites. Recall that the reaction kinetics can be described by the Butler-Volmer equation from Chapter 6 [2,9]:

$$i = i_0\left\{\frac{C_i}{C_0}\exp\left[\frac{\alpha nF(E - E_r)}{RT}\right] - \frac{C_j}{C_n}\exp\left[\frac{-(1-\alpha)nF(E - E_r)}{RT}\right]\right\}$$
$$\tag{8-19}$$

where C_i and C_j are arbitrary concentrations and i_0 is measured as the reference reactant and product concentration values C_0 and C_n. In the high current-density region, the second term in the Butler-Volmer equation drops out, and the expression then becomes:

$$i = i_0\frac{C_i}{C_0}\exp\left[\frac{\alpha nF(E - E_r)}{RT}\right] \tag{8-20}$$

In terms of activation overvoltage:

$$v_{conc} = \frac{RT}{\alpha nF}\frac{C_0}{C_i} \tag{8-21}$$

The ratio can be written as:

$$\frac{C_0}{C_i} = \frac{i_L}{i_L - i} \qquad (8\text{-}22)$$

The fuel cell concentration loss (or mass transport loss) can be written as [2]:

$$v_{conc} = \left(\frac{RT}{nF}\right)\frac{i_L}{i_L - i} \qquad (8\text{-}23)$$

Actual fuel cell behavior frequently has a larger value than what the equation predicts. Due to this, sometimes the equation is obtained empirically. The concentration loss appears at high current density, and is severe. Significant concentration loss limits fuel cell performance.

8.3 Convective Mass Transport in Flow Structures

Fuel cell flow structures are designed to distribute reactants across a fuel cell. The typical fuel cell has a series of small flowfields to evenly distribute reactants, and to keep mass transport losses to a minimum. Flowfield designs are discussed thoroughly in Chapter 12. Some basic equations can be used to obtain an idea of the mass transfer in the flow channels, but to obtain more accurate results, the use of computational fluid dynamics (CFD) software has become a popular method of modeling mass and heat transfer in fuel cell flow channels. Sections 8.3.1 and 8.3.2 present some equations for modeling the mass transport in the flow structures.

8.3.1 Mass transport in flow channels

The mass transport in flow channels can be modeled using a control volume for reactant flow from the flow channel to the electrode layer as shown in Figure 8-2.

The rate of convective mass transfer at the electrode surface ($\dot{m}_s$) can be expressed as [9]:

$$\dot{m}_s = h_m(C_m - C_s) \qquad (8\text{-}24)$$

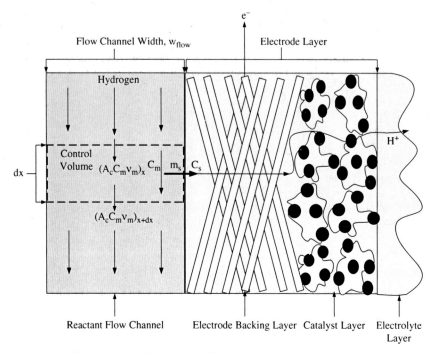

Figure 8-2 Control volume for reactant flow from the flow channel to the electrode layer (Adapted from [9]).

where C_m is the mean concentration of the reactant in the flow channel (averaged over the channel cross-section, and decreases along the flow direction, x), and C_s is the concentration at the electrode surface.

As shown in Figure 8-2, the reactant moves at the molar flow rate, $A_c C_m v_m$ at the position x, where A_c is the channel cross-sectional area and v_m is the mean flow velocity in the flow channel. This can be expressed as [9]:

$$\frac{d}{dx}(A_c C_m v_m) = -\dot{m}_s w_{elec} \qquad (8\text{-}25)$$

where w_{elec} is the width of the electrode surface. If the flow in the channel is assumed to be steady, then the velocity and concentration is constant:

$$\frac{d}{dx}C_m = \frac{-\dot{m}_s}{v_m w_{flow}} \qquad (8\text{-}26)$$

If the current density is small (i < 0.5 i_L), it can be assumed constant. Using Faraday's law, $\dot{m}_s = \frac{i}{nF}$ and integrating, this can be shown as [9]:

$$C_m(x) = C_{m,in}(x) - \frac{\left(\frac{i}{nF}\right)}{v_m w_{flow}} x \qquad (8\text{-}27)$$

where $C_{m,in}$ is the mean concentration at the flow channel inlet.

If the current density is large (i > 0.5 i_L), the condition at the electrode surface can be approximated by assuming the concentration at the surface (C_s) is constant. This can be written as follows [9]:

$$\frac{d}{dx}(C_m - C_s) = \frac{-h_m}{v_m w_{flow}}(C_m - C_s) \qquad (8\text{-}28)$$

After integrating from the channel inlet to location x in the flow channel, equation 8-28 becomes:

$$\frac{C_m - C_s}{(C_m - C_s)_{in}} = \exp\frac{-h_m x}{v_m w_{flow}} \qquad (8\text{-}29)$$

At the channel outlet, x = H, and equation becomes:

$$\frac{C_{m,out} - C_s}{C_{m,in} - C_s} = \exp\frac{-h_m H}{v_m w_{flow}} \qquad (8\text{-}30)$$

where $C_{m,out}$ is the mean concentration at the flow channel outlet.

A simple expression can be derived if the entire flow channel is assumed to be the control volume, as shown in Figure 8-3:

$$\dot{m}_s = v_m w_{flow} w_{elec}(C_{in} - C_{out})$$

$$\dot{m}_s = v_m w_{flow} w_{elec}(\Delta C_{in} - \Delta C_{out}) \qquad (8\text{-}31)$$

If C_s is constant, substituting for $w_{flow} w_{elec}$ [2,9]:

$$\dot{m}_s = A h_m \Delta C_{lm} \qquad (8\text{-}32)$$

where

$$\Delta C_{lm} = \frac{\Delta C_{in} - \Delta C_{out}}{\ln\left(\frac{\Delta C_{in}}{\Delta C_{out}}\right)} \qquad (8\text{-}33)$$

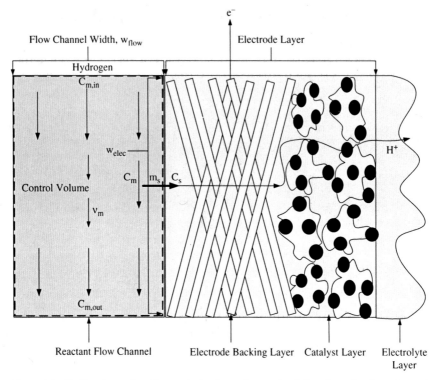

Figure 8-3 Entire channel as the control volume for reactant flow from the flow channel to the electrode layer.

The local current density corresponding to the rate of mass transfer is [9]:

$$i(x) = nFh_m(C_m - C_s)\exp\left(\frac{-h_m x}{v_m w_{flow}}\right) \quad (8\text{-}34)$$

The current density averaged over the electrode surface is:

$$\bar{i} = nFh_m \Delta C_{lm} \quad (8\text{-}35)$$

The limiting current density when C_s approaches 0 is [9]:

$$i_L(x) = nFh_m C_{m,in}\exp\left(\frac{-h_m x}{v_m w_{flow}}\right) \quad (8\text{-}36)$$

$$\bar{i}_L = nFh_m\left[\frac{\Delta C_{in} - \Delta C_{out}}{\ln\left(\frac{\Delta C_{in}}{\Delta C_{out}}\right)}\right] \quad (8\text{-}37)$$

Both the current density and limiting current density decrease exponentially along the channel length.

Example 8-1 A fuel cell operating at 25°C and 1 atm uses bipolar plates with flow fields to distribute the fuel and oxidant to the electrode surface. The channels have a depth of 1.5 mm, with a distance of 1 mm apart. Air is fed parallel to the channels walls for distribution to the cathode electrode. The length of the flow channel is 18 cm, and the air travels at a velocity of 2 m/s. Determine the distribution of the current density due to the limitation of the convective mass transfer $i_L(x)$ and the average limiting current density $\bar{i}_L$.

The Reynolds number can be calculated as follows:

$$Re = \frac{\rho v_m D}{\mu} = \frac{v_m D}{v} = \frac{2 \text{ m/s} * 2 * 1 \times 10^{-3} \text{ m}}{15.89 \times 10^{-6} \text{ m}^2/s} = 251.73$$

Since 251.73 is less than 2000, the flow is laminar.

In order to calculate the limiting current density, the convective mass transfer coefficient, and binary diffusivity coefficient needs to be calculated:

$$D_{i,j}(T) = D_{i,j}(T_{ref}) * \left(\frac{T}{T_{ref}}\right)^{3/2} = D_{O_2-N_2}(273) * \left(\frac{298}{273}\right)^{3/2} = 0.21 \times 10^{-4} \text{ m}^2/s$$

$$h_m = Sh\frac{D_{i,j}}{D_h} = 4.86 * \frac{0.21 \times 10^{-4} \text{ m}^2/s}{2 * 1 \times 10^{-3} \text{ m}} = 0.0105 \text{ m/s}$$

The concentration of O_2 at the channel inlet, with a mol fraction of O_2 is $X_{O_2} = 0.21$ is:

$$C_{O_2,in} = X_{O_2} * \left(\frac{P}{RT}\right) = 0.21 * \frac{101,325}{8.314 \text{ J/molK} * 298 \text{ K}} = 8.588 \text{ mol/m}^3$$

The limiting current density based upon the rate of O_2 transfer:

$$i_L(x) = nFh_m C_{m,in} \exp\left(\frac{-h_m x}{v_m W_{flow}}\right)$$

$$i_L(x) = 4 * 96,487 \frac{C}{molO_2} * 0.0105 \frac{m}{s} * 8.588 \frac{molO_2}{m^3} * \exp\left(\frac{0.0105 \text{ m/s}}{1 * 10^{-3} * 2 \text{ m/s}}\right)$$

$$= 3.4802 * 10^4 \times \exp(-5.25x)$$

The limiting current density will be 3.4802 A/cm² at the channel inlet ($x = 0$) and 3.3022 A/cm² at the channel outlet ($x = 18$ cm). In order to calculate $\bar{i}_L$, the outlet concentration of oxygen needs to be calculated:

$$C_{m,out} = C_{m,in} * \exp\left(\frac{-h_m x}{v_m W_{flow}}\right) = 8.588 \frac{mol}{m^3} * \exp\left(\frac{-0.0105 \text{ m/s} * 0.18 \text{ m}}{1 * 10^{-3} * 2 \text{ m/s}}\right)$$

$$= 3.338 \text{ mol/m}^3$$

Then,

$$\bar{i}_L = nFh_m \left[\frac{\Delta C_{in} - \Delta C_{out}}{\ln\left(\frac{\Delta C_{in}}{\Delta C_{out}}\right)} \right]$$

$$= 4*96{,}487 \frac{C}{molO_2} * 0.0105 \frac{m}{s} \left[\frac{(8.588 - 3.338)\frac{molO_2}{m^3}}{\ln\left(\frac{8.588}{3.338}\right)} \right]$$

$$= 2.251 \frac{A}{cm^2}$$

8.3.2 Pressure drop in flow channels

In many fuel cell types, the flow fields are usually arranged as a number of parallel flow channels; therefore, the pressure drop along a channel is also the pressure drop in the entire flow field. In a typical flow channel, the gas moves from one end to the other at a mean velocity. The pressure difference between the inlet and outlet drives the fluid flow. By increasing the pressure drop between the outlet and inlet, the velocity is increased. The flow through bipolar plate channels is typically laminar, and is proportional to the flow rate. The pressure drop can be approximated using the equations for incompressible flow in pipes:

$$\Delta P = f \frac{L_{chan}}{D_H} \rho \frac{\bar{v}^2}{2} + \Sigma K_L \rho \frac{\bar{v}^2}{2} \qquad (8\text{-}38)$$

where f is the friction factor, L_{chan} is the channel length, m, D_H is the hydraulic diameter, m, ρ is the fluid density, kg/m³, $\bar{v}$ is the average velocity, m/s, and K_L is the local resistance.

The hydraulic diameter can be defined by [3]:

$$D_H = \frac{4 \times A_c}{P_{cs}} \qquad (8\text{-}39)$$

where A_c is the cross-sectional area, and P_{cs} is the perimeter. For the typical rectangular flow field, the hydraulic diameter can be defined as:

$$D_H = \frac{2w_c d_c}{w_c + d_c} \qquad (8\text{-}40)$$

where w_c is the channel width, and d_c is the depth.

The channel length can be defined as:

$$L_{chan} = \frac{A_{cell}}{N_{ch}(w_c + w_L)} \qquad (8\text{-}41)$$

where A_{cell} is the cell active area, N_{ch} is the number of parallel channels, w_c is the channel width, m, and w_L is the space between channels, m.
The friction factor can be defined by:

$$f = \frac{56}{Re} \qquad (8\text{-}42)$$

The velocity at the fuel cell entrance is [3]:

$$v = \frac{Q_{stack}}{N_{cell} N_{ch} A_{ch}} \qquad (8\text{-}43)$$

where v is the velocity in the channel (m/s), Q_{stack} is the air flow rate at the stack entrance, m³/s, N_{cell} is the number of cells in the stack, N_{ch} is the number of parallel channels in each cell, and A_{ch} is the cross-sectional area of the channel.

For the case of the PEMFC fuel cell, the total flow rate at the stack entrance is [3]:

$$Q_{stack} = \frac{I}{4F} \frac{S_{O_2}}{x_{O_2}} \frac{RT_{in}}{P_{in} - \varphi P_{sat}} N_{cell} \qquad (8\text{-}44)$$

where Q is the volumetric flow rate (m³/s), I is the stack current, F is the Faraday's constant, S_{O_2} is the oxygen stoichiometric ratio, x_{O_2} is the oxygen content in the air, R is the universal gas constant, T_{in} is the stack inlet temperature, P_{in} is the pressure at the stack inlet, Φ is the relative humidity, P_{sat} is the saturation pressure at the given inlet temperature, and N_{cell} is the number of cells in the stack.

By combining the previous equations, the velocity at the stack inlet is [3]:

$$v = \frac{I}{4F} \frac{S_{O_2}}{x_{O_2}} \frac{(w_c + w_L)L_{chan}}{w_c d_c} \frac{RT}{P - \phi P_{sat}} \qquad (8\text{-}45)$$

To approximate the pressure drop in the flow channels, it must be determined whether the flow in the channels is laminar or turbulent. This can be accomplished by calculating the Reynolds number. If the Reynold's number is less than 2300, the flow is laminar; if it is greater than 2300, it can be characterized as turbulent. At the channel entrance:

$$Re = \frac{\rho v_m D_{ch}}{\mu} = \frac{v_m D_{ch}}{v} \qquad (8\text{-}46)$$

where v_m is the characteristic velocity of the flow (m/s), D_{ch} is the flow channel diameter or characteristic length (m), ρ is the fluid density (kg/m³), μ is the fluid viscosity (kg/(m∗s or N∗s/m²), and v is the kinematic viscosity (m²/s). It is found that regardless of channel size or flow velocity, f∗Re = 16 for circular channels. Equation 8-48 can be used for circular channels.

For rectangular channels, to compute the effective Reynolds number:

$$Re_h = \frac{\rho v_m D_h}{\mu} \qquad (8\text{-}47)$$

where D_h = 4A/P = 4 ∗ (cross-sectional area)/perimeter.

A relationship in the literature for rectangular channels can be approximated by [3]:

$$f*Re = 24(1 - 1.3553 \times \alpha* + 1.9467 \times \alpha*^2 - 1.7012 \times \alpha*^3 + 0.9564 \times \alpha*^4 - 0.2537 \times \alpha*^5) \qquad (8\text{-}48)$$

where $\alpha*$ is the aspect ratio of the cross-section, and $\alpha* = \frac{b}{a}$ where 2a and 2b are the lengths of the channels' sides. For PEMFCs, the Reynold's number can be calculated by [3]:

$$Re = \frac{\rho v D_H}{\mu} = \frac{1}{\mu} \frac{i}{2F} \frac{S_{O_2}}{x_{O_2}} \frac{(w_c + w_L)L_{chan}}{w_c + d_c} M_{air}$$
$$+ M_{H_2O} \frac{\phi P_{sat(T_{in})}}{P_{in} - \phi P_{sat(T_{in})}} \qquad (8\text{-}49)$$

The flow rate at the stack outlet is usually different than the inlet. Chapter 16 gives more detailed information about calculating flow rates. If it is assumed that the outlet flow is saturated with water vapor, for PEMFCs the flow rate is [3]:

$$Q_{stack} = \frac{I}{4F}\left(\frac{S_{O_2}}{x_{O_2}} - 1\right) \frac{RT_{out}}{P_{in} - \Delta P - \varphi P_{sat(T_{out})}} N_{cell} \qquad (8\text{-}50)$$

where ΔP is the pressure drop in the stack.

The variation in viscosity varies with temperature. For dilute gases, the temperature dependence of viscosity can be estimated using a simple power law:

$$\frac{\mu}{\mu_0} \approx \left(\frac{T}{T_0}\right)^n \qquad (8\text{-}51)$$

where μ_0 is the viscosity at temperature T_0. In these equations, n, μ_0, and T_0 can be obtained from experiments or calculated through kinetic theory.

Fuel cell gas streams are rarely composed of a single species. Usually, they are gas mixtures, such as oxygen and nitrogen from the air. The following expression provides a good estimate for the viscosity of a gas mixture:

$$\mu_{mix} = \sum_{i=1}^{N} \frac{x_i \mu_i}{\sum_{j=1}^{N} x_j \Phi_{ij}} \tag{8-52}$$

where Φ_{ij} is a dimensionless number obtained from:

$$\Phi_{ij} = \frac{1}{\sqrt{8}} \left(1 + \frac{M_i}{M_j}\right)^{-1/2} \left[1 + \left(\frac{\mu_i}{\mu_j}\right)^{1/2} \left(\frac{M_i}{M_j}\right)^{1/4}\right]^2 \tag{8-53}$$

where N is the total number of species in the mixture, x_i and x_j are the mol fractions of species i and j, and M_i and M_j are the molecular weight (kg/mol) of species i and j.

For porous flow fields, the pressure drop is determined by Darcy's Law:

$$\Delta P = \mu \frac{Q_{cell}}{k A_c} L_{chan} \tag{8-54}$$

where μ is the viscosity of the fluid, Q_{cell} is the geometric flow rate through the cell, m³/s, K is the permeability, m², A_c is the cross-sectional area of the flow field, m², and L_{chan} is the length of the flow field.

When using this set of equations, there are a few assumptions that are made that will cause a slight deviation from the actual values:

- The channels are not smooth on both sides. The channels are typically smooth on one side, but the gas diffusion layer or porous electrode side has a rough surface.
- The gas is not simply flowing through the channels. It is also reacting with the catalyst.
- The temperature may not be uniform through the channels.
- There are a number of bends or turns that should be accounted for in the channels.

Example 8-2 Calculate the pressure drop through a PEMFC cathode flow field of a single graphite plate with 100 cm² cell area. The stack operates at 3 atm at 60°C with 100-percent saturated air. The flow field consists of 24 parallel serpentine channels 1 mm wide, 1 mm deep, and 1 mm apart. The cell operates at 0.7 A/cm² at 0.65 V.

The pressure drop is:

$$\Delta P = f\frac{L}{D_H}\rho\frac{\bar{v}^2}{2} + \Sigma K_L \rho\frac{\bar{v}^2}{2}$$

The hydraulic diameter is:

$$D_H = \frac{2w_c d_c}{w_c + d_c} = \frac{2*0.1*0.1}{0.1 + 0.1} = 0.1\ cm$$

The channel length is:

$$L = \frac{A_{cell}}{N_{ch}(w_c + w_L)} = \frac{100}{24(0.1 + 0.1)} = 20.83\ cm$$

The flow rate at the stack entrance is:

$$Q_{stack} = \frac{I}{4F}\left(\frac{S}{x_{O_2}}\right)\frac{RT_{in}}{P_{in} - \varphi P_{sat(T_{in})}}N_{cell}$$

$$= \frac{0.7*100}{4*96,485}*\frac{1}{0.21}*\frac{8.314*333.15}{303,975.03 - 19,944}*1$$

$$= 8.423*10^{-7}\frac{m^3}{s} = 8.423\frac{cm^3}{s}$$

The velocity in a fuel cell channel near the entrance of the cell is:

$$v = \frac{Q_{stack}}{N_{cell}N_{ch}A_{ch}} = \frac{8.423}{1*24*0.1*0.1} = 35.096\ \frac{cm}{s}$$

The Reynolds number at the channel entrance is:

$$Re = \frac{\rho\bar{v}D_H}{\mu}$$

$$\rho = \frac{(P - P_{sat})M_{air} + P_{sat}M_{H_2O}}{RT}$$

$$= \frac{(303,975.03 - 19,944)*29 + 19,944*18}{8314*333.15}$$

$$= 3.10\ kgm^3 = 0.0031\ gcm^3$$

$$\mu = 2*10^{-5}\ kg/ms = 0.0002\frac{g}{cms}$$

$$Re = \frac{\rho\bar{v}D_H}{\mu} = \frac{0.00123*35.096*0.1}{0.0002} = 21.584$$

For rectangular channels:

$$Ref \approx 55 + 41.5\exp\left(\frac{-3.4}{w_c/d_c}\right) = 56$$

$$f = \frac{56}{Re} = \frac{56}{21.584} = 2.594$$

The pressure drop is:

$$\Delta P = f\frac{L}{D_H}\rho\frac{\bar{v}^2}{2} + \Sigma K_{LP}\frac{\bar{v}^2}{2} = 2.594*\frac{0.2083}{0.001}1.23*\frac{0.351^2}{2}$$

$$+ 1.23*\frac{0.351^2}{2} = 33.36 \text{ Pa}$$

Chapter Summary

Mass transport involves the supply of reactants and products in a fuel cell. Inadequate mass transport can result in poor fuel cell performance due to the lack or abundance of fuel cell reactants. Two main mass transport effects are encountered in fuel cells: diffusion in the electrodes, and convection in the flow structures. Depletion of reactants in the fuel cell affects the Nernstian cell voltage and the kinetics at the electrodes. This can be avoided by optimizing the transport in the fuel cell flow channels. Convection in fuel cells can be characterized by the Reynolds number, and the flow is usually laminar. Reactant flow is driven through a flow channel due to a pressure difference. Pressure losses are encountered due to the friction between the fluid and channel wall. Simple mass transport models can provide a lot of information about how reactants are depleted in a flow channel. The mass transport can be improved generally by increasing the reactant flow viscosity, decreasing the channel size or decreasing the diffusion layer thickness.

Problems

1. A fuel cell is operating at 750°C and 1 atm. Humidified air is supplied with the mole fraction of water vapor equal to 0.2 in the cathode. If the channels are rectangular with a diameter of 1.2 mm, find the maximum velocity of air.

2. For the fuel cell in Problem 1, calculate the maximum velocity of air if the channels are circular.

3. A fuel cell is operating at 50°C and 1 atm. The cathode is using pure oxygen, and there is no water vapor present. The diffusion layer is 300 microns with a porosity of 30 percent. Calculate the limiting current density.

4. Calculate the limiting current density for a fuel cell operating at 180°C and 1 atm. The cathode is of the same construction as in Problem 3.

5. Under the conditions from Problem 3, estimate the fuel cell area that can be operated at 0.8 A/cm^2. Assume a stoichiometric number of 2.5, and that the fuel cell is made of a single straight channel with a width of 1 mm and the rib width is 0.8 mm.

Bibliography

[1] Sousa, Ruy Jr., and Ernesto Gonzalez. "Mathematical Modeling of Polymer Electrolyte Fuel Cells." *Journal of Power Sources.* Vol. 147, 2005, pp. 32–45.
[2] O'Hayre, Ryan, Suk-Won Cha, Whitney Colella, and Fritz B. Prinz. *Fuel Cell Fundamentals.* 2006. New York: John Wiley & Sons.
[3] Barbir, Frano. *PEM Fuel Cells: Theory and Practice.* 2005. Burlington, MA: Elsevier Academic Press.
[4] Lin, Bruce. "Conceptual Design and Modeling of a Fuel Cell Scooter for Urban Asia." Princeton University, Masters Thesis. 1999.
[5] Mench, M.M., Z.H. Wang, K. Bhatia, and C.Y. Wang. "Design of a Micro-Direct Methanol Fuel Cell," Electrochemical Engine Center, Department of Mechanical and Nuclear Engineering, Pennsylvania State University. 2001.
[6] Mench, Matthew M., Cao-Yang Wang, and Stephan T. Tynell. "An Introduction to Fuel Cells and Related Transport Phenomena." Department of Mechanical and Nuclear Engineering, Pennsylvania State University. Draft. http://mtrl1.mne.psu.edu/Document/jtpoverview.pdf Last Accessed March 4, 2007.
[7] You, Lixin, and Hongtan Liu. "A Two-Phase Flow and Transport Model for PEM Fuel Cells." *Journal of Power Sources.* Vol. 155, 2006, pp. 219–230.
[8] Springer et al. "Polymer Electrolyte Fuel Cell Model." *J. Electrochem. Soc.*, Vol. 138, No. 8, 1991, pp. 2334–2342.
[9] Li, Xianguo. *Principles of Fuel Cells.* 2006. New York: Taylor & Francis Group.
[10] Springer et al. "Polymer Electrolyte Fuel Cell Model," *J. Electrochem. Soc.*, Vol. 138, No. 8, 1991, pp. 2334–2342.
[11] Rowe, A. and X. Li. "Mathematical Modeling of Proton Exchange Membrane Fuel Cells," *Journal of Power Sources.* Vol. 102, 2001, pp. 82–96.
[12] Mennola, T et al. "Mass Transport in the Cathode of a Free-Breathing Polymer Electrolyte Membrane Fuel Cell." *Journal of Applied Electrochemistry.* 33: 979–987, 2003.
[13] Beale, S.B. "Calculation Procedure for Mass Transfer in Fuel Cells." *Journal of Power Sources.* Vol. 128, 2004, pp. 185–192.
[14] Giles, R.V., J.B. Evett and C. Liu. Fluid Mechanics and Hydraulics. 1995. Third Edition. Schaum's Outline of Theory and Problems. New York: McGraw-Hill.
[15] Schultz, Thorsten and Kai Sundmacher. "Mass, Charge and Energy Transport Phenomena in a Polymer Electrolyte Membrane (PEM) Used in a Direct Methanol Fuel Cell (DMFC): Modeling and Experimental Validation of Fluxes." Journal of Membrane Science. Vol. 276, 2006, pp. 272–285.
[16] Vandersteen, J.D.J., Kenney, B., Pharoah, J.G. and Karan, K. "Mathematical Modelling of the Transport Phenomena and the Chemical/Electrochemical Reactions in Solid Oxide Fuel Cells: A Review." Canadian Hydrogen and Fuel Cells Conference, September 2004.

Chapter 9

Heat Transfer

Temperature in a fuel cell is not always uniform, even when there is a constant mass flow rate in the flow structures. This is due to heat transfer, and for certain fuel cell types, phase change of the reactants or products. In order to accurately predict temperature-dependent parameters as well as rates of reaction and species transport, the temperature and heat distribution need to be determined accurately. Convective heat transfer occurs between the solid surface and the gas streams, and conductive heat transfer occurs in the solid and/or porous structures. The reactants, products, and electricity generated are the basic components to consider in modeling basic heat transfer in a fuel cell as shown in Figure 9-1.

Before heat transfer equations are used to determine the heat gain or loss by the system, an energy balance needs to be completed. The total energy balance around the fuel cell is based upon the power produced, the fuel cell reactions, and the heat loss that occurs in a fuel cell. The energy balance varies for each fuel cell type and system because of the different reactions that occur in different fuel cells, and the system design and components. The general energy balance states that the enthalpy of the reactants entering the cell equals the enthalpy of the products leaving the cell plus the sum of the heat generated, the power output, and the rate of heat loss to the surroundings.

The calculations presented in this chapter give the heat transfer basics for fuel cells. The specific topics to be covered are:

- Fuel Cell Energy Balance
- Heat Generation and Flux in Fuel Cell Layers
- Heat Conduction

152 Chapter Nine

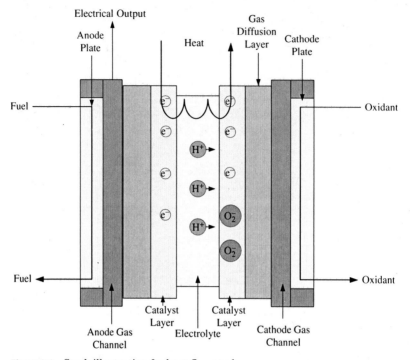

Figure 9-1 Stack illustration for heat flow study.

- Heat Dissipation Through Natural Convection and Radiation
- Fuel Cell Heat Management

The topics covered in this chapter will aid in predicting the temperatures and heat in overall fuel cell stack, and in the stack components.

9.1 Fuel Cell Energy Balance

9.1.1 General energy balance procedure

In order to properly design a fuel cell system, the energy that flows into and out of each process unit in the fuel cell subsystem, and in the fuel cell itself needs to be accounted for in order to determine the overall energy requirement(s) for the process. A typical energy balance calculation determines the cell exit temperature knowing the reactant composition, the temperatures, H_2 and O_2 utilization, the expected power produced, and the percent of heat loss. As a refresher, the procedure for formulating an energy balance is as follows:

1. A flowchart must be drawn and labeled. Enough information should be included on the flowchart to determine the specific enthalpy of

each stream component. This includes known temperatures, pressures, mole fractions, mass flow rates, and phases.

2. Mass balance equations may need to be written in order to determine the flow rates of all stream components (see Chapter 16 for a refresher).
3. The specific enthalpies need to be determined of each stream component. These can be obtained from thermodynamic tables, or can be calculated if this data is not available.
4. The final step is to write the appropriate form of the energy balance equation, and solve for the desired quantity.

An example flowchart in shown in Figure 9-2. The fuel enters the cell at temperature, T, and pressure, P. Oxygen enters the fuel cell from the environment at a certain T, P, x_{O_2} (mole fraction) and m_{O_2} (mass flow rate). The fuel and oxygen react completely in the cell to produce CO_2 and water vapor, which exit in separate streams at T, P, x (mole fraction) and m (mass flow rate). This reaction can be described by:

$$C_aH_b + \left(a + \frac{b}{4}\right)O_2 \rightarrow aCO_2 + \frac{b}{2}H_2O$$

In Figure 9-2, Q is the heat leaving the fuel cell, and W is the work generated through chemical availability. The generic energy balance for the fuel cell in this example is:

$$\frac{W}{m_{fuel}} + \frac{Q}{m_{fuel}} = h_{fuel} + \left(a + \frac{b}{4}\right)h_{O_2} - ah_{CO_2} - \frac{b}{2}h_{H_2O} \quad (9\text{-}1)$$

(The mass balance equations for this example have not been written.)

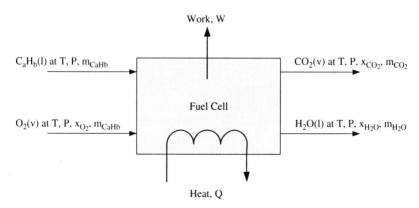

Figure 9-2 Detailed flowchart to obtain energy balance equation.

9.1.2 Energy balance of fuel cell stack

The energy balance on the fuel cell stack is the sum of the energy inputs equals the sum of the energy outputs. The generic heat balance on the stack is [3]:

$$\sum Q_{in} - \sum Q_{out} = W_{el} + Q_{dis} + Q_c \qquad (9\text{-}2)$$

where Q_{in} is the enthalpy (heat) of the reactant gases in, Q_{out} is the enthalpy (heat) of the unused reactants and heat produced by the products, W_{el} is the electricity generated, Q_{dis} the heat dissipated to the surrounding, and Q_c heat taken away from the stack by active cooling. Heat is carried away by the reactants, products, and lost to the surroundings. The remaining heat needs to be taken from the stack through cooling. The heat generation in the fuel cell is associated with voltage losses. Most of the heat is created in the catalyst layers, but there is also heat generation in the membrane, and in the electrically conductive parts of the fuel cell due to ohmic losses [3].

A good estimate for the fuel cell stack energy balance can be obtained by equating the energy of the fuel reacted to the heat and electricity generated [3,26]:

$$\frac{I}{nF} H_{HHV} n_{cell} = Q_{gen} + IV_{cell} n_{cell} \qquad (9\text{-}3)$$

When all of the product water leaves the stack as liquid at room temperature, the heat generated in a fuel cell stack is [26]:

$$Q_{gen} = (1.482 - V_{cell}) I n_{cell} \qquad (9\text{-}4)$$

where Q_{gen} is the heat generated from the stack in Watts, n_{cell} is the number of cells, and V_{cell} is the cell voltage. If all of the product water leaves the stack as vapor, the following equation can be used instead:

$$Q_{gen} = (1.254 - V_{cell}) I n_{cell} \qquad (9\text{-}5)$$

Equations 9-3 to 9-5 are approximations, and do not take into account the heat or enthalpy brought to, or removed from, the stack.

9.1.3 General energy balance for fuel cell

The fuel cell energy balance is the sum of all the energy inputs and the sum of all of the energy outputs [3]:

$$\sum (h_i)_{in} = W_{el} + \sum (h_i)_{out} + Q \qquad (9\text{-}6)$$

The inputs are the enthalpies of the fuel, the oxidant, and the water vapor present (if applicable). The outputs are the electric power produced, enthalpies of the flows out of the fuel cell, and the heat leaving the fuel cell through coolant, convection, or radiation.

The enthalpy (J/s) for each dry gas or mixture of dry gases is [3,16]:

$$h = \dot{m}c_p T \tag{9-7}$$

where $\dot{m}$ is the mass flow rate of the gas or mixture (g/s), c_p is the specific heat (J/[g*K]), and T is the temperature in °C.

If the gas has a high heating value, its enthalpy is then [3,16]:

$$h = \dot{m}(c_p T + h^0_{HHV}) \tag{9-8}$$

where h^0_{HHV} is the higher heating value of that gas (J/g) at 0°C. The heating values are usually reported at 25°C, therefore, the higher heating value may need to be calculated at the chosen temperature. The enthalpy of water vapor is [3,16]:

$$h = \dot{m}_{H_2O(g)} c_{p,H_2O(g)} T + h^0_{fg} \tag{9-9}$$

The enthalpy of liquid water is [16]:

$$h = \dot{m}_{H_2O(l)} c_{p,H_2O(l)} T \tag{9-10}$$

The inputs and outputs of the energy balance can quickly become complex when the heat balance is performed for each individual fuel cell layer and/or stack heating and cooling is involved.

9.1.4 Energy balance for fuel cell components and gases

The overall energy balance for the electrolyte and porous layers (referred to as the MEA in PEMFCs), can be written as [28]:

$$Q_s = Q_c + h_f A_f (T_e - T_{f,av}) + h_a A_a (T_e - T_{a,av}) \tag{9-11}$$

where Q_c is the heat conduction in the solid structure, Q_s is the heat source to account for the electrochemical heat generation, h is the convective heat transfer coefficient, and T is the temperature.

Energy balance in the gaseous phase. Temperature gradients in the gas flows are typically assumed to depend only on convection due to mass and heat transfer from the channel walls to the gases [27]:

$$\sum_i n_i * c_p * \frac{\partial T_g}{\partial x} + \sum_i c_p \frac{\partial n_i}{\partial x} * (T_g - T_s) + h_g * B_g * \frac{1}{a} * (T_g - T_s) = 0 \tag{9-12}$$

where B_g is the ratio between the gas-solid heat exchange area to the cell area, c_p is the specific heat at the constant pressure of species i, T_g is the temperature of the solid, and h_g is the heat transfer coefficient calculated from the Nusselt number:

$$h_g = Nu * \frac{k}{d_h} \qquad (9\text{-}13)$$

The value of the Nusselt number depends upon the channel geometry, the Reynolds number, and the Prandtl number for laminar flow.

Energy balance of a solid structure. The energy balance of the solid structures of the stack describes the unsteady heat conduction. The convective heat, transferred from the gas flow to the solid, and the reaction enthalpies occur as source terms [27]:

$$\frac{1}{s}\sum_g h_g B_g * (T_g - T_s) - \frac{1}{s}\left(\frac{\Delta H}{ne*F} + V\right)I + K_x \frac{\partial^2 T_s}{\partial x^2} = 0 \qquad (9\text{-}14)$$

where K_x is the effective thermal conductivity of the solid structure, accounting for the parallel heat conduction through the parallel layers:

$$K_x = \frac{\sum_h K_h * \delta_h}{\sum_h \delta_h} \qquad (9\text{-}15)$$

Example 9-1 A PEM hydrogen/air fuel cell generates 1 kW at 0.7 V. Air is supplied to the fuel cell stack at 20°C. The mass flow rates for hydrogen going into the cell is 0.02 g/s and air going into the cell is 1.5 g/s. The mass flow rate of the water in air going into the fuel cell is 0.01 g/s. What is the mass flow rate of the air entering the cell? What is the mass flow rate of the water in air leaving the cell? Assume that the heat generated by the fuel cell is negligible. The HHV of hydrogen is 141,900 J/g.

Energy balance is

$$H_{H_2,in} + H_{Air,in} + H_{H_2O\_Air,in} = H_{Air,out} + H_{H_2O\_Air,out} + W_{el}$$

The energy flows are:
Hydrogen in:

$$H_{H_2,in} = m_{H_2,in}(c_{p,H_2}T_{in} + h^0_{HHV})$$

$$h^0_{HHV} = h^{25}_{HHV} - \left(c_{p,H_2} + \frac{1}{2}\frac{M_{O_2}}{M_{H_2}}c_{p,O_2} - \frac{M_{H_2O}}{M_{H_2}}c_{p,H_2O(l)}\right) \cdot 25$$

$$h^0_{HHV} = 141900 - \left(14.2 + \frac{1}{2}*\frac{31.9988}{2.0158}0.913 - \frac{18.0152}{2.0158}4.18\right) \cdot 25$$

$$= 142{,}298 \; J/gK$$

$$H_{H_2,in} = 0.02(14.2 \times 20 + 142{,}298) = 2124.5 \; W$$

Air in:
$$H_{Air,in} = m_{Air,in} c_{p,\,Air} T_{in} = m_{Air,in} \times 1.01 \times 20$$

Water vapor in air in:
$$H_{H_2O\_Air,in} = m_{H_2O\_Air,in} \times (c_{p,H_2O} T_{in} + h^0_{fg})$$
$$H_{H_2O\_Air,in} = 0.01 \times (1.85 \times 20 + 2500) = 25.37$$

Air out:
$$H_{Air,out} = m_{O_2,out} c_{p,O_2} T_{out} + m_{N_2,out} c_{p,N_2} T_{out}$$
$$H_{Air,out} = (0.5*0.913*75) + (1.5*1.04*75) = 151.24$$

The amount of water generated is:
$$I = P/V = 1000 \; W/\, 0.7 \; V = 1428.6$$

$$m_{H_2O,gen} = \frac{I}{2F} M_{H_2O} = \frac{1428.6}{2*96{,}485}*18.015 = 0.133 \; g/s$$

The water mass balance can be written as:
$$m_{H_2O\_Air,in} + m_{H_2O,gen} = m_{H_2O\_Air,out}$$
$$m_{H_2O\_Air,out} = 0.01 \; g/s + 0.133 \; g/s = 0.143 \; g/s$$

Water vapor in air out:
$$H_{H_2O\_Air,out} = m_{H_2O\_Air,out} \times (c_{p,H_2O} T_{out} + h^0_{fg})$$
$$H_{H_2O\_Air,out} = 0.143 \times (1.85 \times 75 + 2500) = 377.34$$

Energy balance is
$$H_{H_2,in} + H_{Air,in} + H_{H_2O\_Air,in} = H_{Air,out} + H_{H_2O\_Air,out} + W_{el}$$
$$2124.5 + H_{Air,in} + 25.37 = 151.24 + 377.34 + 1000$$
$$H_{Air,in} = -621.29 W$$

Substituting this back into the enthalpy calculation for $H_{Air,in}$—
$$H_{Air,in} = m_{Air,in} c_{p,Air} T_{in} = m_{Air,in} \times 1.01 \times 20$$
$$m_{Air,in} = 30.76 \; g/s$$

9.2 Heat Generation and Flux in Fuel Cell Layers

Heat generated by the electrochemical reaction in the anode catalyst layer can be given by [12]:

$$q_{ACL} = i\left(\eta_a - \frac{\Delta H_a - \Delta G_a}{nF}\right) \quad (9\text{-}16)$$

where ΔH_a denotes the anodic reaction enthalpy, and ΔG_a is the anodic Gibbs free energy.

The heat flux can be related to the temperature gradient across the membrane as:

$$q_{ACL} = -\lambda_{mem}\frac{dT}{dx} \quad (9\text{-}17)$$

where λ_{mem} is the effective thermal conductivity of the membrane.

9.3 Heat Conduction

A temperature gradient within a homogenous substance results in an energy transfer through the medium. The rate of heat transfer in the x-direction through a finite cross-sectional area, A, is [25]:

$$q_x = -kA\frac{dT}{dx} \quad (9\text{-}18)$$

where k is the thermal conductivity, W/(m*k). The thermal conductivity of some fuel cell materials is shown in Table 9-1.

TABLE 9-1 Thermal Conductivity of Some Fuel Cell Materials (Adapted from [24])

Material	Thermal conductivity (W/mK) @ 300 K
Aluminum	237
Nickel	90.5
Platinum	71.5
Titanium	22
Stainless Steel 316	13
Graphite	98
$LaCrO_3$	4.3
$La_{0.7}Sr_{0.3}CrO_3$	2.1
$La_{0.8}Ca_{0.2}CrO_3$	1.9
$LaMnO_3$	1.9
Carbon Cloth	1.7
Teflon	0.4

The steady-state heat conduction is governed by the equation:

$$\frac{d^2T}{dx^2} = 0 \qquad (9\text{-}19)$$

When heat is conducted through two adjacent materials with different thermal conductivities, the third boundary condition comes from a requirement that the temperature at the interface is the same for both materials [3]:

$$q = h_{tc} A \Delta T \qquad (9\text{-}20)$$

where h_{tc} is the convective heat transfer coefficient, BTU/hft²F or W/m²K, A is the area normal to the direction of the heat flux, ft² or m², and ΔT is the temperature difference between the solid surface and the fluid F or K.

Internal fuel cell heat generation can be described by the Poisson equation [25]:

$$\frac{d^2T}{dx^2} + \frac{q_{int}}{k} = 0 \qquad (9\text{-}21)$$

where q_{int} is the rate of heat generation per unit volume.

9.4 Heat Dissipation Through Natural Convection and Radiation

The heat lost by the stack through natural convection and radiation to the surroundings is [3]:

$$Q_{dis} = \frac{T_s - T_0}{R_{th}} \qquad (9\text{-}22)$$

where T_s is stack surface temperature, T_0 is the temperature of the surrounding walls, and R_{th} is the thermal resistance, given by [3]:

$$R_{th} = \frac{1}{\dfrac{1}{R_c} + \dfrac{1}{R_R}} \qquad (9\text{-}23)$$

where R_c is the convective thermal resistance:

$$R_c = \frac{1}{hA_s} \qquad (9\text{-}24)$$

where $h = \frac{k}{L}Nu_L$, and R_R is the radioactive thermal resistance defined as [3,25]:

$$R_R = \frac{1}{\sigma F A_s (T_s + T_0)(T_s^2 + T_0^2)} \qquad (9\text{-}25)$$

where σ is Stefan-Boltzmann constant (5.67×10^{-8} W/(m^{2}K^4)), F is the shape factor, and A_s is the stack exposed surface area, m^2.

For small stacks or single cell stacks, the surface area is large and the heat dissipation is greater than the heat generation. Sometimes, small stacks are heated to maintain a certain operating temperature [3].

9.5 Fuel Cell Heat Management

Creating high efficiency fuel cells requires proper temperature control, and heat management to ensure that the fuel cell system runs consistently. Depending upon the fuel cell type, the optimal temperature can range from room temperature to 1000°C, and any deviation from the designed temperature range can result in lowered efficiencies. Higher temperatures also mean faster kinetics and a voltage gain that usually exceeds the voltage loss from the negative thermodynamic relationship between the open-circuit voltage and temperature. Lower temperatures mean shorter warm-up times for the system, and lower thermomechanical stresses. Corrosion and other time- and temperature-dependent processes are retarded, and much less water is required for the saturation of input gases. In the case of PEMFCs, a higher operating temperature means more of the product water is vaporized; thus, more waste heat goes into the latent heat of vaporization, and less liquid water is left to be pushed out of the fuel cell.

Cooling can be achieved through a number of means. Passive cooling of the fuel cell, on the other hand, can be achieved with cooling fins and heat sinks. The fuel cell can also be coupled with subsystems that absorb heat, like turbine reheaters and metal hydride containers (see Chapter 14 and 15). In the case of PEMFCs, the evaporation of some product water at the cathode absorbs some heat, while active cooling with air or liquid coolants can be used to transfer heat to radiators.

Most fuel cell stacks require some type of cooling system to maintain temperature homogeneity throughout the fuel cell stack. Portable fuel cells need to maintain the correct temperature despite temperature, humidity, vibration, and impact changes due to the environment. Very small and micro fuel cells may not require a cooling system, and often can be designed to be self-cooled.

Figure 9-3 shows an example of a simple fuel cell cooling system with cooling plates inserted into the stack or cooling channels machined into the bipolar plates. The coolant is cooled by a heat exchanger, and moved through the system using a pump. Chapter 14 provides more detail about fuel cell subsystem components such as pumps. There are usually thermocouples or temperature sensors in the stack that feed back to a flow controller which adjusts the flow based upon the amount of coolant required to obtain the desired temperature. There are many cooling methods and designs for every type of fuel cell system. Some of the stack cooling options available are:

- Cooling through free convection (air flow)
- Cooling using a condenser
- Cooling using heat spreaders
- Cooling using cooling plates

There are also many other ways to cool the stack. The methods presented in this section are the most commonly used.

Stack cooling using free convection. One of the simplest solutions for cooling a fuel cell stack is through free convection. This method does not require any complicated designs or coolant, and can be suitable for small or low-power fuel cell stacks. Heat dissipation can be achieved through

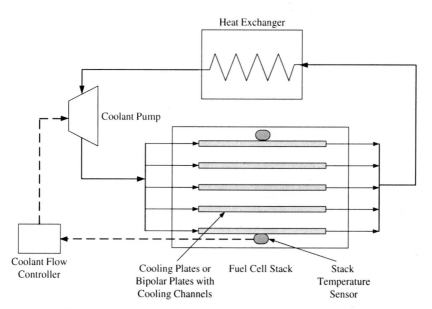

Figure 9-3 Diagram of one type of fuel cell stack cooling system.

manufacturing fins, or through an open cathode flow field design. This cooling method does not have adequate temperature control because the cooling depends largely on the temperature and humidity of the environment. It is also difficult to recover waste heat for power using free convection.

Stack cooling using a condenser. Condenser cooling allows the stack to operate at higher temperatures than other cooling types. For PEM fuel cells, water can be condensed from the exhaust, and then reintroduced into the stack. The condenser must be kept at the water balance temperature so that the overall system will not gain or lose water. Therefore, this system requires precise temperature control like in stack cooling plate systems.

Stack cooling using heat spreaders. Another option for cooling a stack is to use heat spreaders. Heat can obviously be transferred more efficiently outside the stack. Heat spreaders can help transfer heat to the outside of the stack through conduction, and then the heat can dissipate into the surroundings using natural or forced convection. If this method is chosen, high-performance heat spreaders must be used.

Stack cooling using cooling plates. Using cooling plates is a common solution for cooling fuel cell stacks. Thin cooling plates can be manufactured and inserted into the fuel cell, or additional channels can be machined in the bipolar plates to allow air, water, or coolant to flow through the channels to remove heat from the stack. The overall system design can be as simple as only one coolant recirculation path. Precise temperature control is needed to accurately maintain the temperature within a selected range. When cooling plates are used in a PEM fuel cell system, a simultaneous water and heat balance must be conducted to avoid dehydration of the membrane and flooding within the stack.

9.5.1 Heat exchanger model

One method that is used to model the fuel cell is to assume it is a heat exchanger with internal heat generation. Equations 9-26 and 9-29 display the basic heat exchanger equations. The heat, Q, transferred to the cooling fluid is [3]:

$$\frac{dQ_c}{dA_c} = h(T_s - T_c) \qquad (9\text{-}26)$$

where A_c is the entire heat exchange surface. Similar to a heat exchanger:

$$Q_c = UA_c LMTD \qquad (9\text{-}27)$$

where h is the local heat transfer coefficient, $Wm^{2\circ}C$, U is the overall heat transfer coefficient, $Wm^{2\circ}C$, A_c is the heat exchange area (surface area of the cooling channels), m^2, and LMTD is defined as [23,25]:

$$LMTD = \frac{(T_s - T_c)_{in} - (T_s - T_c)_{out}}{\ln\frac{(T_s - T_c)_{in}}{(T_s - T_c)_{out}}} \qquad (9\text{-}28)$$

The temperature difference between the stack body and the cooling fluid can be constant or it can vary, depending upon the stack design. The heat, Q_c, must be absorbed by the cooling fluid and brought to the outside of the stack [3].

$$Q_c = mc_p(T_{c,out} - T_{c,in}) \qquad (9\text{-}29)$$

The temperature difference is a design variable that needs to be optimized depending upon the amount of heat generated and the water that needs to be evaporated. The change in T should be as small as possible to achieve uniform temperatures throughout the stack [3].

9.5.2 Air cooling

As a general rule of thumb, fuel cells with P < 100 W can be cooled effectively using air. One method of achieving cooling with air is to put cooling channels into the bipolar plates as shown in Figure 9-4. With air as the coolant, the flow rate can be found from a simple heat balance. The heat transferred into the air is [3]:

$$Q = m_{coolant}c_p(T_{coolant,out} - T_{coolant,in}) \qquad (9\text{-}30)$$

This equation can be used to estimate the exit temperature of the cooling air. A heat transfer coefficient can be used to estimate the maximum temperature of the channel wall. The Nusselt number is

$$Nu = \frac{hD_h}{k} \qquad (9\text{-}31)$$

where Nu is the Nusselt number, D_h is the hydraulic diameter, h is the convection heat transfer coefficient, and k is the fluid heat conductivity (W/mK). For channels that have a constant heat flux at the boundary, with a high aspect ratio and laminar flow, Incropera and DeWitt [25] report Nu = 8.23.

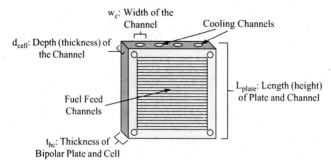

Figure 9-4 A bipolar plate modified for air cooling.

The hydraulic diameter (D_h) can be defined as:

$$D_h = \frac{4A_c}{P_{cs}} \tag{9-32}$$

where A_c and P_{cs} are the cross-sectional area and the perimeter of the cooling channel respectively. Some Nusselt numbers for Reynolds numbers <2300 are shown in Table 9-2.

Recall from Chapter 8, for a circular channel, the Reynolds number must be less than 2300 to insure laminar flow through the channel. The Reynolds number, Re, can be computed using equation 8-48:

$$Re = \frac{\rho v_m D_{ch}}{\mu} = \frac{v_m D_{ch}}{v} \tag{9-33}$$

where v_m is the characteristic velocity of the flow (m/s), D_{ch} is the flow channel diameter or characteristic length (m), ρ is the fluid density (kg/m^3), μ is the fluid viscosity (Ns/m^2), and v is the kinematic viscosity (m^2/s).

This equation is altered slightly for the coolant:

$$Re = \frac{4 m_{coolant}}{\mu_{gas} P_{cs}} \tag{9-34}$$

where μ is the gas or fluid viscosity (Ns/m^2).

TABLE 9-2 Nusselt Numbers for Channel Aspect Ratios for Reynold's Numbers <2300 [17]

Channel aspect ratio	Nusselt number
1	3.61
2	4.12
4	5.33
8	6.49

An empirical correlation from Incropera [25] allows both a Nusselt number and a heat transfer coefficient for air to be determined.

$$Nu = 0.664\, Re^{1/2} * Pr^{1/3} \quad (Pr > 0.6) \qquad (9\text{-}35)$$

The coolant heats up as it travels along the channel, therefore, there is a temperature difference between the inlet and outlet of the flow channel. Assuming a uniform heat flux, the temperature difference is between the solid and gas is [17,23]:

$$Q = L_{plate} P_{cs} h (T_{surface} - T_{gas}) \qquad (9\text{-}36)$$

where L_{plate} is the length of the bipolar plate, $T_{surface}$ is the temperature of the bipolar plate surface, and T_{gas} is the temperature of the gas.

The relationship between the surface temperature and cell edge can be obtained using an energy balance within the bipolar plate, cathode, and anode [17,23]:

$$Q = L_{plate} P_{cs} k_{solid} \frac{(T_{edge} - T_{surface})}{t_{bc}} \qquad (9\text{-}37)$$

where t_{bc} is the thickness of bipolar plate, cathode, anode and electrolyte, k_{solid} is the solid heat conductivity (W/mK), and T_{edge} is the temperature of the cell edge. These equations assume that the temperature difference is constant along the entire channel.

Example 9-2 A 25 watt PEMFC needs to maintain a consistent temperature in order to provide adequate power to the load. The maximum operating temperature that this fuel cell is designed for is 75°C. The fuel cell stack is cooled using natural convection with air at 22°C. The length of the bipolar plate is 5 cm, The cooling channel and cell + bipolar plate thickness are 5 and 0.4 cm respectively. The width of the channel is 0.1 cm. The values for the thermal conductivity of the solid and gas are 20 W/mK and 0.0263 W/mK. The viscosity of the gas is 1.84×10^{-4} g/cms, and the specific heat is 1.0 J/gK. The heat generated per cell is 2 W. What is the air mass flow rate required?

In order to find the temperature of the surface, the channel perimeter needs to be calculated:

$$2d + 2w = 2*0.05 + 2*0.001 = 0.102\ m$$

The solid surface temperature at the cooling channel exit is:

$$Q = L_{plate} P_{cs} k_{solid} \frac{(T_{edge} - T_{surface})}{t_{bc}}$$

$$T_{surface} = 75C - \frac{2W}{0.05} * \frac{1}{0.102} * \frac{0.004}{20\frac{W}{mK}} * \frac{1C}{1K} = 74.9C$$

The hydraulic diameter is given by:

$$D_h = \frac{4A_{cross-section}}{P_{channel}} = \frac{4dw}{2(d+w)}$$

$$D_h = \frac{4}{2}\left(\frac{0.05\ m}{0.05\ m + 0.001\ m}\right)0.001\ m = 0.00196\ m$$

which is close to 2w as d≫w. The heat transfer coefficient is then calculated:

$$h = 8.23\frac{k_{gas}}{D_h} = 8.23*\frac{0.0263\ W/mK}{0.00196\ m} = 110.4\frac{W}{m^2K}$$

The heat transfer coefficient is used to determine the gas exit temperature:

$$T_{gas} = T_{surface} - \frac{Q_{cell}}{hLP_{channel}}$$

$$T_{gas} = 74.9C - \frac{2W}{110.4\ W/m^2K}\left(\frac{1}{0.05\ m}\right)\left(\frac{1}{0.102\ m}\right)\frac{C}{K} = 71.35C$$

The air mass flow rate can be determined by setting $T_{coolant,out} = T_{gas}$.

$$\dot{m}_{coolant} = \frac{Q_{cell}}{C_pT_{coolant,out} - T_{coolant,in}} = \frac{2W}{1.0\frac{J}{gK}(71.35 - 22C)} = 0.0405\ g/s$$

To prove that this is laminar flow, the Reynold's number must be calculated from:

$$Re = \frac{4\dot{m}_{coolant}}{\mu_{gas}P_{channel}} = \frac{4}{1.84*10^{-4}gcm - s}*\frac{0.0405\ g/s}{0.102m}*\frac{m}{100\ cm} = 86.32$$

The total mass flow rate of coolant is equal to the value for one cell multiplied by the number of cells.

9.5.3 Edge cooling

Heat can also be removed from the sides of the cell instead of between the cells. One-dimensional heat transfer can be described in a flat plate with heat generation [3]:

$$\frac{d^2T}{dx^2} + \frac{Q}{kAd_{BP}^{eff}} = 0 \qquad (9\text{-}38)$$

where Q is the heat generated in the cell, W, K is the bipolar plate in-plane thermal conductivity, W/(m∗K), A is the cell active area, m^2, and D_{BP}^{eff} is the average thickness of the bipolar plate in the active area, m.

The solution of the last equation for symmetrical cooling on both sides with T(0) = T(L) = T_0 is [17,23,25]:

$$T - T_0 = \frac{Q}{kAd_{BP}^{eff}} \frac{L^2}{2}\left[\frac{x}{L} - \left(\frac{x}{L}\right)^2\right] \quad (9\text{-}39)$$

where T_0 is the temperature at the edges of the active area, and L is the width of the active area. The maximum temperature difference between the edge and the center is [3]:

$$\Delta T_{max} = \frac{Q}{kAd_{BP}^{eff}} \frac{L^2}{8} \quad (9\text{-}40)$$

The thickness of the plate at the border is d_{BP}. According to Fourier's law, the temperature will be [23]:

$$T_0 - T_b = \frac{Q}{2kA} \frac{L}{d_b} b \quad (9\text{-}41)$$

where t_b is the temperature at the edge of the bipolar plate. The total temperature difference between the center of the plate and the edge of the plate is [3]:

$$\Delta T_{max} = \frac{Q}{kA} L \left(\frac{L}{8d_{BP}^{eff}} + \frac{b}{2d_{BP}}\right) \quad (9\text{-}42)$$

Chapter Summary

Maintaining the design temperature of a fuel cell system is extremely important for proper fuel cell performance. The temperature of the fuel cell system can fluctuate due to heat generated by the fuel cell. In order to predict any heat generated or temperature changes within the fuel cell system, energy balances and heat transfer calculations can be performed. The amount of heat generated and total energy balance around the fuel cell is based upon the power produced, the fuel cell reactions, and the heat loss that occurs in a fuel cell. The energy balance varies for each fuel cell type, size and system because of the different reactions that occur, the thermal conductivity of the components and the plant subsystem. A simple heat transfer model can provide a lot of insight into the phenomena that is occurring in the fuel cell system. Cooling of the stack and/or other fuel cell system components can help to maintain the desired temperature for the system. Common ways of cooling the fuel cell stack include using a condenser, heat spreaders, cooling plates or free convection. The method that is selected depends upon the fuel cell type, size and desired operating temperature.

Problems

1. A fuel cell with a 25 cm^2 active area generates 0.8 A/cm^2 at 0.70 V. The air at the inlet is completely saturated at 80°C and 1 atm. The oxygen stoichiometric ratio is 2.0. Calculate the heat generated, assuming that hydrogen is supplied in a dead-ended mode.

2. Calculate the temperature at the center of a flowfield (2 cm) of a fuel cell operating at 0.65 V and 0.50 A/cm^2. The bipolar plate is made of graphite with k = 22 W/mK; it is 2.2 mm thick in the active area, and 3 mm thick at the border. The border around the active area is 4 mm wide.

3. Calculate the heat generated for a fuel cell with a 100 cm^2 active area that generates 1 A/cm^2 at 0.60 V. The fuel cell operates at 70°C and 3 atm. The oxygen stoichiometric ratio is 2.5.

4. A fuel cell operates at 0.6 V and 0.9 A/cm^2. Calculate the temperature distribution through the gas diffusion layer-bipolar plate on the cathode side. The ionic resistance through the membrane is 0.12 Ohm-cm^2. The heat is removed from the plate by a cooling fluid at 25°C, with a heat transfer coefficient, h = 1600 W/m^{2}k. The electrical resistivity of the gas diffusion layer and the bipolar plate is 0.07 Ohm-cm and 0.06 Ohm-cm, respectively. The contact resistance between the gas diffusion layer and the bipolar plate is 0.006 Ohm-cm. The effective thermal conductivity of the GDL and the bipolar plate is 16 W/mK and 19 W/mK, respectively. The thickness of the GDL and the bipolar plate is 0.30 and 2.5 mm, respectively.

Bibliography

[1] Sousa, Ruy Jr., and Ernesto Gonzalez. "Mathematical Modeling of Polymer Electrolyte Fuel Cells." *Journal of Power Sources*. Vol. 147, 2005, pp. 32–45.
[2] O'Hayre, Ryan, Suk-Won Cha, Whitney Colella, and Fritz B. Prinz. *Fuel Cell Fundamentals*. 2006. New York: John Wiley & Sons.
[3] Barbir, Frano. *PEM Fuel Cells: Theory and Practice*. 2005. Burlington, MA: Elsevier Academic Press.
[4] Lin, Bruce. "Conceptual Design and Modeling of a Fuel Cell Scooter for Urban Asia." Princeton University, Masters Thesis. 1999.
[5] Mench, M.M., Z.H. Wang, K. Bhatia, and C.Y. Wang. "Design of a Micro-direct Methanol Fuel Cell," Electrochemical Engine Center, Department of Mechanical and Nuclear Engineering, Pennsylvania State University. 2001.
[6] Mench, Matthew M., Cao-Yang Wang, and Stephan T. Tynell. "An Introduction to Fuel Cells and Related Transport Phenomena." Department of Mechanical and Nuclear Engineering, Pennsylvania State University. Draft. http://mtrl1.mne.psu.edu/Document/jtpoverview.pdf Last Accessed March 4, 2007.
[7] You, Lixin, and Hongtan Liu. "A Two-Phase Flow and Transport Model for PEM Fuel Cells." *Journal of Power Sources*. Vol. 155, 2006, pp. 219–230.
[8] Springer et al. "Polymer Electrolyte Fuel Cell Model." *J. Electrochem. Soc.* Vol. 138, No. 8, 1991, pp. 2334–2342.
[9] Li, Xianguo. *Principles of Fuel Cells*. 2006. New York: Taylor & Francis Group.
[10] Springer et al. "Polymer Electrolyte Fuel Cell Model." Journal of the Electrochemical Society. Vol 138, No. 8 1991, pp. 2334–2342.
[11] Rowe, A. and X. Li. "Mathematical Modeling of Proton Exchange Membrane Fuel Cells." *Journal of Power Sources*. Vol. 102, 2001, pp. 82–96.

[12] Chen, R. and T.S. Zhao. "Mathematical Modeling of a Passive Feed DMFC with Heat Transfer Effect." *Journal of Power Sources*. Vol. 152, 2005, pp. 122–130.
[13] *Fuel Cell Handbook*, 5th ed. EG&G Services. Parsons Incorporated. Science Applications International Corporation. U.S. Department of Energy, October 2000.
[14] M.W. Chase, Jr. et al. *JANAF Thermochemical Tables*, 3rd ed. American Chemical Society and the American Institute for Physics, Journal of Physical and Chemical Reference Data. Vol. 14, 1985, Supplement 1.
[15] Pitts, Donald and Leighton Sissom. Heat Transfer. 2nd Edition. Schaum's Outline Series. New York: McGraw-Hill, 1998.
[16] Felder, Richard M, and Ronald W. Rousseau. Elementary Principles of Chemical Processes. 2nd edition. New York: John Wiley & Sons, 1986.
[17] Kutz, Myer. Heat Transfer Calculations. New York, McGraw-Hill, 2006.
[18] Oosterkamp, P.F. van den. Critical Issues in Heat Transfer for Fuel Cell Systems. Energy Conversion and management. 47 (2006) 3552–3561.
[19] Kulikovsky, A.A. Heat Balance in the Catalyst Layer and the Boundary Condition for Heat Transport Equation in a Low-Temperature Fuel Cell. Journal of Power Sources. 162 (2006) 1236–1240.
[20] Hwang, J.J. and P.Y. Chen. Heat/Mass Transfer in Porous Electrodes of Fuel Cells. International Journal of Heat and Mass Transfer. 49 (2006) 2315–2327.
[21] Graf, C., A. Vath, and N. Nicolosos. Modeling of Heat transfer in a Portable PEFC System within Matlab-Simulink. Journal of Power Sources. 155 (2006) 52–59.
[22] Faghri, Amir and Zhen Guo. Challenges and Opportunities of Thermal Management Issues Related to Fuel Cell Technology and Modeling. International Journal of Heat and Mass Transfer. 48 (2005) 3891 – 3920.
[23] Middleman, Stanley. An Introduction to Mass and Heat Transfer. Principles of Analysis and Design. New York: John Wiley & Sons, 1998.
[24] "Thermal and Physical Properties of Materials for Fuel Cells." Fuel Cell Group, EEI, AIST. http://unit.aist.go.jp/energy/fuelcells/english/database/thphy1.html Last Accessed March 24, 2007.
[25] Incropera, F. and DdeWitt, D. Fundamentals of Heat and Mass Transfer, 4th ed. New York: Wiley & Sons, 1996.
[26] Larminie, James and Andrew Dicks. Fuel Cell Systems Explained. 2000. John Wiley & Sons, Ltd, Chichester, England.
[27] Srinivasan, Supramaniam. Fuel Cells: From Fundamentals to Applications. 2006. New York: Springer.
[28] Sunden, B. and M. Faghri. Transport Phenomena in Fuel Cells. 2005. Southhampton, UK: WIT Press.

Chapter

10

Fuel Cell Modeling

The last five chapters provided the necessary tools to describe the basic phenomenon that occurs in fuel cells. Fuel cell modeling is helpful for fuel cell developers because it can lead to fuel cell design improvements, as well as cheaper, better, and more efficient fuel cells. Fuel cell models must be robust and accurate and be able to provide solutions to fuel cell problems quickly. A good model should predict fuel cell performance under a wide range of fuel cell operating conditions. Even a modest fuel cell model will have large predictive power. A few important parameters to include in a fuel cell model are the cell, fuel and oxidant temperatures, the fuel or oxidant pressures, the cell potential and the weight fraction of each reactant. Some of the parameters that must be solved for in a mathematical model are shown in Figure 10-1.

The necessary improvements for fuel cell performance and operation demand better design, materials, and optimization. These issues can only be addressed if realistic mathematical process models are available. Many published models for PEM fuel cells and SOFCs exist, but there are only a small number of published models for other fuel cell types. Table 10-1 shows a summary of equations or characteristics of models presented in recent publications on mathematical modeling.

The first column of Table 10-1 shows the number of dimensions the models have in the literature. Most models in the early 1990s were 1D, models in the late 1990s to early 2000s were 2D, and more recently there have been 3D models of certain fuel cell components. The second column specifies that the model can be dynamic or steady-state. Most published models have steady-state voltage characteristics and concentration profiles. The next column of Table 10-1 presents the types of electrode kinetic expressions used. In most models, simple Tafel-type

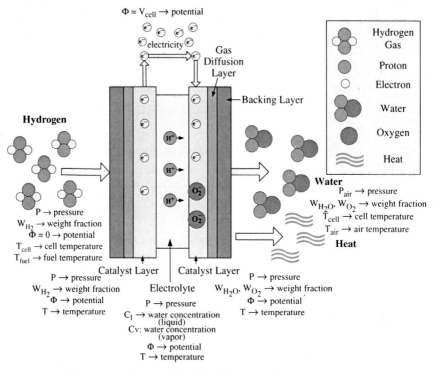

Figure 10-1 Parameters that must be solved for in a mathematical model.

expressions are employed. Certain papers use Butler-Volmer type expressions, and a few other models use more realistic, complex multistep reaction kinetics for the electrochemical reactions. The next column compares the phases used for the anode and cathode structures. It is well-known that there are two phases (liquid and gas) that coexist under a variety of operating conditions. For certain fuel cell types, there is the production of carbon dioxide in the catalyst layer, especially at elevated temperatures on the anode side. For PEMFCs and DMFCs, water may condense inside the cathode structure, and block the way for fresh oxygen to reach the catalyst layer.

An important feature of each model is the mass transport descriptions of the anode, cathode, and electrolyte. Several mass transport models are used in the literature. Simple Fick diffusion models and effective Fick diffusion models typically use experimentally determined effective transport coefficients instead of Fick diffusivities, and do not account for convective flow contributions. Therefore, many models use Nerst-Planck mass transport expressions that combine Fick's diffusion with convective flow. The convective flow is typically calculated from Darcy's Law

TABLE 10-1 Comparison of Recent Mathematical Models

No. of dimensions	Dyn/SS	Anode and cathode kinetics	Anode and cathode phase	Mass transport (anode and cathode)	Mass transport (electrolyte)	Membrane swelling (for PEMFCs and DMFCs)	Energy balance
1, 2, or 3D	Dynamic or steady-state	Tafel type expressions, Butler-Volmer equation, or complex kinetics equations	Gas, liquid, or combination of gas and liquid	Effective Fick's diffusion, Nerst-Plank, Nerst-Plank + Schlogl, or Maxwell-Stefan equation	Nerst-Plank + Schlogl, Nerst-Plank + drag coefficient, or Maxwell-Stefan equation	Empirical or thermodynamic models	Isothermal or full energy balance

using different formulations of the hydraulic permeability coefficient. Some models use Schlogl's formulations for convective flow instead of Darcy's Law, which also accounts for electroosmotic flow, and can be used for mass transport inside the fuel cell. For PEMFCs and DMFCs, a very simple method of incorporating electroosmotic flow in the membrane is by applying the drag coefficient model, which assumes a proportion of water and fuel flow to proton flow. Another popular type of mass transport description is the Maxwell-Stefan formulation for multicomponent mixtures. This has been used for gas-phase transport in many models, but this equation would be better suited for liquid-vapor-phase mass transport. Very few models use this equation for both phases. Surface diffusion models and models derived from irreversible thermodynamics are seldom used. Mass transport models that use effective transport coefficients and drag coefficients usually only yield good approximations to experimental data under a limited range of operating conditions.

The second to last column of Table 10-1 shows that the swelling of polymer membranes is modeled through empirical or thermodynamic models for PEMFCs and DMFCs. Most models assume a fully hydrated PEM. In certain cases, the water uptake is described by an empirical correlation, and in other cases a thermodynamic model is used based upon the change of Gibbs free energy inside the PEM based upon water content.

The last column notes whether the published model includes energy balances. Most models assume an isothermal cell operation, and therefore have no energy balances included. However, including energy balance equations is an important parameter in fuel cell models.

A model is only as accurate as its assumptions allow it to be. The assumption needs to be well understood in order to understand the model's limitations, and to accurately interpret its results. Common assumptions used in fuel cell modeling are:

- Ideal gas properties
- Incompressible flow
- Laminar flow
- Isotropic and homogeneous electrolyte, electrode, and bipolar material structures
- A negligible ohmic potential drop in components
- Mass and energy transport is modeling from a macroperspective using volume-averaged conservation equations

The equations presented in this chapter apply to all fuel cell types, regardless of the fuel cell geometry. Even simple fuel cell models will provide tremendous insight into determining why a fuel cell system performs well or poorly. The physical phenomenon that occurs in a fuel cell can be represented by the solution of the conservation of equations as presented in Sections 10.1–10.5.

10.1 Conservation of Mass

Reactant flow in the fuel cell flow channels is very dependent upon the geometry of the channels (as discussed in Chapter 12). Regardless of the channel geometry, the governing equations are still the same. Equation 10-1 is the conservation of mass in a fuel cell, and is valid for all processes inside the fuel cell, including fluid flow, diffusion, phase change, and electrochemical reactions:

$$\frac{\partial \rho}{\partial t} + \nabla \cdot (\rho v) = S_m \qquad (10\text{-}1)$$

where ρ is the density, kgm^{-3}, v is the velocity vector, ms^{-1}, ∇ is the operator, $\frac{d}{dx} + \frac{d}{dy} + \frac{d}{dz}$, and S_m represents the additional mass sources.

10.2 Conservation of Momentum

The Navier-Stokes equation is sometimes neglected in fuel cell models, but the momentum balance is required to model the fluid velocity and species partial pressures. Momentum conservation is described by Equation 10-2:

$$\frac{\partial (\rho v)}{\partial t} + \nabla \cdot (\rho v) = -\nabla p + \nabla \cdot (\mu_{mix} \nabla v) + S_M \qquad (10\text{-}2)$$

where p is the fluid pressure, Pa, μ_{mix} is the mixture average viscosity, $kgm^{-1}s^{-1}$, and S_M is the external body forces.

The transient term describes momentum with time. For different parts of the fuel cell, the source term is different. For gas channels:

$$S_M = 0 \qquad (10\text{-}3)$$

For backing layers and voids of the catalyst layers:

$$S_M = -\frac{\mu}{K}\varepsilon_{DL} v \qquad (10\text{-}4)$$

where K is the permeability of the gas diffusion layers (or catalyst layer), m^2; and ε_{DL} is the porosity of the diffusion layer. The source term represents a pressure drop from Darcy's drag force imposed by the pore walls on the fluid.

For PEMFCs and DMFCs, an additional source term is electrokinetic permeability for water transport in the polymer phase [24]:

$$S_M = -\frac{\mu}{K_p}\varepsilon_m x_m v + \frac{K_\phi}{K_p} c_f n_f F \nabla \phi_m \qquad (10\text{-}5)$$

where ε_m is the membrane water porosity, x_m is the volume fraction of ionomer in the catalyst layer, K_ϕ is the electrokinetic permeability, K_p is the hydraulic permeability of the membrane, m^2, c_i is the concentration of fixed charge, mol m^{-3}, n_i is the charge number of the sulfonic acid ions, and ϕ_m is the ionomer phase potential.

In the literature, the flow channels are always assumed to be laminar. However, the flow for some fuel cell flow field designs may be turbulent. In these cases, a turbulent model is required.

10.3 Conservation of Energy

Material properties and reaction rates are a strong function of temperature. Therefore, it is important to account for temperature variations within the cell by solving the conservation of energy equation. Conversation of energy for any domain in a fuel cell is described [23,24]:

$$(\rho c_p)_{eff}\frac{\partial T}{\partial t} + (\rho c_p)_{eff}(v \cdot \nabla T) = \nabla \cdot (k_{eff}\nabla T) + S_e \qquad (10\text{-}6)$$

where c_p is the mixture-averaged specific heat capacity, J kg^{-1}K^{-1}, T is the temperature, K, k is the thermal conductivity, W m^{-1}K^{-1}, and S_e is the energy source term. S_e includes the heat from reactions, ohmic heating and heat associated with a phase change. $(\rho c_p)_{eff}$ and k_{eff} can be solved using equations 10-7 and 10-8 [24]:

$$(\rho c_p)_{eff} = (1-\varepsilon)\rho_s c_{p,s} + \varepsilon \rho c_p \qquad (10\text{-}7)$$

$$k_{eff} = -2k_s + \left[\frac{\varepsilon}{2k_s + k} + \frac{1-\varepsilon}{3k_s}\right]^{-1} \qquad (10\text{-}8)$$

where ρ_s, $c_{p,s}$, k_s represent density, specific heat capacity, and thermal conductivity of the solid matrix. This equation balances energy storage, convection, conduction and energy due to species diffusion and a source term, S_e. Therefore, the source term must include the heat from reactions. Equations 10-9–10-11 are applicable for PEMFCs and DMFCs, and can be altered to suit other fuel cell types.

In the gas diffusion layers, the possible heat sources are due to ohmic resistance through solid and phase change in pores, and the gas is not saturated [24]:

$$S_e = \frac{i^2}{\kappa_s^{eff}} - \sigma A_{fg}(x_{sat} - x_{H_2O(g)})(\Delta h_{fg}) \qquad (10\text{-}9)$$

where i is the current density, $A\,m^{-2}$, and κ_s^{eff} is the effective electric conductivity of the gas diffusion layer, $S\,cm^{-1}$.

In the catalyst layers, the source term includes heat released by the electrochemical reaction, heat generated due to ionic and electronic resistance, and the heat of water evaporation [24]:

$$S_e = |i|\left[|\Delta V_{act}| - \frac{T\Delta S}{nF}\right] + \left(\frac{i_m^2}{\kappa_m^{eff}} + \frac{i_e^2}{\kappa_s^{eff}}\right)$$
$$- \sigma A_{fg}(x_{sat} - x_{H_2O(g)})(\Delta h_{fg}) \qquad (10\text{-}10)$$

where i is the transfer current density, $A\,cm^{-3}$, ΔV_{act} is the activation overpotential, V, i_m is the ionic current density, $A\,cm^{-2}$, and κ_m^{eff} is the effective ionic conductivity of ionomer phase in the catalyst layer, $A\,cm^{-1}$.

In the membrane, the only heat source is due to ohmic resistance [24]:

$$S_e = \frac{i_m^2}{\kappa_m} \qquad (10\text{-}11)$$

10.4 Conservation of Species

The species balance equation represents mass conservation for each reactant. Species conservation for the gas phase is [23,24]:

$$\frac{\partial(\varepsilon\rho x_i)}{\partial t} + \nabla \cdot (\upsilon\varepsilon\rho x_i) = (\nabla \cdot \rho D_i^{eff} \nabla x_i) + S_{s,i} \qquad (10\text{-}12)$$

where x_i is the mass fraction of gas species, i = 1,2,3,...N (for example, i = 1 for hydrogen, i = 2 for oxygen, i = 3 for water vapor, and so on), D^{eff} is a function of porosity, and $S_{s,i}$ represents additional species sources. The diffusive flux is sometimes substituted for $\rho D_i^{eff} \nabla x_i$, and it is given in units of $kg/(m^2 * s)$. Often it is more convenient to discuss this in terms of molar diffusive flux, N_i.

The source term for species conservation, $S_{s,i}$, is equal to zero everywhere except in the catalyst layers where the species are consumed or generated in the electrochemical reaction:

$$S_{s,H_2} = -i_a \frac{M_{H_2}}{2F} \qquad (10\text{-}13)$$

$$S_{s,O_2} = -i_c \frac{M_{O_2}}{4F} \qquad (10\text{-}14)$$

For PEMFCs and DMFCs:

$$S_{s,H_2O(g)} = \sigma A_{fg}(x_{sat} - x_{H_2O(g)}) \qquad (10\text{-}15)$$

$$S_{s,H_2O(l)} = -i_c \frac{M_{H_2O}}{2F} - \sigma A_{fg}(x_{sat} - x_{H_2O(g)}) \qquad (10\text{-}16)$$

This equation can be solved for n − 1 species where n is the total number of species present. The last species is solved as a sum of mass fractions equal to one. The flow in the channels is mainly convective, and diffusion in the channels is often ignored to simplify calculations.

10.5 Conservation of Charge

Current transport is described by a governing equation for the conservation of charge [23,24]:

$$\nabla \cdot (\kappa_s^{\text{eff}} \nabla \phi_s) = S_{\phi s} \quad (10\text{-}17)$$

for electrical current, and

$$\nabla \cdot (\kappa_m^{\text{eff}} \nabla \phi_m) = S_{ms} \quad (10\text{-}18)$$

for ionic current, where κ_s^{eff} is the electrical conductivity in the solid phase, S cm^{-1}, κ_m^{eff} is the electrical conductivity in the electrolyte phase, S cm^{-1}, ϕ_s is the solid phase potential, V, ϕ_m is the electrolytic phase potential, V, and S_ϕ is the source term representing volumetric transfer current, at the anode catalyst $S_{\phi s} = -i_a$ and $S_{ms} = -i_a$, at the cathode catalyst layers $S_{\phi s} = -i_s$, $S_{\phi s} = -i_c$, and $S_\phi = 0$ elsewhere.

Early models used correlations to determine many parameters such as mass and heat transfer coefficients or pressure drop. These were based upon many simplifying assumptions. Present-day computing resources allow the full set of governing equations to be solved rapidly. The next section presents the specific equations for the transport phenomena that occur in each fuel cell component.

10.6 The Electrodes

The electrochemical reaction takes place in the electrodes, which serve several functions [23]:

- Transport ions from the reaction site in the electrode to the electrolyte, and from the electrolyte to the reaction site in the opposite electrode
- Transport electrons from the anode sites, and to the cathode sites
- Allow for gas transport/diffusion through the pores to the active sites
- Provide a site for the electrochemical reactions to occur

The electrodes are usually porous structures comprised of two layers: the catalyst layer, where the electrochemical reactions occur; and a porous diffusion or gas diffusion layer that must conduct current and allow for diffusion of the chemical species. The electrode performance depends upon the properties and composition of the materials, microstructural parameters, thickness, and bonding to the adjacent layers. Mathematical modeling is helpful in optimizing and characterizing these parameters.

The reactions that occur in the electrodes involve ions, electrons, and chemical species as shown in Figure 10-2. Therefore, the electrode must account for all of these components.

10.6.1 Mass transport

The first factor to consider in modeling the electrodes is determining the rate at which the species diffuse. Many empirical equations are used in the literature to determine the mass flux and concentration losses. Knowing what empirical equation to use is not as important as determining how to model the system effectively. How to solve the concentration gradients and species distributions requires knowledge of multicomponent diffusion in porous media, and can be a challenging task.

In order to solve the species balance equation, the mass flux, must be determined. The concentration losses are incorporated into a model as the

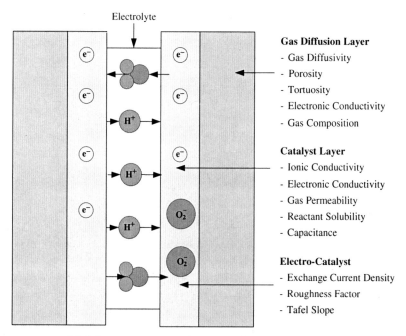

Figure 10-2 Modeling parameters and processes for a PEM electrode layer.

reversible potential decreases due to a decrease in the reactant's partial pressure. There are three basic approaches for determining N [22,23]:

1. The simplest diffusion model is Fick's law, which is used to describe diffusion processes involving two gas species. The binary component form of Fisk's law is shown in equation 10-19.

$$N_i = -cD_{i,j}\nabla X_i \tag{10-19}$$

A multicomponent version of Fick's law is shown in equation 10-20:

$$N_i = -cD_{i,m}\nabla X_i + X_i \sum_{j=1}^{n} N_j \tag{10-20}$$

where c is the total molar concentration. If three or more gas species are present, such as N_2, O_2, and H_2O, a multicomponent diffusion model such as the Stefan-Maxwell equation must be used.

2. The Stefan-Maxwell model is more rigorous, more commonly used in multicomponent species systems, and is employed quite extensively in the literature. It may be used to define the gradient in the mole fraction of components:

$$\nabla y_i = RT \sum \frac{y_i N_j - y_j N_i}{pD_{ij}^{eff}} \tag{10-21}$$

where y_i is the gas phase mol fraction of species i, N_i is the superficial gas phase flux of species i averaged over a differential volume element, which is small with respect to the overall dimensions of the system, but large with respect to the pore size:

$$D_{ij}^{eff} = \frac{a}{p}\left(\frac{T}{\sqrt{T_{c,i}T_{c,j}}}\right)^b (p_{c,i}p_{c,j})^{1/3}(T_{c,i}T_{c,j})^{5/12}\left(\frac{1}{M_i} + \frac{1}{M_j}\right)^{1/2} \varepsilon^{1.5} \tag{10-22}$$

where T_c and p_c are the critical temperature and pressure of species i and j, M is the molecular weight of species, a = 0.0002745 for diatomic gases, H_2, O_2 and N_2, and a = 0.000364 for water vapor, and b = 1.832 for diatomic gases, H_2, O_2, and N_2, and b = 2.334 for water vapor.

3. The Dusty Gas Model is also commonly used, and looks similar to the Stefan-Maxwell equation except that it also takes into account Knudsen diffusion. Knudsen diffusion occurs when a particle's mean-free-path is similar to, or larger in size, than the average pore diameter.

$$-\nabla X_i = \frac{N_i}{D_{i,k}} + \sum_{j=1, j \neq i}^{n} \frac{X_j N_i - X_i N_j}{cD_{i,j}} \tag{10-23}$$

$D_{i,j}$ is the Knudsen diffusion coefficient for species i. The molecular diffusivity depends upon the temperature, pressure, and concentration. The effective diffusivity depends also upon the microstructural parameters such as porosity, pore size, particle size, and tortuosity. The molecular gas diffusivity must be corrected for the porous media. A large portion of the corrections are made using the ratio of porosity to tortuosity (E/T), although in some cases, the Bruggman model is used due to the lack of information for gas transport in porous media:

$$D_{i,j}^{eff} = \left(\frac{\varepsilon}{\tau}\right)D_{i,j} \qquad D_{i,j}^{eff} = \varepsilon^{1.5}D_{i,j} \qquad (10\text{-}24)$$

If the Knudsen number, $Kn = \dfrac{\lambda}{D_{pore}}$, is greater than or equal to 1, Knudsen diffusion is important. The Dusty Gas diffusion model requires a Knudsen diffusivity to be solved, while the other two equations require more work to incorporate Knudsen diffusion.

The Knudsen diffusion coefficient for gas species i can be calculated using

$$D_{i,k} = \frac{2\bar{r}}{3}\sqrt{\frac{8RT}{\pi M_i}} \qquad (10\text{-}25)$$

where M is the molecular mass of species i, λ is the mean-free-path of the gas, and r is the average pore radius.

10.6.2 Electrochemical behavior

The electrochemical reactions in the electrodes create the potential and electrical currents in fuel cells. The Nernst equation from Chapter 5, is used to determine the theoretical electrical potential of the reaction. Equation 10-26 shows the potential for hydrogen electrochemically reacting with oxygen.

$$E = \frac{\Delta G}{2F} + RT\ln\frac{P_{H_2} \cdot P_{O_2}^{1/2}}{P_{H_2O}} \qquad (10\text{-}26)$$

The Nernst equation is used to find the potential at the active locations, and the local potential using the half reactions. To obtain a good approximation of the actual fuel cell potential, the voltage losses described in Chapters 5–8 can be utilized.

The activation overpotential and the rate of species consumption and generation are determined by the electrochemical kinetics and the current density, i. Activation losses and the current density are solved using the appropriate boundary conditions. The reaction rate depends upon the current density, and the mass flux is related to the electric current through Faraday's law (as noted previously in equations 10-13 and 10-14):

$$S_{H_2} = \frac{M_{H_2}}{2F} I \qquad (10\text{-}27)$$

The kinetics can be described by the Butler-Volmer Equation from Chapter 6. The relationship between the current density and the activation losses for the anode is:

$$i_a = i_{0,a} \exp\left(-\alpha_a \frac{Fv_{act,a}}{RT}\right) - \exp\left((1-\alpha_a)\frac{Fv_{act,a}}{RT}\right) \qquad (10\text{-}28)$$

where i_a is the transfer current density (A/m^3), v_{act} is the activation electrode losses, i_o is the exchange current density, and α_a is the anodic charge transfer coefficient. For the cathode:

$$i_c = i_{0,c} \exp\left(-\alpha_c \frac{Fv_{act,c}}{RT}\right) - \exp\left((1-\alpha_c)\frac{Fv_{act,c}}{RT}\right) \qquad (10\text{-}29)$$

From Chapter 6, the exchange current density depends on the local partial pressure of reactants and the local temperature. As the partial pressure of the reactants decreases, the exchange current density will also decrease, which decreases performance. This illustrates how activation and diffusion limitations affect each other, and why the mass flux must be solved precisely. The exchange current density for the anode and cathode are [22,23]:

$$i_{0,c} = i_{0,c}^0 \left(\frac{P_{O_2}}{P_{O_2}^0}\right)^\gamma \exp\left[\frac{-E_{A,c}}{R}\left(\frac{1}{T} - \frac{1}{T^0}\right)\right] \qquad (10\text{-}30)$$

$$i_{0,a} = i_{0,a}^0 \left(\frac{P_{H_2}}{P_{H_2}^0}\right)^{\gamma_1} \left(\frac{P_{H_2O}}{P_{H_2O}^0}\right)^{\gamma_2} \exp\left[\frac{-E_{A,a}}{R}\left(\frac{1}{T} - \frac{1}{T^0}\right)\right] \qquad (10\text{-}31)$$

where $i_{0,c}^0$ and $i_{0,a}^0$ are the reference exchange current density, $\gamma_1 + \gamma_2$ is the reaction order, and E_A is the activation energy.

One assumption in modeling the exchange current density that may be problematic is that the exchange current density is expressed in terms of geometric area. The actual reaction occurs at the active sites, which are a strong function of the shapes and volume of the actual particles. These microstructural parameters are difficult to control and can vary drastically depending on the processing techniques and conditions. Therefore, as one would expect, two electrodes may have vastly different numbers of active sites in the same geometric area. The average current density is the total current generated in a fuel cell divided by the geometric area:

$$i_{avg} = \frac{1}{A}\int_{V_a} i_a dV = \frac{1}{A}\int_{V_c} i_c dV \qquad (10\text{-}32)$$

One of the main challenges in modeling the catalyst layer is finding reliable parameters. The reference exchange current density, the transfer coefficients, and the reaction order are all dependent on the rate determining step(s) of the complex electrochemical reaction, as well as the electrode microstructure. The model becomes more difficult when one has to consider the oxidation of various gases (such as CH_4, CO, and H_2 simultaneously). The reaction order has not been extensively studied, and the experimental kinetic data is still scarce. Thus, there is a need to establish exactly how different fuels are simultaneously oxidized.

10.6.3 Ion/electron transport

Another important component of a fuel cell model is the modeling of the ion and electron transport. A common assumption in fuel cell models is that the electrolyte is the main contributor to the ohmic losses; therefore, the charge transport in the electrodes is neglected. Both electronic and ionic transport are present at the reaction site. Thus, an accurate model must also include this phenomenon in the electrode layer.

The governing equations for the charge transport are Ohm's Law, a charge balance on the electrons and ions, and the conservation of charge equations. Solving these equations gives the local currents and the local losses, and describes the consumption of electronic charge and the formation of ionic charge. For a steady state analysis [23]:

$$i_{el} = -\kappa_s^{eff}\nabla\phi_s \qquad (10\text{-}33)$$

$$i_{io} = -\kappa_m^{eff}\nabla\phi_m \qquad (10\text{-}34)$$

where i_{el} and i_{io} are the local electronic (el) and ionic (io) current densities, ϕ is the potential, and κ^{eff} is the conductivity of the purely electronic or ionic conducting material.

The electronic and ionic currents generated are equal. The total current generated in the anode catalyst layer must be equal to the total current consumed in the cathode catalyst layer as shown by equation 10-35:

$$\nabla \cdot i_{io} = -\nabla \cdot i_{el} \qquad (10\text{-}35)$$

As noted in Section 10.5, the conservation of charge equation for electric current is:

$$\nabla \cdot (\kappa_s^{eff} \nabla \phi_s) = S_{\phi s}$$

and for ionic current:

$$\nabla \cdot (\kappa_m^{eff} \nabla \phi_m) = S_{ms}$$

The most common methods of solving the governing equations for the charge transport processes are the equivalent circuit approach and the charge balance approach. The equivalent circuit approach treats the catalyst layer as a thin film, or uses boundary conditions between the electrode/electrolyte interface and solves for the voltage drop due to electron transport through the electrodes using Ohm's Law. The charge balance method usually creates a more accurate solution since it accounts for local current and losses in the electrode.

10.6.4 Heat transport in the electrodes

In order to accurately predict temperature-dependent parameters as well as rates of reaction and species transport, the temperature and heat distribution needs to be determined accurately. Solving for heat transfer in the electrodes is a challenge because the convective, radial, and conductive heat transfers all exist in the electrodes. The heat transfer equations are similar in the electrodes and flow channels, but there are three main differences with the electrodes:

- Heat transfer in the flow channels is dominated by convection and radiation, whereas conduction can play a dominant role in the electrodes.
- The porous nature of the electrode complicates the heat transfer model. In addition, the heat transfer from the gas phase to the solid phase is difficult to model.
- The heat source in the electrodes is difficult to model in comparison with the flow channels (where the heat source is large and a known parameter).

The conservation of energy equation is the most useful for modeling the porous media. Heat is generated in the electrodes by several different methods:

- Ohmic heat is generated due to the irreversible resistance to current flow.
- Heat is generated due to activation and transport losses. The energy not transformed into current ends up as heat, and this heat is released in the electrodes.

- The change in entropy because of the electrochemical reaction generates heat. Entropic heat effects can be endothermic or exothermic and are generated at the two electrodes in unequal amounts.
- If there is internal reforming in the fuel cell, the chemical reactions also result in the release or absorption of energy. This process is usually modeled as a heat sink.

A critical parameter in modeling the heat transfer in fuel cells is determining where the heat is released in the electrode. Most researchers ignore radiative transfer, but it is known that the electrodes absorb, emit, and transmit radiation. Modeling this radiative transfer will aid in developing better fuel cells in the future.

10.7 The Electrolyte

In order to model the electrolyte accurately, the transport of both charge and energy must be included in the model. Equations 10-17, 10-18, 10-33, and 10-34 must be solved in the electrolyte for ion transport. Contact resistance between the electrode and the electrolyte can also be significant and should be incorporated into the model.

For a PEMFC or a DMFC, the two important fluxes are the proton flux and the water flux as described in Chapters 7 and 8. Figure 10-3 shows the water flux in a polymer electrolyte. The membrane needs to stay hydrated in order to ionically conduct hydrogen, therefore, the water

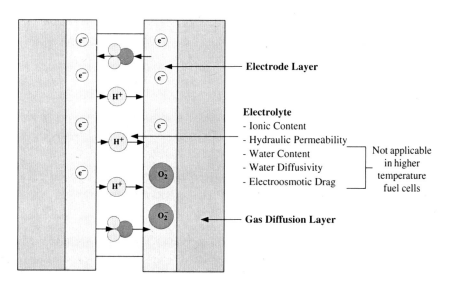

Figure 10-3 Modeling parameters and processes for the electrolyte layer.

profile must be calculated in the electrolyte. In the Nafion membrane, two types of water flux are present: back diffusion and electroosmotic drag. From Chapter 7, both fluxes can be accounted for by the following equation [4]:

$$J_{H_2O}^M = 2n_{drag}\frac{i}{2F}\frac{\lambda}{22} - \frac{\rho_{dry}}{M_m}D_\lambda\frac{d\lambda}{dz} \qquad (10\text{-}36)$$

The water content is not constant in this equation. By obtaining the water content, the resistance of the electrolyte can be estimated.

The conservation of species describes the transport in the membrane, which includes the net affect of the electroosmotic drag, diffusion due to the concentration gradient and the convection due to a pressure gradient. This can be described by a form of the Nernst-Planck equation:

$$N_i = -z_i\frac{F}{RT}D_iC_i\frac{d\Phi_m}{dx} - D_i\frac{dC_i}{dx} + vC_i \qquad (10\text{-}37)$$

where Ni is the superficial flux of species i, z_i is the charge number of species i, C_i is the concentration of species i, and D is the diffusion coefficient of species i, and Φ_m is the electrical potential in the membrane, v is the velocity of H_2O, which is generated by the electric potential and pressure gradient, and can be described by Schogl's equation:

$$v = \frac{k_\Phi}{\mu}z_fc_fF\frac{d\Phi_m}{dx} - \frac{k_p}{\mu}\frac{dp}{dx} \qquad (10\text{-}38)$$

where k_Φ is the electrokinetic permeability, μ is the pore-fluid viscosity, z_f is fixed-site charge, c_f is fixed-charge concentration and k_p is the hydraulic permeability. For all fuel cell types, mass transport in the electrolyte can also be modeled using the Stefan-Maxwell equation (10-21) or the Nernst-Planck equation (10-37) combined with the drag coefficient.

Chapter Summary

Much progress has been made in fuel cell modeling during the last decade, but a great deal of work is still needed. For continued progress in this field, models must be based on an accurate description of the fundamental principles underlying the various processes occurring at the micro-scale level. Constant improvements in mathematical modeling will allow models to be very useful in making design decisions and performance predictions, and will help commercialize fuel cells. Many types of software can aid in modeling fuel cells, such as MATLAB, FEMLAB, FLUENT, and CFD Research (ESI Group). The equations presented in this chapter can be put into any of these programs and solved simultaneously to obtain a good idea of the phenomena occurring in a fuel cell stack.

Problems

1. Is it always necessary to include Knudsen diffusion in a mathematical model under all of the operating conditions?

2. Where is the heat released in the electrode and electrolyte?

3. What are the half-cell potentials and the change in entropy for the half-cell reactions?

4. What is the role of radiation heat transfer in the electrodes, and in the electrolyte?

5. Is radiation scatter important? What is the role of radiation with participating media?

Bibliography

[1] Cha, S.W., R. O'Hayre, Y. Saito, and F.B. Prinz. "The Scaling Behavior of Flow Patterns: A Model Investigation." *Journal of Power Sources*. Vol. 134, 2004, pp. 57–71.
[2] U.S. Patent 6,551,736 B1 Fuel Cell Collector Plates with Improved Mass Transfer Channels. Vladimir Gurau, Frano Barbir, and Jay K. Neutzler. Teledyne Energy Systems, Inc, Hunt Valley, MD. April 22, 2003.
[3] Nguyen, Nam-Trung and Siew Hwa Chan. "Micromachined Polymer Electrolyte Membrane and Direct Methanol Fuel Cells – A Review." *J. Micromech. Microeng.* Vol. 16, 2006, pp. R1–R12.
[4] Ren, X and S. Gottesfeld. "Electro-osmotic Drag of Water in Poly (perfluorosulfonic acid) Membranes." *Journal of Electrochemical Society*. Vol. 148, No. 1, 2001, pp. A87–A93.
[5] M.M. Mench, Z.H Wang, K. Bhatia, and C.Y. Wang. "Experimental Study of a Direct Methanol Fuel Cell." Electrochemical Engine Center, Department of Mechanical and Nuclear Engineering, Pennsylvania State University, 2002.
[6] C.Y. Wang, M.M. Mench, S. Thynell, Z.H. Wang, and S. Boslet. "Computational and Experimental Study of Direct Methanol Fuel Cells." *Int. J. Transport Phenomena*. Vol. 3, August 2001.
[7] Sundmacher, T., T. Schultz, S. Zhou, K. Scott, M. Ginkel, and E.D. Giles. "Dynamics of the direct methanol fuel cell (DMFC): experiments and model-based analysis." *Chem. Eng. Sci.* Vol. 56, No. 2, 2001, pp. 333–341.
[8] Zhou, S., T. Schultz, M. Peglow, and K. Sundmacher. "Analysis of the nonlinear dynamics of a direct methanol fuel cell" *Phys. Chem. Chem. Phys.* Vol. 3, No. 3, 2001, pp. 347–355.
[9] Scott, K., W.M. Taama, J. Cruickshank. "Performance and modelling of a direct methanol solid polymer electrolyte fuel cell." *Journal of Power Sources*. Vol. 65, 1997, pp. 159–171.
[10] Scott, K., P. Argyropoulos, and K. Sundmacher. "A model for the liquid feed direct methanol fuel cell". *J. Electroanal. Chem.* Vol. 477, Issue 2, 1999, pp. 97–110.
[11] K. Sundmacher and K. Scott. "Direct methanol polymer electrolyte fuel cell: Analysis of charge and mass transfer in the vapour–liquid–solid system." *Chem. Eng. Sci.* Vol. 54, 1999, pp. 2927–2936.
[12] Kulikovsky, A.A. Analytical model of the anode side of DMFC: the effect of non-Tafel kinetics on cell performance. *Electrochem. Commun.* Vol. 5, Issue 7, 2003, pp. 530–538.
[13] Jeng, K.T. and C.W. Chen. "Modeling and simulation of a direct methanol fuel cell anode." *Journal of Power Sources*. Vol. 112, Issue 2, 2002, pp. 367–375.

[14] Siebke, A., W. Schnurnberger, F. Meier, and G. Eigenberger. "Investigation of the Limiting Processes of a DMFC by Mathematical Modeling." *Fuel Cells*. Vol. 3, Issue 1-2, 2003, pp. 37–47.
[15] Meyers, J. P. and J. Newman, "Simulation of the Direct Methanol Fuel Cell." *J. Electrochem. Soc.* Vol. 149, No. 6, 2002, pp. A710–A717.
[16] Meyers, J. P. and J. Newman, "Simulation of the Direct Methanol Fuel Cell." *J. Electrochem. Soc.* Vol. 149, No. 6, 2002, pp. A718–A728.
[17] Meyers, J. P. and J. Newman, "Simulation of the Direct Methanol Fuel Cell." *J. Electrochem. Soc.* Vol. 149, No. 6, 2002, pp. A729–A735.
[18] Murgia, G., L. Pisani, A.K. Shukla, and K. Scott. "A Numerical Model of a Liquid-Feed Solid Polymer Electrolyte DMFC and Its Experimental Validation." *J. Electrochem. Soc.* Vol. 150, No. 9, 2003, pp. A1231–A1245.
[19] Argyropoulos, P., K. Scott, A.K. Shukla, and C. Jackson. "A semi-empirical model of the direct methanol fuel cell performance: Part I. Model development and verification." *J. Power Sources*. Vol. 123, Issue 2, 2003, pp. 190–199.
[20] Guo, H. and C. Ma. "2D analytical model of a direct methanol fuel cell" *Electrochem. Commun.* Vol. 6, Issue 3, 2004, pp. 306–312.
[21] Schultz, Thorsten and Kai Sundmacher. "Rigorous Dynamic Model of a Direct Methanol Fuel Cell Based on Maxwell-Stefan Mass Transport Equations and a Flory-Huggins Activity Model: Formulation and Experimental Validation." *Journal of Power Sources*. Vol. 145, 2005, pp. 435–462.
[22] Schultz, Thorsten and Kai Sundmacher. "Mass, Charge and Energy Transport Phenomena in a Polymer Electrolyte Membrane (PEM) Used in a Direct Methanol Fuel Cell (DMFC): Modeling and Experimental Validation of Fluxes." *Journal of Membrane Science*. Vol. 276, 2006, pp. 272–285.
[23] Vandersteen, J.D.J., Kenney, B., Pharoah, J.G. and Karan, K. "Mathematical Modelling of the Transport Phenomena and the Chemical/ Electrochemical Reactions in Solid Oxide Fuel Cells: A Review." *Canadian Hydrogen and Fuel Cells Conference*, September 2004.
[24] Barbir, Frano. PEM Fuel Cells: Theory and Practice. 2005. Burlington, MA: Elsevier Academic Press.

Chapter

11

Fuel Cell Materials

Fuel cells are made of an ion-conductive electrolyte layer, and two electrode layers. The electrolyte layer is an ion exchange material or fluid that is an excellent proton conductor, and differs depending upon the fuel cell type. The electrolyte must be relatively impermeable to fuel and oxidizer to minimize reactant crossover, stable in oxidizing and reducing environments over time, and maintain its structural integrity at operating conditions. Reactants mixing or permeating into the electrolyte results in mixed potentials at the electrodes, reduced performance, and possibly degradation of the catalyst. The catalyst layer must be low-cost and easily applied to the gas diffusion layer, porous electrode or electrolyte. The gas-diffusion or porous electrode layer must be a thin, conductive, porous layer that can withstand the chemical environment and be a compatible material with the electrolyte and catalyst layers. The requirements for the electrode include low activation losses, long-term stability, and acceptable ionic/electronic conductivity. These components are illustrated in Figure 11-1 with a PEM fuel cell.

This chapter explains the physical characteristics, properties, and manufacturing processes of the popular components used for PEMFCs, DMFCs, SOFCs, MCFCs, PAFCs, and AFCs. The most common materials used for each fuel cell type are listed in Table 11-1.

Specific topics covered in this chapter include

- The Electrolyte Layer
- Catalyst/Electrode Layer

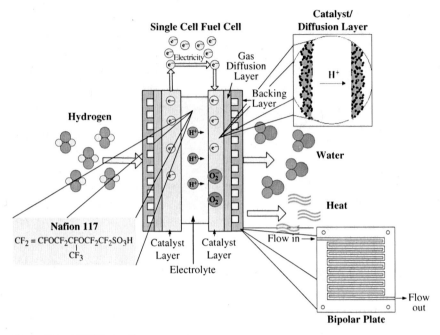

Figure 11-1 PEM fuel cell components.

- Fuel Cell Processing Techniques
- Detailed Method for Building a Fuel Cell

The basic fuel cell materials will be described, along with a detailed summary of the fuel cell processing technologies and a method for building a PEM fuel cell.

11.1 Electrolyte Layer

The electrolyte layer is essential for a fuel cell to work properly. In low temperature fuel cells, when the fuel in the fuel cell travels to the catalyst layer, the fuel molecule gets broken into protons (H^+) and electrons. The electrons travel to the external circuit to power the load, and the hydrogen proton (ions) travel through the electrolyte until it reaches the cathode to combine with oxygen to form water. In high temperature and alkaline fuel cells, the oxygen reacts at the cathode to produce either hydroxide (OH^-), a carbonate ion (CO_3^{2-}), or an oxygen ion (O^{-2}). The ion travels through the electrolyte to react with hydrogen at the cathode. Depending upon the fuel cell type, the electrons are produced

TABLE 11-1 Common Materials Used for Each Fuel Cell Type

Fuel cell/component	Proton exchange membrane fuel cell (PEMFC)	Direct methanol fuel cell (DMFC)	Solid oxide fuel cell (SOFC)	Alkaline fuel cell (AFC)	Phosphoric acid fuel cell (PAFC)	Liquid molten carbonate fuel cell (MCFC)
Most Common Electrolyte	Perflourosulfonic acid membrane (Nafion by DuPont)	Perflourosulfonic acid membrane (Nafion by DuPont)	Yttria stabilized zirconia (YSZ) (8 mol% Y)	Potassium hydroxide	Liquid phosphoric acid	Liquid molten carbonate
Electrolyte Thickness	~50 to 175 μm	~50 to 175 μm	~25 to 250 μm	N/A	N/A	0.5 to 1 mm
Ion Transferred	H^+	H^+	O^{2-}	OH^-	H^+	CO_3^{2-}
Most Common Anode Catalyst	Pt	Pt/Ru	Nickel/YSZ	Pt/Pa	Pt	Nickel
Anode Catalyst Layer Thickness	~10 to 30 μm	~10 to 30 μm	~25 to 150 μm	N/A	~10 to 30 μm	0.20 to 1.5 mm

at either the cathode or anode. Regardless of the fuel cell type, the electrolyte must meet the following requirements:

- High ionic conductivity
- Present an adequate barrier to the reactants
- Be chemically and mechanically stable
- Low electronic conductivity
- Ease of manufacturability/availability
- Preferably low-cost

Finding a material that meets all of these requirements is tough. The toughest requirements are high ionic conductivity, and a material that is stable in both an oxidizing and reducing environment.

11.1.1 PEMFCs and DMFCs

The standard electrolyte material presently used in low-temperature fuel cells is a fully fluorinated Teflon-based material (perfluorosulfonic acid [PFSA]), produced by DuPont for space applications in the 1960s. This membrane is a PTFE-based structure, and is relatively strong and stable in both oxidative and reductive environments, and has high protonic conductivity (0.2 S/cm) at typical PEMFC and DMFC operating temperatures. Figure 11-2 illustrates the chemical structure [21-23].

The DuPont electrolytes have the generic brand name Nafion, and the specific type used most often is number 117. The Nafion membranes, are stable against chemical attack in strong bases, strong oxidizing and reducing acids, H_2O_2, Cl_2, H_2, and O_2 at temperatures up to 125°C. Similar materials have been developed for PEMFC and DMFC by Dupont, Gore and Associates, Asahi Glass, Asahi Chemical, and Pall, as illustrated in Table 11-2 [18].

The proton-conducting membrane usually consists of a PTFE-based polymer backbone, to which sulfonic acid groups are attached. The proton conducting membrane works well for fuel cell applications because the H^+ jumps from SO_3 site to SO_3 site throughout the material. The H^+ emerges on the other side of the membrane. The membrane must remain hydrated to be proton-conductive. This limits the operating temperature of PEM fuel cells to under the boiling point of water and makes water management a key issue in PEM fuel cell development. Figure 11-3 illustrates the SO_3 sites in the Nafion membrane.

$$CF_2 = CFOCF_2CFOCF_2CF_2SO_3H$$
$$|$$
$$CF_3$$

Figure 11-2 The chemical structure of Nafion.

TABLE 11-2 Properties of Commercial Ion-Exchange Membranes [Adapted from 18]

Membrane	Membrane chemistry	IEC (mequiv/g)	Thickness (mm)	Conductivity (S/cm) @ 30°C and 100% RH
Asahi Chemical K-101	Sulfonated polyarylene	1.4	0.24	0.0114
Asahi Glass CMV	Sulfonated polyarylene	2.4	0.15	0.0051
Asahi Glass DMV	Sulfonated polyarylene	—	0.15	0.0071
DuPont Nafion–117	Perfluorinated	0.9	0.2	0.0133
DuPont Nafion–901	Perfluorinated	1.1	0.4	0.01053
Ionac 61AZL386	—	2.3	0.5	0.0081
Ionac 61CZL386	—	2.7	0.6	0.0067
Pall RAI R-1010	Perfluorinated	1.2	0.1	0.0333

Figure 11-3 A pictorial illustration of Nafion.

Nafion membranes come in various thicknesses, and can be cut to any size. Nafion membranes are available in 25.4 μm (Nafion NRE-211), 50.8 μm (Nafion NRE-212), 127 μm (Nafion 115), 183 μm (Nafion 117) and 254 μm (Nafion NE-1110). It is a clear membrane that has to be carefully handled to avoid tears or defects. Figure 11-4 shows a PEM fuel cell with a Nafion membrane.

PFSA membranes, such as Nafion, have a low cell resistance (0.05 Q cm^2) for a 100 μm thick membrane with a voltage loss of only 50 mV at 1 A/cm^2 [22,23]. There are also several disadvantages of PFSA membranes, such as material cost, supporting structure requirements, and temperature-related limitations. The plant components required to keep a PFSA membrane hydrated also adds considerable cost and complexity to the fuel cell system. The fuel cell efficiency increases at higher temperatures, but issues with the membrane, such as membrane dehydration, reduction of ionic conductivity, decreased affinity for water, loss of mechanical strength via softening of the polymer backbone, and increased parasitic losses through high fuel permeation become worse. PFSA membranes must be kept hydrated to retain proton conductivity, but the operating temperature must be kept below the boiling point of water. The largest challenge in finding a replacement for PFSA membranes is low-cost materials.

Hydrocarbon alternative membranes may provide some advantages over PFSA membranes, such as cost, commercial availability, and high water uptakes over a wide temperature range, with the absorbed water restricted to the polar groups of polymer chains. Five main categories of membranes are currently being researched: (1) perfluorinated, (2) partially fluorinated, (3) non-fluorinated (including hydrocarbon), (4) non-fluorinated (including hydrocarbon) composite, and (5) others. There is a wide range of material properties between the membranes

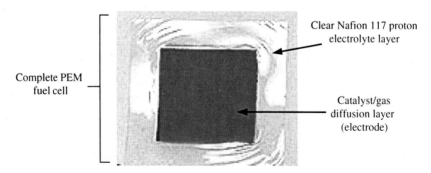

Figure 11-4 PEM fuel cell with Nafion 117 proton electrolyte layer and electrode.

TABLE 11-3 Examples of Alternate Polymer Membranes Being Researched for Low-Temperature Fuel Cells (Adapted from [21-24])

Chemical classification	Membrane type	Performance
Perfluorinated	<ul><li>Perfluorosulfonic acid</li><li>Perfluorocarboxylic acid</li><li>Gore-Select</li><li>Bis(perfluoroalkylsulfonyl) imide</li></ul>	<ul><li>Good proton conductivities and resistance</li><li>Very durable membrane (> 60,000 hrs)</li></ul>
Partially flourinated	<ul><li>a, β, β – trifluorostyrene grafted onto poly (tetrafluoroethylene-ethylene) with post sulfonation</li><li>Styrene grafted and sulfonated poly(vinylidenefluoride)</li></ul>	<ul><li>Less durable and lower performance perfluorinated then</li></ul>
Non-flourinated	<ul><li>Methylbenzensulfonated polybenzimidazoles [MBS-PBI]</li><li>Naphthalenicpolyimide</li><li>Sulfonatedpolyetherketone</li></ul>	<ul><li>Good water absorption</li><li>Some types have good conductivity; others poor</li></ul>
Non-fluorinated composite	<ul><li>Acid-doped polybenzimidazoles</li><li>Base-doped S-polybenzimidazoles</li></ul>	<ul><li>Good proton conductivity.</li><li>Durability needs to be further tested.</li></ul>

in each category. Table 11-3 shows some examples of the polymer membranes currently being researched. Most membranes have degradation temperatures ranging from 250 to 500°C, water uptake from 2.5 to 27.5 H_2O/SO_3H, and conductance from 10 to 10 –S/cm [21-24].

11.1.2 PAFCs

Phosphoric acid fuel cells use a phosphoric acid electrolyte to tolerate the carbon dioxide in reactant gas streams and because of the low rate of electrolyte loss due to evaporation. Other reasons include high oxygen solubility, and good ionic conductivity at high temperatures. When phosphoric acid is above 150°C, phosphoric acid is in the polymeric state as pyrophosphoric acid ($H_4P_2O_7$) [9]. Usually 100-percent phosphoric acid is used in PAFCs, which have a solidification temperature of 42°C [42]. The solidification results in a volume increase, which can damage the porous electrode and matrix structures, and thus lower cell performance and shorten lifetime [9]. PAFC stacks must then be maintained above 45°C at all times. The solidification temperature is dependent upon the phosphoric acid concentration. It is the highest for pure phosphoric acid solution, and then it decreases with decreasing concentration [13].

The acid loss due to evaporation becomes an issue over four to five years. The PAFC operating temperature range of 150 to 220°C dissociates into $H_4P_2O_7$, as shown in the following equation:

$$2H_3PO_4 \rightarrow H_4P_2O_7 + H_2O$$

The phosphoric acid is contained in a matrix structure of silicon carbide (SiC) powder bound with a small amount of PTFE. The phosphoric acid is held by capillary pressure in the pores, and is typically 0.1 to 0.2 mm thick [9,13].

11.1.3 AFCs

The electrolyte used for alkaline fuel cells is a potassium hydroxide (KOH)/water solution with normalities ranging from 6 to 9 [9]. The solution must be as pure as possible to prevent the impurities from contaminating the catalyst. The electrolyte can be in a mobile or immobile form. The mobile electrolyte is pumped through the cells, and removes water and waste heat from the fuel cell. These cells typically have large 2 to 3 mm flow channels to allow rapid flow [13]. The large thickness increases the ohmic polarization which is a major design consideration for AFCs [42].

In AFCs that use immobile electrolyte, the KOH/H_2O solution is held in an asbestos matrix. The electrolyte layer can be as thin as 0.05 mm; therefore, ohmic polarization is not as much of an issue in this type of AFC. The electrolyte typically used is 30-percent potassium hydroxide, which yields the optimal ionic conductivity when AFCs are operated at 60 to 80°C [9,13].

Increasing the KOH concentration helps AFC performance, but it is not practical and feasible to use high concentrations of KOH in water due to the nonuniformity of KOH concentrations in operating cells. The oxygen reduction reaction may reduce the water concentration near the cathode and then the electrolyte solution may solidify, thus preventing reactant transport.

11.1.4 MCFCs

MCFCs use an electrolyte which is a mixture of lithium carbonate (Li_2CO_3) and potassium carbonate (K_2CO_3) with smaller amounts of sodium bicarbonate (Na_2CO_3) and the carbonates of alkaline metals. Optimization of electrolyte composition and cell operating temperature is extremely important for minimizing ohmic and cell polarization, as well as nickel oxide solubility. Most developers have been using 62 percent Li_2CO_3/38 percent K_2CO_3 eutectic since 1975 [9]. A summary of the basic properties for carbonate electrolyte/matrix is presented in Table 11-4.

TABLE 11-4 Basic Properties for Molten Carbonate Electrolyte/Matrix (Adapted from [9])

Component/property	Current status
Electrolyte Support Matrix	
Material	γ– $LiAlO_2$
Thickness	1.8 mm (hot pressed), 0.5 mm (tape cast)
Pore Size	0.5 to 0.8 microns
Surface Area	0.1 to 12 m^2/g
Electrolyte-Filled Matrix	
Composition	40 to 50% $LiAlO_2$ by weight (34 to 42% by volume), 50 to 60% carbonates

A porous electrolyte matrix is used to hold molten carbonate in place using the capillary effect. The porous matrix is made of ceramic powder, such as lithium aluminate ($LiAlO_2$). It typically contains about 40 percent $LiAlO_2$ and 60 percent carbonate by weight. The carbonate must maintain about 60 percent of the weight to remain paste-like in nature. The matrix is typically made using tape-casting or hot pressing methods commonly used in the ceramics and electronics industry. The hot pressing technique involves the pressing of the $LiAlO_2$ carbonate matrix at pressures around 5000 psi, and temperatures just below the melting point of the carbonates. This process is simple, but it usually leads to many issues with the electrolyte such as high porosity (>4 percent), poor microstructure and a high thickness (1 to 2 mm) [9]. The process that is used more frequently is the tape-casting process, which involves dispersing the ceramic materials in a solvent, and then the material is cast in a thin film over a moving surface. The thickness is obtained by shearing with an adjustable blade device. The material is heated, and any remaining organic binder is burnt off at 250 to 300°C [9,13]. Tape casting is an effective method of producing large-area components.

The ohmic resistance of the MCFC electrolyte has an important effect on the fuel cell voltage as in other fuel cells. Most of the ohmic losses are a result of the electrolyte, and are dependent upon the thickness of the electrolyte according to [9]:

$$\Delta V = 0.533t \qquad (11\text{-}1)$$

where t is the thickness in centimeters. A fuel cell with an electrolyte structure of 0.025 cm thickness can operate at a cell voltage 82 mV higher than an identical cell with a structure of 0.18 thickness [9]. Current electrolytes can be made as thin as 0.25 to 0.5 mm, which is helpful in reducing ohmic thickness. However, there is a trade-off for utilizing very thin electrolytes because of the stability obtained with thicker materials [42].

When assembling the MCFC electrolyte, the electrodes, the matrix, the current collectors, and the bipolar plates are slowly heated up to the cell temperature of at least 450°C while it is being assembled together [9,13]. As the carbonate is heated up, it is absorbed into the matrix, which creates shrinkage of the stack components. A reducing gas must also be supplied to the anode to ensure that the nickel anode remains in the reduced state while it is being heated. Because of this conditioning process, the MCFC must warm up for at least 14 hours. Every time the stack is cooled below the electrolyte melt temperature, stresses are set up through the electrolyte, and the resulting damage may cause fuel crossover. The anode compartment also needs to be protected as the fuel cell shuts down by flushing the anode compartment with inert gas [9,13,42].

11.1.5 SOFCs

The most effective and commonly used electrolyte for SOFCs is a zirconia doped with 8 to 10 mol% yttria (yttria-stabilized zirconia [YSZ]). Zirconia is chemically stable, conducts O = ions through the fluorite structure of zirconia where the Zr^{4+} ions are replaced with Y^{3+} ions [42]. When the ion exchange mechanism is occurring, a number of oxide-ion sites become vacant because three O = ions replace four O = ions. The ionic conductivity of YSZ is comparable with liquid electrolytes and can be made very thin (25 to 50 microns) [13]. The structure of YSZ is shown in Figure 11-5. There are several ways to improve the mechanical stability of the electrolyte layer. One method is to add alumina particles, while another popular method is to use the tetragonal phase of the zirconia to strengthen the electrolyte [9].

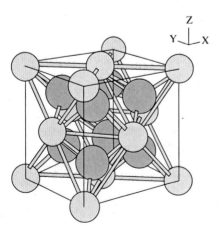

Figure 11-5 Yttria-stabilized zirconia crystal lattice structure [17].

The most popular techniques for creating the yttria electrolyte layers are electrochemical vapor deposition and tape casting. The electrochemical vapor deposition process was originated by Siemens Westinghouse to produce layers of oxides for their tubular SOFC design. This technique is now only used for the electrolyte layer of the SOFC. This process creates a dense and uniform layer from a metal chloride vapor and an oxygen/steam mixture [42].

Zirconia-based electrolytes have pure ionic conductivity. Other material types have exhibited higher oxygen ionic conductivities, but they are less chemically stable, which increases material defects, and internal electrical current. A few electrolyte layers that look hopeful for the future are the LaSrGaMgO (LSGM), which provides performance at 800°C, adding zinc to lanthanum-doped Bi_2O_3, and stabilized ceria with gadolinium [13]. A comparison of zirconia electrolytes is shown in Table 11-5.

11.2 Fuel Cell Electrode Layers

The fuel cell electrode layer is the other critical component of a fuel cell. The electrode layer is made up of the catalyst and porous electrode or gas diffusion layer. When the fuel in the flow channels meets the electrode layer, it diffuses into the porous electrode or gas diffusion layer as described in Chapter 8. The reactant travels to the catalyst layer where it is broken into ions, electrons and other molecules or recombined into new molecules depending upon the fuel cell type and whether the reaction is occurring at the anode or cathode. The electrons travel to the external circuit to power the load, and the ions travel through the electrolyte until it reaches the other electrode to form water and other molecules. Regardless of the fuel cell type, the catalyst layer must be very effective at breaking molecules into protons and electrons, have high surface area and is preferably low-cost. The gas diffusion or porous electrode layer must meet the following requirements:

- High electronic conductivity
- Be chemically and mechanically stable
- Ease of manufacturability/availability
- Preferably low-cost
- Must thermally, chemically and physically interact well with the surrounding fuel cell components
- Adequate porosity

TABLE 11-5 Comparison of Zirconia Electrolytes (Adapted from [9,13,42])

	Stabilization (mol %)	Ionic conductivity (850°C) (S/m)	Coefficient of linear expansion (20 – 1000°C) (10^{-6} 1/K)	Bending strength (N/mm^2)	Density (g/cm^3)	Gas permeability	Activation energy (eV)	Thermal conductivity (W/(m*K))
3YSZ	Yttria, 3	>6	≈11	>1000	>98% th. D	Gas proof	~0,80	~3
5ScO.5CeSZ	Scandia, 5	>7	≈11	—	>98% th. D	Gas proof	~0,80	—
8YSZ	Yttria, 8	>10	≈11	>265	>98% th. D	Gas proof	~0,80	—
10Sc1CeSZ	Scandia, 10	>12	≈11	>250	>98% th. D	Gas proof	~0,7	~2

Finding a material that meets all of these requirements is difficult. The toughest requirements to meet is finding a low-cost catalyst that is effective at breaking the reactants into protons and electrons.

11.2.1 PEMFC, DMFC, and PAFC catalysts

The fuel cell electrode is a thin, catalyst layer where electrochemical reactions take place. The electrodes are usually made of a porous mixture of carbon-supported platinum and ionomer. In order to catalyze reactions, catalyst particles must have contact to both protonic and electronic conductors. Furthermore, there must be passages for reactants to reach catalyst sites and for reaction products to exit. The contact point of the reactants, catalyst, and electrolyte is traditionally referred to as the three-phase interface. In order to achieve acceptable reaction rates, the effective area of active catalyst sites must be several times higher than the geometric area of the electrode. Therefore, the electrodes are made porous to form a three-dimensional network, in which the three-phase interfaces are located. An illustration of the catalyst, electrolyte, and gas diffusion layer is shown in Figure 11-6.

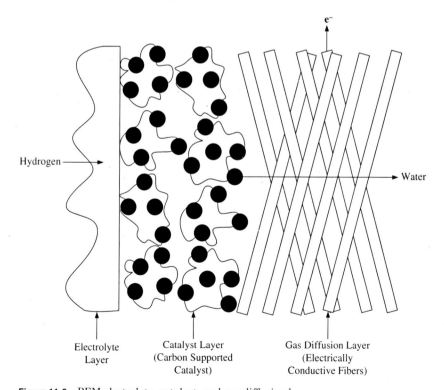

Figure 11-6 PEM electrolyte, catalyst, and gas diffusion layer.

Most PEM fuel cell developers have chosen the thin-film approach, in which the electrodes are manufactured directly on the membrane surface. The benefits of thin-film electrodes include lower price, better use of catalyst and improved mass transport. The thickness of a thin-film electrode is typically 5 to 15 microns, and the catalyst loading is between 0.1 to 0.5 mg/cm^2 for the anode and cathode [23,24].

The catalyst surface area matters more than the weight, so it is important to have small platinum particles (4 nm or smaller) with a large surface area finely dispersed on the surface of catalyst support, which is typically carbon powders with a high mesoporous area (>75 m^2/g). The typical support material is Vulcan XC72R, Black Pearls BP 2000, Ketjen Black International, or Chevron Shawinigan [3]. In order to designate the particle size distribution, the platinum particle surface area on a per-unit mass basis can be calculated by assuming all of the platinum particles are spherical [3]:

$$A_s = \frac{\int f(D)\pi D^2 dD}{\int f(D)\rho_{Pt}\left(\frac{\pi D^3}{6}\right)dD} = \frac{6}{\rho_{Pt}D_{32}} \qquad (11\text{-}2)$$

where ρ_{Pt} is the density of the platinum black and D_{32} is the volume-to-surface area mean diameter of all the particles. The active area per unit mass can be estimated from the mean D_{32}, and a typical value is 28 m^2/g Pt.

This layer should be created reasonably thin to minimize cell potential losses due to the rate of proton transport and reactant gas permeation in the depth of the electrocatalyst layer. The metal active surface area should be maximized; therefore, higher Pt/C ratios should be selected (>40 percent by weight). It has been noted in the literature that the cell's performance remained unchanged as the Pt/C ratio was varied from 10 to 40 percent with a Pt loading of 0.4 mg/cm^2 [22-24]. When the Pt loading was increased beyond 40 percent, the cell performance actually decreased. Fuel cell performance can be increased by better platinum utilization in the catalyst layer, instead of increasing the Pt loading.

The type of catalyst needed in a PEMFC or DMFC is dependent upon the type of fuel used. Tolerance to carbon monoxide is an important issue, especially when methanol is supplied to the fuel cell by steam reforming. Methanol reformate can contain as much as 25 percent carbon dioxide (CO_2), along with a small amount (1 percent) of carbon monoxide (CO). Fuel cell performance drops with very small amounts of CO concentration (several parts per million), due to the strong chemisorption force of CO onto the catalyst. Two methods of resolving CO poisoning is fuel reforming or using more than one catalyst in the anode.

If fuel reforming is being used to provide fuel to the fuel cell, the CO concentration must be reduced to at least 100 ppm if the fuel cell type is a PEMFC or PAFC. Some of the methods used for removing CO from the fuel include [20-24]:

- **Selective oxidation:** The reformer fuel is typically mixed with hydrogen and oxygen before the fuel cell is fed into the stack itself. Sometimes this method is used before the fuel is fed to the stack, or within the stack itself. Selective oxidation technologies can reduce CO levels to less than 10 ppm, but this small level of CO is difficult to maintain during actual operating conditions.
- **Catalysis:** The CO level in the fuel cell can be significantly reduced by passing reforming methanol and oxygen over a Pt aluminum catalyst.
- **Hydrogen peroxide bleeding:** The use of hydrogen peroxide in an anode humidifier has reduced CO to 100 ppm in a hydrogen feed. This has been accomplished through the decomposition of H_2O_2 in the humidifier.

The other method of reducing CO in the fuel system is to combine one or two other catalysts to the base catalyst. Table 11-6 provides a list of the most researched types. In the literature, seven catalyst types claim to have equal performance to the typical Pt/C catalyst. These are: Pt-Ru/C, Pt-Mo/C, Pt-W/C, Pt-Ru-Mo/C, Pt-Ru-W/C, Pt-Ru-Al4, and Pt-Re-(MgH). Iwase and Kawatsu [25] investigated ten of the catalysts listed in Table 11-6: Pt-Ru/C, PtIr/C, Pt-V/C, Pt-Rh/C, Pt-Cr/C, Pt-Co/C, Pt-Ni/C, Pt-Fe/C, Pt-Mn/C, and Pt-Pd/C [20-24]. The catalyst combinations were made 20-wt. percent alloy on carbon with a Pt loading rate of 0.4 mg/cm^2 in a 5-wt. percent PFSA solution. The results of this study indicated that only the Pt-Ru catalyst showed a cell performance equivalent to the single Pt/C catalyst when exposed to 100 ppm CO. The results of this research also found that the Pt/Ru catalyst absorbs water, and facilitates the oxidation of CO. Although Pt/Ru works effectively over the Ru range of 15 to 85 percent, the optimum ratio found was 50:50. Pt/Ru is typically used for DMFCs because methanol is fed into the cell instead of pure hydrogen. The Pt/Ru combination in DMFCs helps to prevent catalyst poisoning and deactivation.

TABLE 11-6 Recently Researched Anode Catalyst Materials

Single metal catalyst	Binary catalysts	Tertiary catalysts
Pt/C	Pt-Co/C , Pt-Cr/C, Pt-Fe/C, Pt-Ir/C, Pt-Mn/C, Pt-Mo/C, Pt-Ni/C, Pt-Pd/C, Pt-Rh/C, Pt-Ru/C, Pt-V/C, Au-Pd/C	Pt-Ru-Al4, Pt-Ru-Mo/C, Pt-Ru-Cr/C, Pt-Ru-lr/C, Pt-Ru-Mn/C, Pt-Ru-Co, Pt-Ru-Nb/C, Pt-Ru-Ni/C, Pt-Ru-Pd/C, Pt-Ru-Rh/C, Pt-Ru-W/C, Pt-Ru-Zr/C, Pt-Re-(MgH$_2$)

Other researchers have found that Pt-Mo/C and a non-Pt-based alloy Au-Pd/C are able to achieve low CO levels. Pt-Mo/C can achieve CO levels as low as <20 ppm, but for allowable higher levels, the benefit of this catalyst is lessened. Other researchers have found that the traditional Pt/Ru still outperforms these catalysts. Lawrence Berkeley researchers [33] have developed a nonplatinum-based binary catalyst, and have reported a three-fold improvement in electro-oxidation of CO/H_2 with their Au-Pd catalyst, as compared to a Pt-Ru catalyst [20-24].

Tertiary catalysts are typically based on a Pt-Ru alloy. A large number of alternative tertiary catalysts have been investigated by scientists at ECI Laboratories [29], and include Pt-Ru alloys with Ni, Pd, Co, Rh, Ir, Mn, Cr, W, Zr, and Nb. Of the catalysts that were investigated, the binary catalysts Pt053-Ru047 and Pt082-W018 were superior to pure platinum in the presence of CO. Pt-Ru performed better in the low potential region, and Pt-W at high current densities. The tertiary catalysts performed even better than these. Other tertiary catalysts that performed well with CO gas concentrations of 100 ppm include Pt-Ru-Al_4 with no carbon support by Denis et al. [3], and PtRe-(MgH_2) without carbon support by Dodelet et al. [31].

The catalyst for the cathode has not been investigated as extensively because these do not have to be CO-tolerant. Cathode catalyst types that have been investigated include Pt- Ni/C, Pt-Co/C, and a nonplatinum-based catalyst produced by pyrolysis of iron acetate, adsorbed on perylenetetracarboxylic dianhydride in Ar:H_2:NH_3 under ambient conditions.

Since relatively high platinum loading is necessary, the catalyst cost may be very high for many commercial applications. If a PEM fuel cell is operated at $E = 0.6$ V and $I = 500$ mA/cm^2, the power density is [9]:

$$P'_{stack} = EI = 0.6V \times 0.5 A/cm^2 = 0.3 W/cm^2 \qquad (11\text{-}3)$$

on a per-electrode surface-area basis. Since each cell requires a total of $m_{Pt} = 8$ mg Pt/cm^2 (anode and catalyst combined), of catalyst loading, and the automobile may require P'$_{stack}$ = 50 kW for the stack power output, and the platinum price is $600/oz.t. the cost of platinum can be calculated as follows [9]:

$$Platinum\_Cost = \frac{P_{stack}}{P'_{stack}} \times m_{Pt} \times Price \qquad (11\text{-}4)$$

$$Platinum\_Cost = \frac{50,000 \ W}{0.3 \ W/cm^2} \times 8 \ mgPt/cm^2 \times \frac{1 oz.t.}{31,103 \ mg} = \$25,721$$

which is too expensive for a platinum catalyst alone. Due to the costs of platinum, alternative fuel cell catalysts will continue to be researched in order to make the cost of fuel cells comparable with traditional combustion engines, batteries, and other commonly used sources of energy.

11.2.2 PEMFC, DMFC, and PAFC gas diffusion layers

The gas diffusion layer is between the catalyst layer and the bipolar plates. In a PEMFC, DMFC or PAFC, the fuel cell layers (MEA) are sandwiched between flow field plates. On each side of the catalyst layer, there are gas diffusion backings. They provide electrical contact between the electrodes and the bipolar plates, and distribute reactants to the electrodes. They also allow reaction product water to exit the electrode surface and permit the passage of water between the electrodes and the flow channels.

Gas diffusion backings are made of a porous, electrically conductive material (usually carbon cloth or carbon paper). An illustration of carbon cloth and Toray paper is shown in Figure 11-7. The substrate can be treated with a fluoropolymer and carbon black to improve water management and electrical properties. These material types promote effective diffusion of the reactant gases to the membrane/electrode assembly. The structure allows the gas to spread out as it diffuses to maximize the contact surface area of the catalyzed membrane. The thicknesses of various gas diffusion materials vary between 0.0017 to 0.04 cm, density varies between 0.21 to 0.73 g/cm^2, and the porosity varies between 70 and 80 percent [20-24].

The GDL also helps with managing water in PEMFCs and DMFCs because it only allows an appropriate amount of water vapor to contact the membrane electrode assembly to keep the membrane humidified. In addition, it promotes the exit of liquid water from the cathode to help eliminate flooding. This layer is typically wet-proofed to insure the pores in the carbon cloth or paper do not become clogged with water. The most common wet-proofing agent is PTFE.

A limited amount of research has been conducted regarding the best GDL layer for water management. Properties of some of the commercially

Figure 11-7   Carbon cloth and toray paper (picture courtesy of *Fuel Cell Scientific*) [38].

TABLE 11-7 Properties of Commercially Available Carbon Papers Used as Substrates in PEMFC Electrodes [19]

Carbon paper	Thickness (mm)	Porosity (%)	Density (g/cm^3)
Toray TGPH-090	0.30	77	0.45
Kureha E-715	0.35	60 to 80	0.35 to 0.40
Spectracarb 2050A-1041	0.25	60 to 90	0.40

available carbon papers are shown in Table 11-7. Ralph et al. found that carbon cloth offered a distinct advantage at high current densities in Ballard Mark V cells. The slope of the pseudolinear region of the cell potential versus current density plot was lowered from 0.27 to 0.21 Ω cm^2 and the limiting current was substantially raised by the use of the carbon cloth [22,23]. Also, the cloth was found to enhance mass transport properties at the cathode derived from improved water management and enhanced oxygen diffusion rates. The surface porosity and hydrophobicity of the cloth substrate are also more favorable for the movement of the liquid water.

Many treatments exist for the gas diffusion layer. Most of these treatments are used to make the diffusion media hydrophobic to avoid flooding in the fuel cell. Either the anode or the cathode diffusion media, or both can be PTFE-treated. The diffusion material is dipped into a 5 to 30 percent PTFE solution, followed by drying and sintering. The interface with the catalyst layer can be fitted with a coating or microporous layer to ensure better electrical contacts and efficient water transport in and out of the diffusion layer. This layer consists of carbon or graphite particles mixed with PTFE binder. The resulting pores are between 0.1 and 0.5 microns, and are, therefore, much smaller than the pore size of the carbon fiber papers [20-24].

The porosity of gas diffusion media is between 70 and 80 percent porous. Porosity of the gas diffusion layer may be calculated from its real weight, thickness, and the density of the solid phase. The porosity also depends upon the compressed thickness:

$$\varepsilon = 1 - \frac{W_A}{\rho_{real} d} \quad (11\text{-}5)$$

where W_A is the real weight (g/cm^2), ρ is the solid phase density, and D is the thickness (either compressed or uncompressed).

One of the functions of the gas diffusion layer is to electrically connect the catalyst layer with the bipolar plates. Both the through plane and the in-plane resistivities of the gas diffusion material are important. The gas diffusion layer is typically compressed to minimize the contact resistance losses. The diffusion coefficients include the effects of material porosity and tortuosity. The Darcy coefficient relates to pressure drop and is proportional to the volumetric flow rate [3]:

$$Q = K_D \frac{A}{\mu l} \Delta P \qquad (11\text{-}6)$$

where Q is the volumetric flow rate, m^3/s, K_D is the Darcy coefficient, m^2, A is the cross-sectional area perpendicular to flow, m^2, M is the gas viscosity, kg/(m∗s), L is the length of the path (thickness of diffusion media), m, and ΔP is the pressure drop, Pa.

11.2.3 AFC electrodes

The AFC electrodes can be hydrophobic or hydrophilic. The hydrophobic electrodes are carbon-based with PTFE, while the hydrophilic electrodes are usually made of metallic materials such as nickel and nickel-based alloys. The electrodes usually have several layers with different porosities for the liquid electrolyte, fuel, and oxidant. AFCs can use both precious and nonprecious metal catalysts. The precious metal catalysts used are platinum or platinum alloys that are deposited on carbon supports, or manufactured on nickel-based metallic electrodes. The catalyst loading is typically 0.25 mg Pt/cm^2 and up [9,13]. The most commonly used nonprecious metal catalysts are Raney nickel for the anodes at a loading of 120 mg Ni/cm^2, and silver-based powders for the cathodes with a loading between 1.5 to 2 mg Ag/cm^2 [9,13,42]. Descriptions of nickel powder and raney metal catalysts are given in the next couple of paragraphs.

Sintered nickel powder. When the alkaline fuel cell was originally designed, it was made to use precious metal catalysts and employ low-cost materials. The electrodes were made of porous powdered nickel that was sintered to make it a rigid structure. To ensure good three-phase contact between the reactant gas, the liquid electrolyte, and the solid electrode, the nickel was made of two layers of different sizes of nickel powders. This structure gave good results, but it was hard to optimize the fuel cell catalyst layer at the time due to the lack of today's current coatings and instrumentation to investigate the properties of materials. This structure is sometimes used today with and without additional catalysts [9,13].

Raney metals. Raney metals can be used to achieve very active and porous forms of a metal. These are prepared by mixing the active metal (nickel) with an inactive metal such as aluminum. The mixing is accomplished so that the two metals are not mixed to form an alloy, but the regions and properties of both metals are maintained. A strong alkali is then applied to the mixture to dissolve the aluminum. This leaves a porous material with a very high surface area. This process allows the pore size to be changed by altering the degree of mixing between the two metals [9,13].

11.2.4 MCFC electrodes

Since molten carbonate is extremely corrosive, the MCFC cathode only uses noble metals for the catalyst material. The currently used cathode material is lithiated nickel oxide (NiO) with a porosity of 70 to 80 percent. The porosity is then reduced by in-situ oxidation to between 55 and 65 percent. The mean pore size is 5 to 7 microns. The smaller pores are filled with electrolyte. The larger pores allow for the diffusion of gas into the interior of the electrode. Polarization is affected by cathode thickness. Ohmic and gas diffusional losses increase with thickness, but losses due to liquid diffusion resistance and kinetic activation decrease as the small pore size increases [9]. The optimal cathode thickness that decreases polarization is currently from 0.4 to 0.8 mm [9]. Alternative cathode materials are being explored because of the limiting factor of cell lifetime.

Many metals are suitable for the electrocatalyst for hydrogen oxidation, such as nickel, cobalt, and copper, which can be used as catalysts when in the form of powdered alloys or composite oxides. MCFC anodes are currently fabricated from porous sintered nickel that contains a few percent of chrominum or aluminum, which forms submicron particles of $LiCrO_2$ or $LiAlO_2$ on the surface of the Ni particles. Sintering increases particle size and decreases the ability of the anode to hold electrolyte [13,42]. The basic properties for molten carbonate electrodes are shown in Table 11-8.

11.2.5 SOFC electrodes

The anode of the current SOFCs are made of metallic nickel and a YSZ structure. The zirconia inhibits the sintering of metal particles and provides a thermal expansion coefficient similar to the electrolyte. The anode has a high porosity (20 to 40 percent) to promote the mass transport of gases. There is some ohmic polarization loss between the anode and the electrolyte, and several companies are working to resolve this

TABLE 11-8 Basic Properties for Molten Carbonate Electrodes (Adapted from [9,13])

Component/property	Anode	Cathode
Material	Ni with 2% to 20% Cr/Ni-Al	Lithiated NiO (NiO with 1% to 2 wt% Li)
Thickness	0.5 to 1.5 mm	0.4 to 0.75 mm
Porosity	50 to 70%	70 to 80%
Pore size	3 to 6 um	7 to 15 um
Surface area	0.1 to 1 m^2/g	0.15 m^2/g Ni pretest; 0.5 m^2/g Ni postest

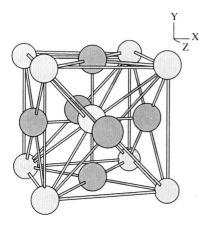

Figure 11-8 Sr-doped lanthanum manganite (LSM) [17].

using a two-layer anode. This improves the anode to withstand temperature changes in the chemical environment. The stability of the anode can also be managed by carefully controlling the particle size of the YSZ. One area currently being researched is novel ceramic anodes that promote the direct oxidation of methane and other fuels. Some of these material types are Gd-doped ceria mixed with Zr and Y and various TiO_2 systems [9,13].

The cathode is also a porous structure that must allow rapid mass transport of reactant and product gases. The most commonly used cathode material is strontium-doped lanthanum manganite ($La_{0.84}Sr_{0.16}MnO_3$), but many other materials can be used in SOFCs that operate around 650°C [42]. Figure 11-8 shows an example of the structure of strontium-doped lanthanum manganite. Table 11-9 summarizes the basic properties of SOFC electrodes.

TABLE 11-9 Basic Properties of SOFC Electrodes [Adapted from 9]

Component/property	Anode	Cathode
Material for High Temperature SOFCs	Cermet made of Ni/YSZ	Strontium-doped lanthanum manganite
Material for Low Temperature SOFCs	Samaria-doped Ceria NiO, Ni/CeO_2 cermet, Copper/ceria cermet	Lanthanum strontium cobalite (LSCo)
Thickness	25 μm – 150 μm	25 μm – 1.4 mm
Porosity	20 – 40%	20 – 40%
Thermal Expansion Coefficient (high temperature materials)	1.25×10^{-5} cm/(cm*°C)	1.2×10^{-5} cm/(cm*°C)

11.3 Low-Temperature Fuel Cell Processing Techniques

The two most common methods of assembling the fuel cell layers (MEA) in low temperature fuel cells are: (1) applying the catalyst layer to the gas diffusion layer (GDL) and then adding the membrane, or (2) applying the catalyst layer to the membrane followed by the GDL addition. The catalyst layer can be applied in one or two steps. For the first method, there are five common ways to prepare and apply the catalyst for the GDL/catalyst assembly [20-24]:

1. **Spreading:** This method consists of preparing a catalyzed carbon and PTFE mixture by mixing and spreading it on using a heavy stainless steel cylinder on a flat surface. This leads to a thin and uniform layer where the Pt loading is related to the thickness [43].

2. **Spraying:** The electrolyte is suspended in a mixture of water, alcohol, and colloidal PTFE. This mixture is sprayed onto a wet-proofed carbon cloth, and the electrode is sintered between spraying to prevent the components from redissolving in the next layer. The electrode is then rolled to produce a thin layer of uniform thickness and porosity on the GDL/catalyst assembly [43].

3. **Catalyst powder deposition:** The Vulcan XC-72, PTFE powder, and a variety of Pt/C loadings are mixed in a fast-running knife mill under forced cooling. This is then applied to a wet-proofed carbon cloth. Applying a layer of carbon/PTFE also evens the surface of the paper and improves the gas and transport properties within the MEA.

4. **Ionomer impregnation:** The catalytically active side of the GDL is painted with a mixture of PFSA in a mixture of lower aliphatic alcohols and water. The catalyst and ionomer are mixed before the catalyst layer is deposited, in order to create a more reproducible uniform layer [46].

5. **Electro-deposition:** Electrodeposition impregnates the porous carbon structure with ionomer, an exchange of the cations in the ionomer by a cationic complex of platinum and electrodeposition of platinum from this complex onto the carbon support. This results in deposition of platinum only at sites accessed effectively by both carbon and ionomer [45].

For the second method, there are five common ways to prepare and apply the catalyst for the GDL/catalyst assembly [20-24]:

1. **Impregnation reduction:** The membrane is ion-exchanged to the Na form and equilibrated with an aqueous solution of $(NH_3)_4PtCl_7$ and a co-solvent of H_5O/CH_3OH. One side of the membrane is then

exposed to dried PFSA in the H form, and the other side to aqueous reductant $NaBH_4$ [44].

2. **Evaporative deposition:** $(NH_3)4PtCl_2$ is deposited onto a membrane through evaporation of an aqueous solution. Metallic platinum is produced by immersion of the entire membrane in a solution of $NaBH_4$. This method produces metal loadings of the order of < 0.1 mg Pt/cm^2 on the membrane/catalyst assembly [44].

3. **Dry spraying:** The reactive materials (Pt/C, PTFE, PFSA powder, and/or filler materials) are mixed in a knife mill, and the mixture is atomized and sprayed in nitrogen through a slit nozzle directly onto the membrane. The adhesion of this layer is good, but to improve the electric and ionic contact, the layer is then hot-rolled or pressed [47].

4. **Catalyst decaling:** Platinum ink is thoroughly prepared by mixing the catalyst and solubilized PFSA. The protonated form of PFSA in the ink is then converted to the TBA+ (tetrabutylammonium) form by the addition of TBAOH in methanol. The stability and spreadability of the ink is improved by adding glycerol to the mixture. The ink is then cast onto PTFE blanks for transfer to the membrane by hot pressing. When the PTFE blank is peeled away, a thin casting layer of catalyst is left on the membrane. In the last step, the catalyzed membranes are rehydrated and ion-exchanged to the H form by immersing them in lightly boiling sulfuric acid, followed by rinsing them in deionized water [46].

5. **Painting:** Pt ink is prepared in the same manner as the catalyst decaling method. A layer of ink is painted directly onto a dry membrane in the Na form and baked to dry the ink. The solvent is removed through drying in a heated vacuum chamber. The catalyzed membranes are rehydrated and ion-exchanged to the H form by immersing them in lightly boiling sulfuric acid, followed by rinsing them in deionized water [46].

In addition to the methods previously listed, sputtering can also be used as a single step for catalyst preparation and application. Platinum can be sputter deposited onto one or both sides of the GDL. To enhance the performance, various coatings can be brushed on to the catalyzed surfaces of the membrane/catalyst assembly, such as a mixture of PFSA solution, carbon powder, and isopropyl alcohol. The solvent is removed by a vacuum chamber [20-24].

The final step for both methods is the addition of the membrane and GDL. Hot pressing is normally used for applying both of these layers. During hot-pressing, the membrane will dry out, but it becomes rehydrated after insertion into the stack. It has been suggested in the literature that the membrane be treated with an H_2O/H_2O_2 solution heated to the boiling point, rinsed in deionized water, immersed in hot dilute sulfuric acid, and treated several times in boiling water. This process removes impurities and traces of acid from the finished MEA. This method is described in more detail in the "Method for Building a Fuel Cell" section at the end of this chapter.

The most important factor in assembling the MEA is to achieve good contact between the membrane, GDL and catalyst layers. Good contact insures maximum utilization of the catalyst during cell operation.

11.4 SOFC manufacturing method

The three different SOFC configurations all have thin layers for the electrodes, electrolyte, and cell interconnection (see Chapter 12), but the fabrication techniques vary with the SOFC stack and developer. The three types of SOFC stacks are the tubular SOFC, the monolithic SOFC and the planar SOFC. The manufacturing process for each type is described in the next few paragraphs.

In a tubular SOFC, the cathode tubes are made of calcia-stabilized zirconia, and fabricated by extrusion from a plastic mass, and then sintering. A slurry coating of Sr-doped $LaMnO_3$ powder is deposited on the support tube. The SOFC tube has a porosity of 30 to 40 percent to allow the transport of reactants and products to the cathode/electrolyte interface where the electrochemical reactions occur [42]. The cell interconnection is then deposited on the support tube by electrochemical vapor deposition (EVD). The electrolyte is then applied to the cathode tube also by electrochemical vapor deposition (EVD). Zirconium and yttrium-chloride vapor is applied to one side of the cathode tube, and O_2/H_2O is applied to the other side [42]. The gases on both sides of the tube act to form a galvanic couple, which results in a dense, uniform oxide layer. The deposition rate of this layer is controlled by the diffusion rate of ions on the surface, and the concentration of electronic charge carriers. The anode is made of Ni and YSZ. A nickel powder slurry coating is deposited on the electrolyte layer, and then yttria-stabilized zirconia is deposited by EVD to form the nickel-cement anode layer [9]. YSZ has a coefficient of thermal expansion that is similar to other materials, and also inhibits the sintering of the metal particles. The anode has a porosity of 20 – 40 percent to allow the mass transport of reactants and products [9,13,42].

In the monolithic SOFC stack, the layers are fabricated to support each other without any separate support layers. The tape casting or tape calendaring process is typically used. Tape casting is a commonly used technique in the ceramics industry. The process involves dispersing ceramic powder in a solvent which contains plasticizers, dissolved binders, and additives. This is cast over a moving, smooth substrate, and the required thickness is established with a doctor blade device. In the tape calendaring process, three flat tapes for the electrolyte and the anode and cathode are made by rolling through a two-roll mill. These tapes are then laminated together by rolling through a second two-roll mill. This tape is then corrugated through molding, put into the fuel cell stack, and then sintered into one piece at the desired temperature [9,13,42].

For the planar SOFC configuration, each fuel cell component is tape cast and sintered separately at the desired temperature [9]. These components are then assembled into the fuel cell stack. The manufacturing process for planar SOFC stacks is more time-consuming compared with the other stacks. Chapter 13 describes stack configurations in more detail.

11.5 Method for Building a Fuel Cell

The following example presents a traditional method for making a PEMFC MEA. Many of the materials can be purchased from a major chemical company, laboratory supply, or a fuel cell company, such as Clean Fuel Cell Energy, LLC. The following materials and equipment are required:

- A proton exchange membrane, such as Nafion 1135, 115 or 117
- Five-percent Nafion solution
- Tetra butyl ammonium hydroxide
- 1 M methanol
- Furan decals
- Teflon release agent
- Camel hair brush
- Cotton gloves
- Six glass beakers

- Carbon cloth or paper for the GDL
- 20-wt. percent platinum carbon catalyst
- Glycerol
- DI water
- Three-percent H_2O_2 solution
- One-percent NaOH solution
- 0.5 M H_2SO_4 solution
- Heated vacuum table
- Heated laboratory press
- Plastic storage bags

11.5.1 Preparing the polymer electrolyte membrane

The proton exchange membrane should be placed on a clean surface and handled using clean cotton gloves to avoid contaminating the sheet. The appropriate-sized PEM pieces should be cut according to your fuel cell design. A proton exchange membrane is illustrated in Figure 11-9.

The PEM film is then prepared for catalyst application by dipping it in six different heated solutions in glass beakers. The solutions are all

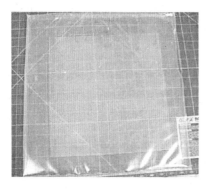

Figure 11-9 Nafion 115 membrane (courtesy of *Fuel Cell Scientific* [38]).

held at about 80 to 100°C using heating plates. Each beaker holds the PEM film for one hour in sequence.

1. Hydrate the membrane and dissolve surface contaminants with 100 mL of distilled (DI) water.
2. Remove organic contaminants from the PEM surface with 100 mL of 3-percent hydrogen peroxide solution (USP). Rinse with DI water.
3. Hydrate the membrane with 100 mL of distilled (DI) water.
4. Exchange membranes to Na^+ form by boiling in a 1-percent NaOH solution for one hour. Rinse with DI water.
5. Hydrate PEM in 100 mL DI water.

While the film is in the beakers, it should remain submerged at all times. A thermocouple or thermometer should be kept in each beaker to make sure the temperature is 80 to 100°C. Store the membranes in DI water until needed for hot pressing. Before hot pressing, place the wet membranes on a heated vacuum table at room temperature. Heat the vacuum table to 130°C for five to ten minutes, cool to room temperature, and remove and store in plastic bags for hot pressing.

11.5.2 Catalyst/electrode layer material

The catalyst/electrode layer is made from a mixture of platinum and carbon powder bonded to a conductive carbon fiber cloth. An example of a PEMFC electrode is shown in Figure 11-10. Each fuel cell MEA (membrane electrode assembly) requires two pieces of catalyst/electrode material. The carbon fiber cloth is the substrate for a gas diffusion catalyst holder. The cloth is often wet-proofed on one side (coated with Teflon) to help keep the water management in a fuel cell stack under control. The catalyst can be applied by any one of several methods, such as painting, screen-printing, sputter diffusion, electrochemical

Fuel Cell Materials

Figure 11-10 A PEMFC electrode (courtesy of *Fuel Cell Scientific* [38]).

deposition, electroless deposition, mechanical deposition, and so on. For simplification, the painting method will be used in this example.

1. Measure the desired amount of 5-percent Nafion solution (1100 EW) into a small beaker or vial.
2. Add 20-wt. percent Pt carbon catalyst, and stir for no less than one hour.

$$2.5 * \frac{1.0\,g}{20} = 0.125\ g$$

3. Add half the weight amount of the 5-percent Nafion solution of glycerol to the solution, and stir for no less than 15 minutes.
4. Add 50 mg of tetra butyl ammonium hydroxide in 1 M methanol per gram of 5-percent Nafion solution. Stir for no less than one hour.
5. Add half the weight amount of the 5-percent Nafion solution of glycerol to the solution, and stir for no less than 24 hours.
6. Clean and then spray the furon decals with the Teflon release agent. Record the weights of each decal, and then paint a coat of catalyst ink on one side with a camel-hair brush. Put the decals into an oven set at 140°C for no less than one hour.
7. Remove and weigh each decal after baking. Repeat the painting in a 90° direction, and the baking in the oven until the desired loading is achieved. The final weighing should be taken after 24 hours of drying.

11.5.3 Hot-pressing the MEA

The two catalyst layers and polymer electrolyte membrane need to be fused together by temperature and pressure for proper mass transfer. This is usually accomplished with a heated laboratory press like the one

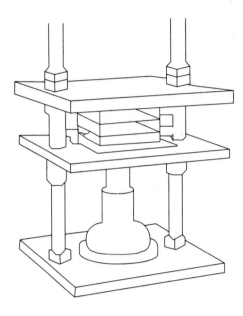

Figure 11-11 A miniature laboratory hot press.

illustrated in Figure 11-11. The three layers (catalyst–PEM–catalyst) are then set on top of the lower heating plate. The upper heating plate was placed on top of the layers.

1. The painted side of the two decals are placed facing each other on either side of a dried Na+ membrane. The decal/membrane assembly is put between two furan sheets with a thin, high durometer silicon outside one of the furan sheets to ensure even pressure.

2. The furan/membrane assembly is placed in a 210°C hot press and compressed to 110 lbs/cm^2 for five minutes. The decals are then peeled away from the membrane, leaving the catalyst electrodes fused to the membrane.

3. Boil the PEM with electrodes in 0.5 M H_2SO_4 for one hour.

4. Rinse the PEM with electrodes with DI H_2O and then boil in DI H_2O for one hour.

5. Dry the protonated PEM on a heated vacuum table at 60°C for 15 minutes, cool to room temperature in a vacuum, remove, and store it in a plastic bag.

The TBA+ exchanged ionomer (Nafion) in the electrodes becomes thermoplastic during hot pressing, in order to make contact with the catalyst particles and create a good bond with the membrane. The completed PEMFC MEA should look like the MEA illustrated in Figure 11-12.

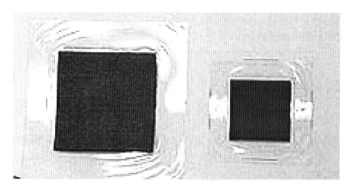

Figure 11-12 A completed PEMFC MEA.

Chapter Summary

The typical fuel cell has three to five layers, which includes: an ion-conductive electrolyte layer, and two electrode layers. The electrolyte layer must be a good proton conductor, chemically stable and able to withstand the temperatures and compression forces of the fuel cell stack. The catalyst layer must be low-cost, easily applied to the gas diffusion layer or electrolyte, and most importantly, be effective at removing the protons and electrons from the reactants. The gas-diffusion or porous electrode layer must be a thin, conductive porous layer that can withstand the chemical environment and be a compatible material with the electrolyte and catalyst layers. The requirements for the electrode include low activation losses, long-term stability, and acceptable ionic/electronic conductivity. There are many choices for each of these materials in the fuel cell, and the decision regarding each component must depend upon many factors including most importantly, cost and mass manufacturing capabilities.

Problems

1. Calculate the resistance of the Nafion 115 membrane if the Nafion conductivity is 0.2 S/cm when 100-percent saturated.

2. Calculate the resistance of the Nafion 115 membrane if the Nafion conductivity is 0.2 S/cm when 25-percent saturated.

3. Calculate the permeation rate through Nafion 117 if the hydrogen pressure is 200 kPa.

4. Calculate the ionic resistance and hydrogen crossover rate for Nafion 1135 if the fuel cell operates at 60°C and 1 atm.

5. Two 100 cm^2 fuel cells are operating at 75°C and 2 atm with 100-percent humidity. They are both generating the same current, and the only difference between the two cells is the catalyst loading, which is 0.35 mg/cm^2 Pt in the first cell. What is the platinum loading of the second cell?

Bibliography

[1] Sousa, Ruy Jr., and Ernesto Gonzalez. "Mathematical Modeling of Polymer Electrolyte Fuel Cells." *Journal of Power Sources.* Vol. 147, 2005, pp. 32–45.
[2] O'Hayre, Ryan, Suk-Won Cha, Whitney Colella, and Fritz B. Prinz. *Fuel Cell Fundamentals.* 2006. New York: John Wiley & Sons.
[3] Barbir, Frano. *PEM Fuel Cells: Theory and Practice.* 2005. Burlington, MA: Elsevier Academic Press.
[4] Lin, Bruce. "Conceptual Design and Modeling of a Fuel Cell Scooter for Urban Asia." Princeton University, Masters Thesis. 1999.
[5] Mench, M.M., Z.H. Wang, K. Bhatia, and C.Y. Wang. "Design of a Micro-direct Methanol Fuel Cell." Electrochemical Engine Center, Department of Mechanical and Nuclear Engineering, Pennsylvania State University. 2001.
[6] Mench, Matthew M., Cao-Yang Wang, and Stephan T. Tynell. "An Introduction to Fuel Cells and Related Transport Phenomena." Department of Mechanical and Nuclear Engineering, Pennsylvania State University. Draft. http://mtrl1.mne.psu.edu/Document/jtpoverview.pdf Last Accessed March 4, 2007.
[7] You, Lixin, and Hongtan Liu. "A Two-Phase Flow and Transport Model for PEM Fuel Cells." *Journal of Power Sources.* Vol. 155, 2006, pp. 219–230.
[8] Springer et al. "Polymer Electrolyte Fuel Cell Model." *J. Electrochem. Soc.* Vol. 138, No. 8, 1991, pp. 2334–2342.
[9] Li, Xianguo. *Principles of Fuel Cells.* 2006. New York: Taylor & Francis Group.
[10] Springer et al. "Polymer Electrolyte Fuel Cell Model." *J. Electrochem. Soc.* Vol. 138, No. 8, 1991, pp. 2334–2342.
[11] Rowe, A. and X. Li. "Mathematical Modeling of Proton Exchange Membrane Fuel Cells." *Journal of Power Sources.* Vol. 102, 2001, pp. 82–96.
[12] Chen, R. and T.S. Zhao. "Mathematical Modeling of a Passive Feed DMFC with Heat Transfer Effect." *Journal of Power Sources.* Vol. 152, 2005, pp. 122–130.
[13] Larminie, James and Andrew Dicks. *Fuel Cell Systems Explained*, 2nd ed. 2003. West Sussex, England: John Wiley & Sons.
[14] Souzy, R., B. Ameduri, B. Boutevin, G. Gebel, and P. Capron. "Functional Fluoropolymers for Fuel Cell Membranes." *Solid State Ionics.* Vol. 176, 2005, pp. 2839–2848.
[15] Mogensen, Mogens, Nigel M. Sammes, and Geoff A. Tompsett. "Physical, Chemical, and Electrochemical Properties of Pure and Doped Ceria." *Solid State Ionics.* Vol. 129, 2000, pp. 63–94.
[16] Kharton, V.V., F.M.B. Marques, and A. Atkinson. "Transport Properties of Solid Electrolyte Ceramics: A Brief Review." *Solid State Ionics.* Vol. 174, 2004, pp. 135–149.
[17] Crystal Lattice Structures Web page, http://cst-www.nrl.navy.mil/lattice/, provided by the Center for Computational Materials Science of the United States Naval Research Laboratory. Last Updated March 13, 2007. Last Accessed March 20, 2007.
[18] Smitha, B., S. Sridhar, and A.A. Khan. "Solid Polymer Electrolyte Membranes for Fuel Cell Applications – A Review." *Journal of Membrane Science.* Vol. 259, 2005, pp. 10–26.
[19] Hinds, G. "Preparation and Characterization of PEM Fuel Cell Electrocatalysts: A Review." *National Physical Laboratory.* NPL Report DEPC MPE 019, October 2005.
[20] Lister, S. and G. McLean. "PEM Fuel Cell Electrode: A Review." *Journal of Power Sources.* Vol. 130, 2004, pp. 61–76.

[21] Hinds, G. "Performance and Durability of PEM Fuel Cells: A Review." *National Physical Laboratory*. NPL Report DEPC MPE 002, September 2004.
[22] Haile, Sossina M. "Fuel Cell Materials and Components." *Acta Materialia*. Vol. 51, 2003, pp. 5981–6000.
[23] Mehta, Viral and Joyce Smith Copper. "Review and Analysis of PEM Fuel Cell Design and Manufacturing." *Journal of Power Sources*. Vol. 114, 2003, pp. 32–53.
[24] Antoli, E. "Recent Developments in Polymer Electrolyte Fuel Cell Electrodes." *Journal of Applied Electrochemistry*. Vol. 34, 2004, pp. 563–576.
[25] Silva, V.S., J. Schirmer, R. Reissner, B. Ruffmann, H. Silva, A. Mendes, L.M. Madeira, and S.P. Nunes. "Proton Electrolyte Membrane Properties and Direct Methanol Fuel Cell Performance." *Journal of Power Sources*. Vol. 140, 2005, pp. 41–49.
[26] Yamada, Masanori, Itaru Honma. "Biomembranes for Fuel Cell Electrodes Employing Anhydrous Proton Conducting Uracil Composites." *Biosensors and Bioelectronics*. Vol. 21, 2006, pp. 2064–2069.
[27] Coutanceau, C., L. Demarconnay, C. Lamy, and J.M. Leger. "Development of Electrocatalysts for Solid Alkaline Fuel Cell (SAFC)." *Journal of Power Sources*. Vol. 156, 2006, pp. 14–19.
[28] Morita, H., M. Komoda, Y. Mugikura, Y. Izaki, T. Watanabe, Y. Masuda, and T. Matsuyama. "Performance Analysis of Molten Carbonate Fuel Cell Using a Li/Na Electrolyte." *Journal of Power Sources*. Vol. 112, 2002, pp. 509–518.
[29] Wong, C.W., T.S. Zhao, Q. Ye, and J.G. Liu. "Experimental Investigations of the Anode Flow Field of a Micro Direct Methanol Fuel Cell." *Journal of Power Sources*. Vol. 155, 2006, pp. 291–296.
[30] Liu, J.G., T.S. Zhao, Z.X. Liang, and R. Chen. "Effect of Membrane Thickness on the Performance and Efficiency of Passive Direct Methanol Fuel Cells." *Journal of Power Sources*. Vol. 153, 2006, pp. 61–67.
[31] Hahn, R., S. Wagner, A. Schmitz, and H. Reichl. "Development of a Planar Micro Fuel Cell with Thin Film and Micro Patterning Technologies." *Journal of Power Sources*. Vol. 131, 2004, pp. 73–78.
[32] Wang, Cheng, Mahesh Waje, Xin Wang, Jason M. Trang, Robert C. Haddon, and Yushan Yan. "Proton Exchange Membrane Fuel Cells with Carbon Nanotube-Based Electrodes." *Nano Letters*. Vol. 4, No. 2, 2004, pp. 345–348.
[33] Li, Wenzhen, Changhai Liang, Weijiang Zhou, Jieshan Qui, Zhenhua Zhou, Gongquan Sun, and Qin Xin. "Preparation and Characterization of Multiwalled Carbon Nanotube-Supported Platinum for Cathode Catalysts of Direct Methanol Fuel Cells." *J. Phy. Chem. B*. Vol. 107, 2003, pp. 6292–6299.
[34] Girishkumar, G., K. Vinodgopal, and Prashant V. Kamat. "Carbon Nanostructures in Portable Fuel Cells: Single-Walled Carbon Nanotube Electrodes for Methanol Oxidation and Oxygen Reduction." *J. Phys. Chem. B*. Vol. 108, 2004, pp. 19960–19966.
[35] Matsumoto, Taketoshi, Toshiki Komatsu, Kazuya, Arai, Takahisa Yamazaki, Masashi Kijima, Harukazu, Shimizu, Yosuke Takasawa, and Junji Nakamura. "Reduction of Pt Usage in Fuel Cell Electrocatalysts with Carbon Nanotube Electrodes." *Chem. Commun*. 2004, pp. 840–841.
[36] Kim Chan, Yong Jung Kim, Yoong Am Kim, Takashi Yanagisawa, Ki Chul Park, and Morinobu Endo. "High Performance of Cup-Stacked Type Carbon Nanotubes as a Pt-Ru Catalyst Support for Fuel Cell Applications." *Journal of Applied Physics*. Vol. 96, No. 10, November 2004.
[37] Xin, Wenzhen, Xin Wang, Zhongwei Chen, Mahesh Waje, and Yushan Yan. "Carbon Nanotube Film by Filtration as Cathode Catalyst Support for Proton Exchange Membrane Fuel Cell." *Langmuir*. Vol. 21, 2005, pp. 9386–9389.
[38] *Fuel Cell Scientific*. http://www.fuelcellscientific.com/. Provider of fuel cell materials and components.
[39] US Patent 6,696,382 B1. Catalyst Inks and Method of Application for Direct Methanol Fuel Cells. Zelenay, Piotr, John Davey, Xiaoming Ren, Shimshon Gottesfeld, and Sharon Thomas. The Regents of the University of California. February 24, 2004.
[40] US Patent 5,234,777. Membrane Catalyst Layer for Fuel Cells. Wilson, Mahlon. The Regents of the University of California. August 10, 1993.

[41] US Patent 5,211,984. Membrane Catalyst Layer for Fuel Cells. Wilson, Mahlon. The Regents of the University of California. May 18, 1993.
[42] The Fuel Cell Handbook. EG&G Technical Services. 7th ed. U.S. Department of Energy, Nov. 2004.
[43] Srinivasan, A. Ferreira, R. Mosdale, S. Mukerjee, J. Kim, S. Hirano, S. Lee, F. Buchi, A. Appleby. Proceedings of the Fuel Cell—Program and Abstracts on the Proton Exchange Membrane Fuel Cells for Space and Electric Vehicle Application, 1994. pp. 424–427.
[44] Foster, S., P. Mitchell, R. Mortimer. Proceedings of the Fuel Cell—Program and Abstracts on the Development of a Novel Electrode Fabrication Technique for Use in Solid Polymer Fuel Cells, 1994. pp. 442–443.
[45] Taylor, E., E. Anderson, N. Vilambi. Preparation of High Platinum Utilization Gas Diffusion Electrode for Proton Exchange Membrane Fuel Cells. J. Electrochem. SOc. 139, 1992. pp. 442–443.
[46] Gottesfeld, S. and T. Zawodzinski. Polymer Electrolyte Fuel Cells. Adv. Electrochem. Sci. Eng. 5, 1997. pp. 195–301.
[47] Gulzow, E., M. Schulze, N. Wagner, T. Kaz, R. Reissner, G. Steinhilber, A. SChneider. Dry Layer Preparation and Characterization of Polymer Electrolyte Fuel Cell Components. J. Power Sources. 86, 2000. pp. 352–362.

Chapter 12

Fuel Cell Stack Components and Materials

After the actual fuel cell layers have been assembled, the cell(s) must be placed in a fuel cell stack to evenly distribute fuel and oxidant to the cells, and collect the current to power the desired devices. This chapter covers the most commonly used fuel cell stack components and materials, including:

- Bipolar plates (design and materials)
- Coatings for bipolar plates
- Bipolar plate manufacturing
- Gaskets
- End plates

The most common materials used for each fuel cell type is shown in Table 12-1. The design and optimization of the bipolar plates or cell interconnects is necessary for creating the most efficient fuel cell stack possible for the desired application.

12.1 Bipolar Plates

In a fuel cell with a single cell, there are no bipolar plates (only single-sided flow-field plates), but in fuel cells with more than one cell, there is usually at least one bipolar plate. Bipolar plates perform many roles in fuel cells. They distribute fuel and oxidant within the cell, separate the individual cells in the stack, collect the current, carry water away from each cell, humidify gases, and keep the cells cool. In order to simultaneously perform these functions, specific plate materials and designs

TABLE 12-1 Fuel Cell Stack Materials

Fuel cell component	Proton exchange membrane fuel cell (PEMFC)	Direct methanol fuel cell (DMFC)	Solid oxide fuel cell (SOFC)	Alkaline fuel cell (AFC)	Phosphoric acid fuel cell (PAFC)	Liquid molten carbonate fuel cell (MCFC)
Bipolar plate/ interconnect/ current collector material	Graphite, titanium, stainless steel, and doped polymers	Graphite, titanium, stainless steel, and doped polymers	Doped $LaCrO_3$, $YCrO_3$, Inconel alloys, and stainless steel	Nickel mesh	Graphite, titanium, stainless steel, and polymer carbon	Stainless steel
Gaskets/ spacers	Silicon, EPDM rubber, PTFE, PTFE-coated fiber thread, EPDM/ PTFE	Silicon, EPDM rubber, PTFE, PTFE-coated fiber thread, EPDM/ PTFE	Glass, glass-ceramic seals, metal brazes, mica, and hybrid mica seals	PTFE	PTFE	Electrolyte contact with bipolar plate
End plates	Graphite, titanium, stainless steel, aluminum, nickel and polymers	Graphite, titanium, stainless steel, aluminum, nickel and polymers	High-temperature compatible ceramic materials, stainless steel	Stainless steel	Graphite, titanium, stainless steel	Stainless steel

are used. Commonly used designs can include straight, serpentine, parallel, inter-digitated, or pin-type flow fields. Materials are chosen based upon chemical compatibility, resistance to corrosion, cost, density, electronic conductivity, gas diffusivity/impermeability, manufacturability, stack volume/kW, material strength, and thermal conductivity. The materials most often used are stainless steel, titanium, $LaCrO_3$, $YCrO_3$, nonporous graphite, and doped polymers. Several composite materials are currently being researched and are beginning to be used for bipolar plates. Bipolar plates also have reactant flow channels on both sides, forming the anode and cathode compartments of the unit cells on the opposing sides of the bipolar plate.

Most PEMFC and DMFC bipolar plates are made of resin-impregnated graphite. Solid graphite is highly conductive, chemically inert, and resistant to corrosion, but expensive and costly to manufacture. Flow channels are typically machined or electrochemically etched on the bipolar plate surfaces. These methods are not suitable for mass production, which is why new bipolar materials and manufacturing processes are being researched.

Fuel Cell Stack Components and Materials

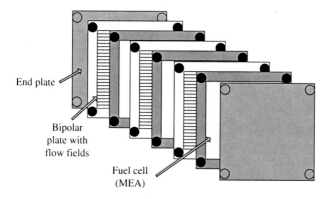

Figure 12-1 An exploded view of a fuel cell stack.

Figure 12-1 shows an exploded view of a fuel cell stack. The stack is made of repeating units of fuel cells and bipolar plates. Increasing the number of cells in the stack increases the voltage, while increasing the surface area increases the current. Fuel cell bipolar plates account for most of the stack weight and volume; therefore, it is desirable to produce plates with the smallest dimensions possible (<3 mm width) for portable and automotive fuel cells [3].

12.1.1 Bipolar plate materials for low and medium temperature fuel cells

Graphite has been traditionally used for bipolar plates in low temperature fuel cells due to its chemical stability in a fuel cell environment. The two commonly used types of materials for bipolar plates are metallic and graphite plates.

Many types of metallic plates have been used for fuel cells, including aluminum, steel, titanium, and nickel. Metallic plates are suitable for mass production, and also can be made in very thin layers, which results in lightweight and portable stacks. The bipolar plates are exposed to a corrosive environment, and dissolved metal ions can diffuse into the electrolyte, which lowers ionic conductivity and reduces fuel cell life. A coating is needed to prevent corrosion while promoting conductivity. Some commonly used coatings are graphite, gold, silver, palladium, platinum, carbon, conductive polymer, and other types. Some of the issues with protective coatings include (1) the corrosion resistance of the coating, (2) micropores and microcracks in the coating, and (3) the difference between the coefficient of thermal expansion and the coating.

Graphite-carbon composite plates have been made using thermoplastics or thermosets with conductive fillers. These materials are usually chemically stable in fuel cells, and are suitable for mass production techniques, such as compression molding, transfer molding, or injection molding. Often, the construction and design of these plates are a trade-off between manufacturability and functional properties. Important properties that need to be considered when designing these plates are tolerances, warping, and the skinning effect. Some issues associated with these plates are that they are slightly brittle and bulky. Although the electrical conductivity is several orders of magnitude lower than the conductivity of the metallic plates, the bulk resistivity losses are only on the order of a magnitude of several millivolts.

One of the most important parameters of the fuel cell stack is the electrical conductivity. The contact resistance is more important than bulk resistance in the fuel cell. Bulk resistivity is not a significant source of voltage loss in fuel cells. Greater resistivity losses result from interfacial contacts between the bipolar plate and the gas diffusion layer. The interfacial contact resistivity losses can be determined by sandwiching a bipolar plate between two gas diffusion layers, and then passing an electrical current through the sandwich and measuring voltage drop. The total voltage drop is a strong function of clamping pressure. Bulk resistance of the bipolar plate and the gas diffusion media should be independent of the clamping force, but the contact resistance is a strong function of the clamping force.

Interfacial contact resistance depends not only upon the clamping pressure, but the surface characteristics and the two surfaces in contact. The relationship between the contact resistance and the clamping pressure between a soft material (such as the GDL) and a hard material (such as a bipolar plate) is as follows [2]:

$$R = \frac{A_a K G^{D-1}}{\kappa L^D} \left[\frac{D}{(2-D)p*} \right]^{D/2} \quad (12\text{-}1)$$

where R is the contact resistance, Ωm^2, A_a is the apparent contact area at the interface, m^2, K is the geometric constant, G is the topothesy (fractal parameter) of a surface profile, m, D is the fractal dimension of a surface profile, L is the scan length, m, p* is the dimensionless clamping pressure (ratio of actual clamping pressure and comprehensive modulus of gas diffusion layer), and κ is the effective electrical conductivity of two surfaces, S/m, described by:

$$\frac{1}{\kappa} = \frac{1}{2}\left(\frac{1}{\kappa_1} + \frac{1}{\kappa_2}\right) \quad (12\text{-}2)$$

12.1.2 Coated metallic plates

The most common materials for fuel cell bipolar plates include aluminum, stainless steel, titanium, and nickel. Bipolar plates have to be chemically resistant, and if they are not properly designed, corrosion or dissolution will occur. Due to these issues, metallic bipolar plates use protective coatings. If the metal plate begins to dissolve, the metal ions will diffuse into the electrolyte, and lower the ionic conductivity. The formation of corrosion layers on the surface of the plate increases the electrical resistance, and decreases the efficiency of the cell. Coatings for bipolar plates should be corrosion-resistant and protect the material from the operating environment. Borup and Vanderborgh [6] present an overview of carbon-based and metallic bipolar plate coating materials in the literature. Carbon-based coatings include graphite, conductive polymer, diamond-like carbon, and organic self-assembled monopolymers, while metal-based coatings consist of noble metals, metal nitrides, and metal carbides. Table 12-2 lists bipolar plate coatings suggested by Borup, and Vanderborgh [6], and others [4,5].

When selecting a coating, some important considerations are the conductivity, corrosion resistance, thermal expansion, and the absence of micropores and microcracks. Temperature differentials that the metal plates may be exposed to should be considered when selecting the coating and metallic plate type because the two metals may expand and contract at different rates.

Micropores and microcracks may lead to failure if the base metal becomes exposed to the acidic fuel cell environment. Thermal expansion differences, as well as microcracks and micropores, can be minimized by adding intermediate coating layers between that of adjacent layers.

Many methods are used for depositing coatings onto metallic bipolar plates. Processes include physical vapor deposition techniques like electron beam evaporation, sputtering and glow discharge decomposition, the chemical vapor deposition technique, and liquid phase chemical

TABLE 12-2 Common Materials Used for Bipolar Plates (Adapted from [4])

Material/resin	Filler/fiber
Stainless steel	N/A
Graphite	N/A
Aluminum	N/A
Titanium	N/A
Nickel	N/A
Poly (vinylidene fluoride)	Carbon/graphite particles
Polypropylene	Carbon black graphite powder
Phenyl-aldehyde resol or phenyl aldehyde novolac	Graphite powder or coke-graphite particles /cellulose fibers (not rayon and cellulose acetate)
Phenol resin	Graphite powder/carbon fibers
Vinyl ester	Graphite powder/cotton flock

techniques like electro- and electroless deposition, chemical anodization/ oxidation overcoating, and painting [4]. A summary of coating techniques for bipolar plate material types can be found in Table 12-3.

12.1.3 Composite plates

New fuel cell materials have been under investigation during the past decade that are low-cost and easy to manufacture. Many of these new plates are metal or carbon-based. One interesting design, developed by Los Alamos National Laboratory [23], combines porous graphite, polycarbonate plastic, and stainless steel to utilize the beneficial properties of each material type. The stainless steel provides rigidity to the structure while the graphite resists corrosion. The polycarbonate provides chemical resistance and can be molded to any shape to provide for gaskets and manifolding. This layered plate seems to be a good alternative from a chemical, physical, electrical, and cost point-of-view [4].

An extensive amount of research has been conducted on the use of carbon-based composite bipolar plates in fuel cells. Some of the carbon composite bipolar plates have been made using thermoplastic (polypropylene, polyethylene, and poly[vinylidene fluoride]) or thermosetting resins

TABLE 12-3 Coating Materials for Metallic Bipolar Plates (Based on [4])

Coating type	Coating process	Bipolar plate materials
Gold (usually needs an intermediate layer, such as Ni)	Pulse current electrodeposition	Aluminum, titanium, nickel, stainless steel
Graphite foil layer (usually needs an intermediate layer, such as graphite particles in an organic solvent)	Painting or pressing	Aluminum, titanium, nickel
Graphite topcoat layer (usually needs an intermediate layer, such as titanium, Cr, or a combination of various layers)	Physical vapor deposition (PVD) or a chemical anodization/oxidation overcoating	Aluminum, titanium, nickel, stainless steel
Indium tin oxide	Electron beam evaporation	Stainless steel
Lead oxide (usually needs an intermediate layer, such as lead)	Vapor deposition and sputtering	Stainless steel
Stainless steel (usually needs an intermediate layer, such as nickel phosphorous or titanium nitride)	Physical vapor deposition or chemical vapor deposition (CVD), electroless deposition	Aluminum, titanium, stainless steel

(phenolics, epoxies, and vinyl esters) with fillers and with or without fiber reinforcement. Recent bipolar plate manufacturers are producing bipolar plates with thermoplastic resins because they are easily injection-moldable and recyclable.

12.2 Flow-Field Design

In fuel cells, the flow field should be designed to minimize pressure drop (reducing parasitic pump requirements), while providing adequate and evenly distributed mass transfer through the diffusion layer to the catalyst surface for reaction. The three most popular channel configurations for bipolar stack designs are serpentine, parallel, and inter-digitated flow. Serpentine, parallel, and inter-digitated flow channels are shown in Figures 12-2 through 12-6. Some small-scale fuel cells do not use a flow field to distribute the hydrogen and/or air, but rely on diffusion processes from the environment. Since the hydrogen reaction is not rate limiting, and water blockage in the humidified anode can occur, a serpentine arrangement is typically used for the anode in smaller fuel cells.

As shown in Figure 12-2, serpentine flow path is continuous from start to finish. An advantage of the serpentine flow path is that any obstruction in the path will not block all downstream activity of the obstruction. A disadvantage of serpentine flow is the fact that the reactant is depleted through the length of the channel, so an adequate amount of the gas must be provided to avoid excessive polarization losses. When air is used as an oxidant, problems usually arise with the cathode gas flow distribution and the cell water management. When the fuel cell operates for long periods of time, the water formed at the cathode accumulates in the

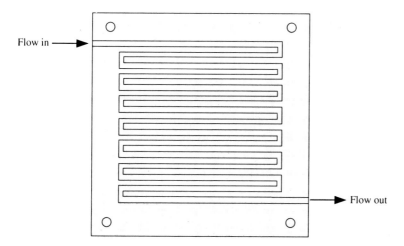

Figure 12-2 A serpentine flow field design.

228   Chapter Twelve

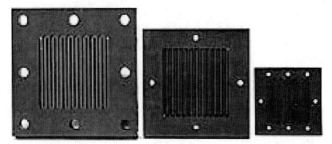

Figure 12-3 Graphite bipolar plates with a serpentine design.

cathode. A force is required to move the water out of the channels. This design is relatively effective at providing flow distribution across the electrode surface of the fuel cell. However, this design may cause high pressure loss due to the relatively long flow path. For high current density operation, very large plates, or when air is used as an oxidant, alternate designs have been proposed based upon the serpentine design [3,4].

Several continuous flow channels can be used to limit the pressure drop, and reduce the amount of power used for pressurizing the air

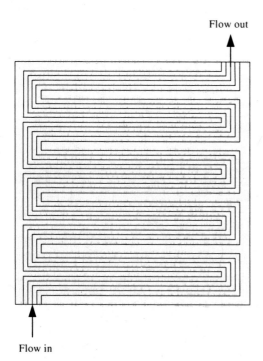

Figure 12-4 Multiple serpentine flow channel design.

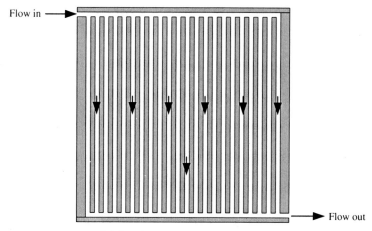

Figure 12-5 A parallel flow field design.

through a single serpentine channel as shown in Figure 12-4. This design allows no stagnant area formation at the cathode surface due to water accumulation. The reactant pressure drop through the channels is less than the serpentine channel, but still high due to the long flow path of each serpentine channel [3].

Although some of the reactant pressure losses can increase the degree of difficulty for hydrogen recirculation, for PEMFCs they are helpful in removing the product water in vapor form. The total reactant gas

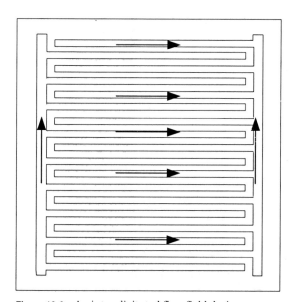

Figure 12-6 An inter-digitated flow field design.

pressure is $P_T = P_{vap} + P_{gas}$, where P_{vap} and P_{gas} are the partial pressures of the water vapor and reactant gas in the reactant gas stream respectively. The molar flow rate of the water vapor and reactant can be related as follows:

$$\frac{N_{vap}}{N_{gas}} = \frac{P_{vap}}{P_{gas}} = \frac{P_{vap}}{P_T - P_{vap}} \qquad (12\text{-}3)$$

The total pressure loss along a flow channel will increase the amount of water vapor that can be carried and taken away by the gas flow if the relative humidity is maintained [3]. This can help remove water in both the anode and cathode flow streams.

In the parallel flow field design configuration (see Figure 12-5), the flow channels require less mass flow per channel, and provide more uniform gas distribution with a reduced pressure drop. If air is used as the oxidant, it is found that low and unstable cell voltages may occur after long periods of operation due to water buildup and cathode fuel distribution. When the fuel cell is operated continuously, the water accumulates in the flow channels. The disadvantage of the parallel flow configuration is that an obstruction in one channel results in flow redistribution among the remaining channels, and thus a dead zone downstream of the blockage. The amount of water in each channel could vary, which leads to uneven gas distribution. This problem can also occur in the pin-type flow field discussed later. Another issue with this design is that the channels are short, and have few directional changes. As a result, the pressure drop in the channels are low, but the pressure drop in the piping system and stack distribution manifold may not be. The first few cells near the manifold inlet have a greater amount of flow than the cells toward the end of the manifold.

The reactant flow for the inter-digitated flow field design is parallel to the electrode surface. Often, the flow channels are not continuous from the plate inlet to the plate outlet. The flow channels are dead-ended, which forces the reactant flow, under pressure, to go through the porous reactant layer to reach the flow channels connected to the stack manifold. This design can remove water effectively from the electrode structure, which prevents flooding and enhances performance. The inter-digitated flow field is good since the gas was pushed into the active layer of the electrodes where forced convection avoids flooding and gas diffusion limitations. This design is sometimes noted in the literature as outperforming conventional flow field design, especially on the cathode side of the fuel cell. Most studies in the literature indicate that the traditional serpentine design outperforms all other designs. The inter-digitated design is shown in Figure 12-6. A variation of the interdigitated design–the spiral interdigitated flow design is shown in Figure 12-7.

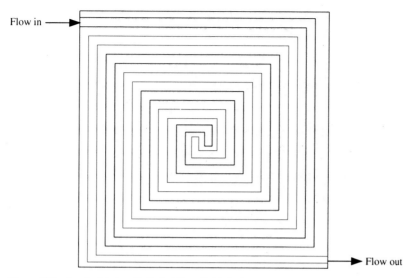

Figure 12-7 A spiral inter-digitated flow field design.

Instead of the traditional flow fields that are typically machined, pressed, or injection-molded, Figure 12-8 shows a flow field made of pins that can be of various geometric shapes. The most common shapes for the pins are rectangular or tubular pins, which protrude from the plates. The gases/fluids flow through the grooves formed by the pins. Pin flow fields result in a low pressure drop; however, there is a chance of

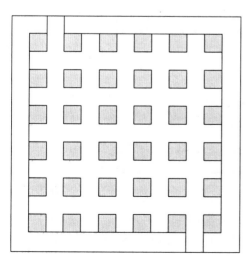

Figure 12-8 Pin-type flow field design.

stagnant reactant areas, uneven reactant distribution, and poor fuel cell performance. The Reynolds number of this plate type can range from a few tens to the low hundreds [3].

12.3 Materials for SOFCs

SOFCs plates must be physically stable, and resist corrosion from the electrolyte at high temperatures. There are two categories of materials used for SOFC plates: high temperature perovskite materials (900 to 1000°C), and metallic alloys for lower temperature operation (600 to 900°C). The high temperature perovskite materials commonly used in SOFCs are primarily doped lanthanum and yttrium chromites (dopants typically include Mg, Sr, Ca, Ca/Co). These materials have good electronic conductivity, which increases with temperature. The type of dopant alters the material properties, and determines the compatibility with the fuel cell layers. Ceramic plates or interconnects are chemically and physically stable in SOFCs for long periods of time, but because the plates are made of ceramic, they do not have good flexibility for sealing the cell, and are rigid and weak. To remedy this issue, conductive pastes or a contact felt is sometimes used. These help with the flexibility and sealing, but the life of these components are not as good as the ceramic plate [25,26].

When thin-electrolyte anode-supported SOFCs were developed within the last decade, this enabled the SOFC to be operated at lower temperatures (650 to 800°C), which enabled the use of some metallic components in the cell. Even at these "lowered" temperatures, it is difficult to find reliable metals that can be used as bipolar plates. Metals that will be used as SOFC interconnects at these temperatures must have the following characteristics:

- They must be able to maintain uniform temperature across the plate without deformation.
- They should have sufficient creep strength.
- They have to tolerate a corrosive environment.
- They should maintain a high conductivity as discussed in section 12.1.

Many high-chrome interconnects have been researched and developed, such as $Cr_5Fe_1Y_2O_3$, which operates at around 900°C. The chrome is problematic because it can poison the electrode, and it has very high material costs. If the operating temperatures were reduced, ferritic steels could be used, which would reduce the material and processing costs. The following materials are commonly used as plate materials in fuel cells:

- Lanthanum chromite
- $Cr_5Fe_1Y_2O_3$
- Avesta 600 (Fe-28/Cr-4/Ni-2/Mo)
- Inconel 601 (NiCr + Al)
- SS 310
- SS 316L
- SS 310S

In order for many metals to be noncorrosive, they have to be coated (like other fuel cell types), which means that the coating also needs to have the same properties as the ideal interconnect. The most commonly used coating is a stable oxide (chromia) which leads to evaporation and electrode poisoning. Two coatings that minimize evaporation and insure good contact resistance are interconnect coatings of strontium-doped lanthanum cobaltite or manganite. These coatings perform well in the intermediate temperature range, but better coatings and interconnect materials still need to be developed in the future for commercially viable products. Some of the characteristics of the materials and coatings for SOFCs that need to be improved include the conductivity, corrosion resistance, mechanical stability, creep strength, and cost. Chapter 13 provides further information on SOFC interconnects.

12.4 Materials for MCFCs

The MCFC bipolar plate includes a separator, current collectors, and the wet seal. The most common materials for MCFC bipolar plates are alloys such as Incoloy 825, 310S, or 316L stainless steel that are coated on one side with a Ni layer. The plates are typically 15 mm thick, and the nickel-coated side is exposed to fuel gases in the anode compartment, and provides a conductive surface coating with low contact resistance. Corrosion of plates are reduced by applying a coating at the vulnerable locations on the bipolar plate. A thin aluminum coating can form a protective $LiAlO_2$ after reaction of Al with Li_2CO_3. This coating is not useful for areas that need to be conductive since it is an insulating protective layer. Materials that are commonly used to promote conductivity and resist corrosion are 316 stainless steel and Ni plated stainless steels. Type 310 and 446 stainless steels have better corrosion resistance than Type 316 in corrosion tests.

The separator and current collector is a Ni-coated 310S/316L and the wet seal is formed by aluminization of the metal. A low oxygen partial pressure on the anode side of the bipolar plate prevents oxidation. Single alloy bipolar current collector plates need to be developed that work on both the cathode and anode sides of the fuel cell. The primary construction materials are stainless steels (austenitic stainless steels). Nickel-based alloys also resist corrosion well. Certain nickel-based coatings can protect the anode side. Electroless nickel coatings are thick and uniform, but expensive, and contain large amounts of impurities. Electrolytic nickel coatings are not sufficiently dense or uniform in thickness.

12.5 PAFC Materials and Design

PAFCs also use bipolar plates for reactant distribution and current collection. Graphite resin mixtures are typically carbonized at low temperatures (~900°C/1,652°F), and are not suitable due to their rapid degradation in PAFC operating environments. Corrosion resistance of these plates can be improved by heat treatment at 2,700°C (4,892°F). Pure graphite bipolar plates are corrosion-resistant for a projected life of 40,000 hours in PAFCs, but like in other low-temperature fuel cells—are costly and expensive to manufacture. Most bipolar plate designs in PAFCs are similar to low-temperature fuel cells, but another type of commonly used design is a multicomponent bipolar plate, which has a thin, impervious plate that separates the reactant gases in adjacent cells, and a separate porous pate with ribbed channels that is used for gas flow. The impervious plate is subdivided into two parts, and each joins one of the porous plates. Figure 12-9 shows two pieces of porous, ribbed substrate and one sheet of conductive, impermeable material such as carbon. The porous ribbed flow field plate hold phosphoric acid electrolyte.

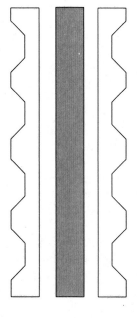

Figure 12-9 PAFC cell interconnections made from ribbed substrates.

12.6 Channel Shape, Dimensions, and Spacing

Fluid flow channels are typically rectangular in shape, but other shapes such as trapezoidal, triangular, and circular have been demonstrated. The change in channel shape can have an affect upon the water accumulation in the cell, and, therefore, the fuel and oxidant flow rates. For instance, in rounded flow channels, the condensed water forms a film at the bottom of the channel, and in tapered channels, the water forms small droplets. The shape and size of the water droplets is also determined by the hydrophobicity and hydrophilicity of the porous media and channel walls. Channel dimensions are usually around 1 mm, but a large range exists for micro- to large-scale fuel cells (0.1 to 3 mm). Simulations have found that optimal channel dimensions for macro fuel cell stacks (not MEMs fuel cells) are 1.5, 1.5, and 0.5 mm for the channel depth, width, and land width (space between channels), respectively. These dimensions depend upon the total stack design and stack size. The channel dimensions affect the fuel and oxidant flow rates, pressure drop, heat and water generation, and the power generated in the fuel cell. Wider channels allow greater contact of the fuel to the catalyst layer, have less pressure drop, and allow more efficient water removal. However, if the channels are too wide, there won't be enough support for the fuel cell. If the spacing between flow channels is also wide, this benefits the electrical conductivity of the plate but reduces the area exposed to the reactants, and promotes the accumulation of water [2,3,6].

12.7 Bipolar Plate Manufacturing

When designing and building a fuel cell stack, the materials selected, the methods of assembly and the manufacturing of fuel cell components are important considerations because the ultimate goal is to not only have a highly efficient fuel cell stack—but also a highly reproducible stack that can be mass produced. The materials and designs currently used for most fuel cell plates could not be mass produced. Therefore, in the process of selecting and designing a bipolar plate, it may be beneficial for the researcher to consider alternative methods, materials, and manufacturing methods that would be useful for producing more than a few units. Sections 12.7.1–12.7.3 describe common fabrication methods for traditional fuel cell bipolar plate materials.

12.7.1 Nonporous graphite plate fabrication

One of the modern alternatives to the traditional solid graphite bipolar plates are graphite mixtures that contain crystalline graphite with additives and/or binders, are compression molded, and subjected to a heat treatment in the absence of oxygen. The graphite mixtures are typically made with additives like aluminum oxide, zircon dioxide, silicon dioxide, titanium dioxide, silicon carbide, and powdered coke, and include binders such as fructose, glucose, galactose, and mannose, as well as oligosaccharides such as sucrose, maltose, and lactose [4].

12.7.2 Coated metallic plate fabrication

The fabrication of coated metallic plates includes the formation of the base plate, surface preparation and cleaning operations, and coating processes. A standard method for forming solid metallic bipolar plate designs is machining or stamping. Other metal forming processes currently being researched are cold closed die forging, die-casting, investment casting, powder metal forging, and electroforming. A summary of the metal forming processes are given in Table 12-4. These machining processes are for macro-sized fuel cell plates (>1 cm^3). Techniques not compatible with smaller-faced "macro" fuel cell designs are investment casting and powder metal forging, and often die-casting. For MEMs fuel cells, (<1 cm^3), there are numerous methods of fabrication that are not covered in this text, but the fabrication basics for these fuel cells can be studied in any microfabrication or micromachining text.

TABLE 12-4 Process Options for Bipolar Plates (Adapted from [4])

Material	Process options	Plate size(s)
Aluminum	Machining Cold closed die forging Stamping Die-casting Investment casting Powder metal forging	Machining and cold closed die forging can be used for almost any size. Investment casting and powder metal forging can only be used for larger-faced, thinner plates.
Stainless steel	Machining Cold closed die forging Stamping Investment casting Powder metal forging	Machining and cold closed die forging can be used for almost any size. Investment casting and powder metal forging can only be used for larger-faced, thinner plates.
Titanium	Machining Cold closed die forging Stamping Investment casting Powder metal forging	Machining, stamping, and cold closed die forging can be used for almost any size. Investment casting and powder metal forging can only be used for certain-sized larger-faced, thinner plates.
Nickel	Machining Cold closed die forging Stamping Investment casting Powder metal forging Electroforming	Machining and cold closed die forging can be used for almost any size. Investment casting and powder metal forging can only be used for larger-faced, thinner plates. Electroforming should only be used for smaller-faced, thicker plates.
Carbon composite	Compression molding Injection molding Transfer molding Reaction injection molding	These processes should only be used for smaller-faced, thicker plates.

12.7.3 Composite plate fabrication

The layered composite plate developed at Los Alamos [23] uses a multistep process in its fabrication. The stainless steel layer is first produced using shearing and stamping methods. The graphite powder and resin are mechanically mixed and molded to the required shape by conventional compression or injection molding. The graphite plate is then baked in an oven. A conductive adhesive is then applied to the graphite plate using a screen printing method. Afterward, a hot press is used to join the stainless steel and graphite plates [4].

Several processes are used for carbon-based composite bipolar plates, including mold fabrication, fiber preparation, composite formation, and carbonization. Common post-processing steps are carbonization and graphitization. In carbonization, the resin is converted to carbon through the decomposition of the C–H bonds [4]. Graphitization is the process of making the material more dense.

12.8 Gaskets and Spacers

Gas leaks lead to unused reactant and poor fuel cell performance. Gas tightness can be improved by increasing stack compression and/or by finding a better gasket material. Many types of gasket materials can be used in fuel cell stacks. Some of the considerations when selecting the appropriate gasket material should be:

- Long-term chemical stability
- Prevent mixing of fuel and oxidant
- Prevent mixing the reactants with the environment
- Vibration and shock resistance
- Electrical insulation between components
- Minimal mechanical and thermal mismatch stresses
- Stability over the required temperature range
- Low-cost
- Excellent sealing capability
- Low-cost stack manufacturing methods
- Prevent mechanical bonding of components (for certain fuel cell types)

These requirements are often tough to meet simultaneously, and are often much more of a challenge for high temperature fuel cells. Some commonly used gasket materials for low temperature fuel cells include silicon, EPDM rubber, and PTFE. Gasket materials for higher temperature fuel cells are more of a challenge because it is tougher to seal ceramic components to prevent gas leakage.

12.8.1 PEMFCs/DMFCs/AFCs

The following materials listed in Table 12-5 are commonly used as gasket materials in low-temperature fuel cells.

All of the material types listed in Table 12-5 are relatively inexpensive and easy to shape into gaskets. Of the materials listed in Table 12-5, EPDM is the best material for gas tightness, but thin EPDM sheets may not be readily available. PTFE is a little more difficult to make gas-tight than EPDM. The combination of EPDM and PTFE gaskets enables the designer to obtain the correct thickness. PTFE-coated glass fiber thread is the easiest to handle, and is available in several thicknesses, but requires more compression in order to be gas-tight. Gas tightness can be improved by bonding the gaskets to the fuel cell (MEA) holder (usually Mylar) using a suitable adhesive.

In small stacks at ambient pressure, soft gasket materials such as silicon work very well for sealing the cell. A general rule of thumb in designing PEM or DMFCs is that the gasket material is approximately the same thickness as the fuel cell (MEA). However, in large or pressurized stacks, the soft gasket materials deform under compression, and may block gas flow. This can be prevented in several ways: (1) using more rigid gasket materials, (2) covering the flow channels with bridge pieces, or (3) machining the flow channels to a deeper width and depth.

12.8.2 SOFC Seals

There are many challenges associated with sealing high temperature fuel cells such as SOFCs. Custom seal designs are often made for particular cell and stack designs. Certain tubular and monolithic designs (see Chapter 13) require no seals at all. Many planar designs require many seals per unit, and the designs can vary greatly for a given cell area. The two main types of seals that are being developed for SOFCs are bonded and compressive seals [25,26].

TABLE 12-5 Process Options for Gasket Materials

Gasket type	Advantages/disadvantages
Silicon	▪ Good gas tightness
	▪ Available in several thicknesses
	▪ May deform and obstruct gas flow
EPDM rubber	▪ Good gas tightness
	▪ Thin sheets may not be readily available
PTFE	▪ Difficult to make gas-tight
PTFE-coated glass fiber thread	▪ Easy to handle
	▪ Available in several thicknesses
	▪ Requires more compression for gas tightness
EPDM/PTFE	▪ Available in several thicknesses

Bonded seals. Bonded seals are either rigid or flexible. The rigid type of seal must have a close thermal expansion coefficient match to the other components around it. If the seal is flexible, the thermal expansion coefficient can vary to a much larger degree. The bonding temperature for a flexible seal should be between the operating temperature and the stability limit for other cell materials. One of the issues with SOFCs is the method of sealing the ceramic components to obtain no gas leakage. The current approach is to use glasses and ceramics that have transition temperatures close to the operating temperature of the cell (650 to 800°C) [25]. This type of seal performs well because as the cell is heated up, the material softens and forms a tight seal. Glass and glass ceramic seals also have a wide range of compositions which can be custom-made to have the desired properties. This type of seal is also low-cost, and easy to manufacture and apply. These seals also have some disadvantages, such as brittleness, and possible cell failure during cool down. In larger cells, the thermal expansion coefficient does not have a close enough match, despite the customability of the glass ceramic material. There is also the issue of the migration of silica from glasses, which causes degradation in cell performance. These seal types sometimes can also volatilize during operation, which can negatively interact with other cell components [25,26].

Another type of sealing material are metal brazes, which have a molten metal filler that insures sealing. This seal type facilitates hermetic sealing, has customizable properties and are easy to fabricate. A few of the issues with this type of seal are that they are electrically conductive, and often incompatible with SOFC operating conditions. Bonded seals result in compact structures, therefore, no additional pressure is required.

Compressive seals. Compressive seals must be elastic over the operating temperature range, and soft enough to fill the surface roughness between the parts to be sealed. Some advantages of compressive seals over bonded seals include: a reduction of thermal stress, a greater range of thermal expansion coefficients can be used, and certain types are low- cost and easy to fabricate. Some of the issues with this seal type are finding materials that can work within the desired temperature range, and the pressure required to keep the stack components together. With this seal type, a load-frame is required, which can be bulky and expensive, and complicate the stack design. Another challenge is balancing the pressure between all of the components, since different parts may require different pressures to operate properly [25,26].

Compressive seals that are used for SOFCs include mica and hybrid mica seals. These seals have the ability to withstand thermal cycling, but have high leak rates. This can be sometimes remedied by placing a thin layer of glass on either side of the seal. Although this process is encouraging, there is still a lot of improvement needed for future seal types.

12.9 End Plates

The traditional fuel cell stack is designed with surfaces (bipolar and end plates) that are parallel to one another with a high degree of precision, as shown in Figure 12-10. The end plate must be mechanically sturdy enough to support the fuel cell stack, and be able to uniformly distribute the compression forces to all of the major surfaces of each fuel cell of the fuel cell stack. In fuel cell stacks containing a large number of fuel cells, the tolerance accumulation can result in substantial non-parallelism at the terminal end cells of the fuel cell stack. The material selection for end plates is quite wide for low-temperature stacks, but the larger the stack becomes, the more important it is to uniformly transmit the compressive forces to the major surfaces of the end cells of the fuel cell stack. Some fuel cell stacks are designed with the last bipolar plate doubling as the end plate (flow fields on one side, and smooth on the other side), while other stacks use a separate plate next to the last bipolar plate as the end plate. Obviously, the material selection for the flow-field/end-plate combination is much more limited than having separate bipolar and end plates. Some considerations when selecting the appropriate end plate material should be:

- High compressive strength
- Vibration and shock resistance
- Stability over the required temperature range
- Low-cost
- Mechanical stability (providing support for the stack)

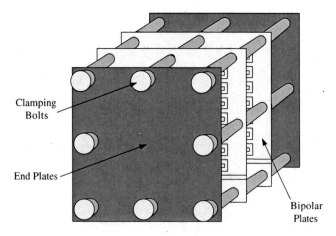

Figure 12-10 A two cell stack with end plates, clamping bolts, and bipolar plates.

The following materials are commonly used as end-plate materials in fuel cells:

- Graphite
- Stainless steel
- Aluminum
- Titanium
- Nickel
- Metal foams
- PVC
- Polycarbonate
- Polyethylene
- Various other polymers

In addition to the characteristics previously listed, the endplate must be easily machinable because it often has the connections or ports for receiving sources of air, fuel, water, and any other materials needed for the fuel cell to function properly. Such ports may also be connected to each other via external conduits. These connections or ports may also link to conduits connected to valves, heat exchangers, and any other desired balance-of-plant components to connect to the fuel cell.

12.10 Constructing the Fuel Cell Bipolar Plates, Gaskets, End Plates, and Current Collectors

After the fuel cell (MEA) has been manufactured (see Chapter 11), the reactant flow field design needs to be chosen, and the appropriate material selected for the flowfield plates, gasket, end plates, and current collectors. In this example, the materials and designs are chosen for a standard temperature and pressure, portable, air-breathing, single fuel cell PEM stack. The stack in this example is then put together in Chapter 13.

12.10.1 Bipolar plate design

The choice of material to construct the bipolar plates from should be inexpensive and easy to manufacture. Any material such as graphite or other non-corrosive materials with high conductivity and a rigid structure can be used. The flow fields must be designed to accommodate the fuel cell (MEA) and fuel cell stack size must be appropriate for the application. As discussed in this chapter, the most common flow field design for bipolar plates is the serpentine design. Many machine shops and fuel cell companies will machine the plates from a variety of materials if provided

242 Chapter Twelve

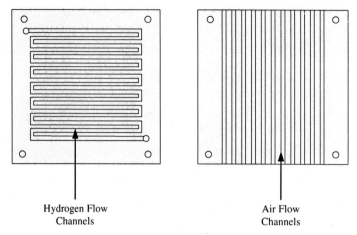

Figure 12-11 Bipolar plate design for hydrogen and air flow channels for a PEMFC.

with a mechanical drawing (like Clean Fuel Cell Energy, LLC). The mechanical drawing must include the flow channel design, and the desired holes surrounding the flow design for the bolts or other clamping device that will be used to hold the plates and fuel cell stack together. For an air-breathing hydrogen PEM fuel cell stack, Figure 12-11 presents an example of the bipolar plate design used with graphite plates.

12.10.2 Gasket selection

For the small air-breathing stack presented in Figure 12-12, 0.010 or 0.020-inch silicone gaskets can be used (depending upon the fuel cell

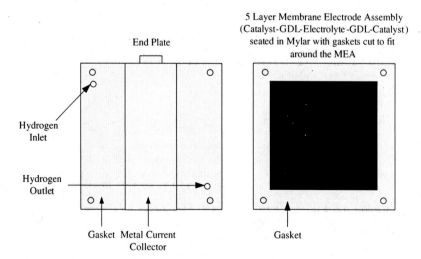

Figure 12-12 Gaskets, end plate, and MEA for PEMFC.

[MEA] thickness). Another popular option is using a 0.010-inch thick fiberglass reinforced silicone rubber anode gasket, and a 0.010-inch thick furan cathode gasket. The gasket is placed around the flow fields next to the electrode/diffusion layers to create a seal to prevent gas leakage. These pieces should be cut and made to fit around the MEA and flow fields, as shown in Figure 12-12.

12.10.3 End plates

End plates can be made of metal, polymer, or a number of other materials to hold the stack in place when it is clamped together by nuts and bolts or some other clamping device. The gaskets and metal current collectors can be placed/connected to the top of the end plate, as shown in Figure 12-13. The end plates in most portable, air breathing fuel cell stacks are next to the gaskets/current collectors and the conductive flow plate as shown in Figure 12-13.

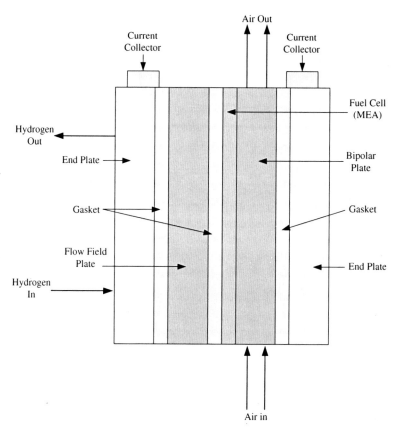

Figure 12-13 End and flowfield plates, gaskets, current collectors, and MEA for a single-cell PEMFC stack.

12.10.4 Current collectors

The metal current collectors can be made from any type of conductive metal, and are seated on the end plates to collect the electrons. Commonly used materials include tin, brass, aluminum, and many other metal types. An example of one type of current collector design is shown in Figures 12-12 and 12-13.

Chapter Summary

After the actual fuel cell (membrane electrode assembly [MEA]) has been fabricated, the cell must be integrated into the fuel cell stack. Bipolar plates have multiple jobs, such as evenly distributing fuel and oxidant to the cells, collecting the current to power the desired devices, and evenly distributing or discarding heat and water products. Bipolar plates are dominated by the management of cost, stack mass, chemical resistance, and high temperature and pressures in certain fuel cell environments. Currently, no bipolar plate meets all of the desired criteria for fuel cells. The selection of the appropriate bipolar plates can be a challenge in order to select the most conductive material balanced with the most chemically inert, and the most easily manufactured. The materials typically used for bipolar plates are graphite, coated metallic plates, composite plates, and high temperature ceramics. The plate that performs the best in low-temperature fuel cells is nonporous graphite. This material provides high corrosion resistance without the need for a protective coating, and good thermal and electrical conductivity at a low density. Graphite plates are still the preferred material for most PEMFC and DMFC applications, although they are lower in electrical conductivity, compressive strength, and recyclability than metallic plates. They also cannot be mass-produced efficiently; therefore, they are more expensive. Composite plates have the advantage of being flexible for incorporating seals, manifolding, cooling systems, and other features—although processing can include many steps. However, composite plates are usually manufacturable compared with graphite plates. The material used for flow fields/current collectors/interconnects in high-temperature fuel cells vary depending upon the temperature that the fuel cell operates at. Several ceramics and stainless steels work well for these cells. However, high temperature fuel cells face the same dilemas with plate/interconnect materials as low temperature fuel cells.

The flow field design is also very important for insuring even distribution of the reactants and products throughout the cell. Commonly used designs range from the traditional serpentine design to parallel, interdigitated, and pin-type, as well as combinations of these flow field types. Many process options are available to manufacture the bipolar plates. The option selected should be carefully considered to insure it will perform the best, and will enable mass production of the fuel cell type. The design of end plates also offers many options. If the end plates are not part of the fuel flow fields, they can be made of almost any sturdy material that will resist the stack compression. However, if the end plates have flow fields, or cooling channels, the material will probably be the same as for the bipolar plates, or it, too, will have to be carefully selected.

Problems

1. A fuel cell has a 25 cm^2 active area and a current density of 0.7 A/cm^2 with nine parallel channels on the cathode. Each channel is 1 mm wide and 1 mm deep with 1 mm of spacing between channels. Air at the inlet is 100 percent humidified at 75°C. The pressure is 2 atm, and there is a 0.2 atm pressure drop through the flow field. The oxygen stoichiometric ratio is 2.0. Calculate the velocity and Reynolds number at the air inlet and outlet.

2. Calculate the pressure drop through a PEMFC cathode flow field of a single graphite plate with 250 cm^2 cell area. The stack operates at 1 atm at 70°C with 100 percent saturated air. The flow field consists of 24 parallel serpentine channels 0.8 mm wide, 1 mm deep, and 1 mm apart.

3. A fuel cell has a 250 cm^2 active area and a current density of 0.6 A/cm^2 with 25 parallel channels on the cathode. Each channel is 1.5 mm wide and 1.5 mm deep with 1 mm spacing between channels. Air at the inlet is 100 percent humidified at 80°C. The pressure is 3 atm, and there is a 0.3 atm pressure drop through the flow field. The oxygen stoichiometric ratio is 3.0. Calculate the velocity, Reynolds number, and pressure drop at the air inlet and outlet.

Bibliography

[1] O'Hayre, Ryan, Suk-Won Cha, Whitney Colella, and Fritz B. Prinz. *Fuel Cell Fundamentals*. 2006. New York: John Wiley & Sons.
[2] Barbir, Frano. *PEM Fuel Cells: Theory and Practice*. 2005. Burlington, MA: Elsevier Academic Press.
[3] Li, Xianguo, and Imran Sabir. "Review of Bipolar Plates in PEM Fuel Cells: Flow-Field Designs." *International Journal of Hydrogen Energy*. Vol. 30, 2005, pp. 359–371.
[4] Mehta, Viral and Joyce Smith Copper. "Review and Analysis of PEM Fuel Cell Design and Manufacturing." *Journal of Power Sources*. Vol. 114, 2003, pp. 32–53.
[5] Hermann, Allen, Tapas Chaudhuri, and Priscila Spagnol. "Bipolar Plates for PEM Fuel Cells: A Review." *International Journal of Hydrogen Energy*. Vol. 30, 2005, pp. 1297–1302.
[6] Cha, S.W., R. O'Hayre, Y. Saito, and F.B. Prinz. "The Scaling Behavior of Flow Patterns: A Model Investigation." *Journal of Power Sources*. Vol. 134, 2004, pp. 57–71.
[7] U.S. Patent 6,551,736 B1. Fuel Cell Collector Plates with Improved Mass Transfer Channels. Gurau, Vladimir Gurau, Frano Barbir, and Jay K. Neutzler. Teledyne Energy Systems, Inc, Hunt Valley, MD. April 22, 2003.
[8] Nguyen, Nam-Trung and Siew Hwa Chan. "Micromachined Polymer Electrolyte Membrane and Direct Methanol Fuel Cells – A Review." *J. Micromech. Microeng*. Vol. 16, 2006, pp. R1–R12:

[9] Mench, M.M., Z. H Wang, K. Bhatia, and C. Y. Wang. "Experimental Study of a Direct Methanol Fuel Cell." Electrochemical Engine Center, Department of Mechanical and Nuclear Engineering, Pennsylvania State University, 2002.
[10] Wang, C. Y., M.M. Mench, S. Thynell, Z.H. Wang, and S. Boslet. "Computational and Experimental Study of Direct Methanol Fuel Cells." *Int. J. Transport Phenomena*. Vol. 3, August 2001.
[11] Hsieh, ShouShing, Ching-Fang Huang, Jenn-Kun Kuo, Huang-Hsiu Tsai, and Sheng-Huang Yang. "SU-8 Flow Field Plates for a Micro PEMFC." *Journal of Solid State Electrochemistry*. Vol. 9, 2005, pp. 121–131.
[12] Gulzow, E., M. Schulze, and U. Gerke. "Bipolar Concept for Alkaline Fuel Cells." *Journal of Power Sources*. Vol. 156, 2006, pp. 1–7.
[13] Muller, M. C. Muller, F. Gromball, M. Wolfle, and W. Menz. "Micro-Structured Flow Fields for Small Fuel Cells." *Microsystem Technologies*. Vol. 9, 2003, pp. 159–162.
[14] Muller, M.A., C. Muller, R. Forster, and W. Menz. "Carbon Paper Flow Fields made by WEDM for Small Fuel Cells." *Microsystem Technologies*. Vol. 11, 2005, pp. 280–281.
[15] Feindel, Kirk, W. Logan, P.A. LaRocque, Dieter Starke, Steven H. Bergens, and Roderick E. Wasylishen. *J. Am. Chem. Soc.* Vol. 126, 2004, pp.11436–11437.
[16] He, Suhao, Matthew M. Mench, and Srinivas Tadigadapa. "Thin Film Temperature Sensor for Real-Time Measurement of Electrolyte Temperature in a Polymer Electrolyte Fuel Cell." *Sensors and Actuators A*. Vol. 12, 2006, pp. 170–177.
[17] Chen, X., N.J. Wu, L. Smith, and A. Ignatiev. "Thin Film Heterostructure Solid Oxide Fuel Cells." *Applied Physics Letters*. Vol. 84, No. 14, April 2004.
[18] Motokawa, Shinji, Mohamed Mohamedi, Toshiyuki Momma, Shuichi Shoji, and Tesuya Osaka. "MEMS-Based Design and Fabrication of a New Concept Micro Direct Methanol Fuel Cell." *Electrochem. Comm.* Vol. 6, 2004, pp. 562–565.
[19] Lee, Shuo-Jen, Yu-Pang Chen, and Ching-Han Huang. "Electroforming of Metallic Bipolar Plates with Micro-Featured Flow Field." *Journal of Power Sources*. Vol. 145, 2005, pp. 369–375.
[20] Hsieh, Shou-Shing, Sheng-Huang Yang, Jenn-Kun Kuo, Chin-Feng Huang, and Huang-Hsiu Tsai. "Study of Operational Parameters on the Performance of Micro PEMFCs with Different Flow Fields." *Energy Conversion and Management*. Vol. 47, 2006, pp. 1868–1878.
[21] Change, Min-Hsing, Falin Chen, and Nai-Siang Fang. "Analysis of Membraneless Fuel Cell Laminar Flow in a Y-Shaped Microchannel." *Journal of Power Sources*. Vol. 159, 2006, pp. 810–816.
[22] Choban, E.R., J.S. Spendelow, L. Gancs, A. Wieckowski, and P.J.A. Kenis. "Membraneless Laminar Flow-Based Micro Fuel Cells Operating in Alkaline, Acidic, and Acidic/Alkaline Media." *Electrochimica Acta*. Vol. 50, 2005, pp. 5390–5398.
[23] Cohen, Jamie L, Daron A. Westly, Alexander Pechenik, and Hector D. Abruna. "Fabrication and Preliminary Testing of a Planar Membraneless Microchannel Fuel Cell." *Journal of Power Sources*. Vol. 139, 2005, pp. 96–105.
[24] Cohen, Jamie L., David J. Volpe, Daron A. Westly, Alexander Pechenik, and Héctor D. Abruña. "A Dual Electrolyte H_2/O_2 Planar Membraneless Microchannel Fuel Cell System with Open Circuit Potentials in Excess of 1.4 V." *Langmuir*. Vol. 21, 2005, pp. 3544–3550.
[25] The Fuel Cell Handbook. EG&G Technical Services. 7th ed. U.S. Department of Energy, Nov. 2004.
[26] Larminie, James and Andrew Dicks. Fuel Cell Systems Explained. 2000. John Wiley & Sons, Ltd, Chichester, England.

Chapter 13

Fuel Cell Stack Design

As demonstrated in Chapters 11 and 12, there are many parameters that must be considered when designing a fuel cell. A very important design consideration is the overall stack design and configuration. The most commonly used stack configuration is the bipolar configuration, which has been described in previous chapters and is shown in Figure 13-1. This chapter also presents many alternative stack configurations. For all of these configuration types, the materials, designs, and methods of fabricating the components differ only slightly. When considering the design of a fuel cell stack, several limitations should be considered. Some of these limitations may include:

- Size, weight, and volume at the desired power
- Cost
- Operating temperature
- Humidification and water management (if applicable)
- Fuel and oxidant pressures
- Fuel type(s) and storage

These limitations then translate into design requirements, which helps one to design the entire fuel cell system. Table 13-1 presents a summary of many of the basic design parameters and requirements.

There are two Appendices devoted to helping the designer to create a fuel cell design. Along with the technical details of fuel cell design, a comprehensive plan should be written on how the fuel cell will meet all of the customer requirements. Appendix H shows a brief list of Product Design Specifications. Depending upon the size of the company or project, whole reports can be written for some of the entries in this table.

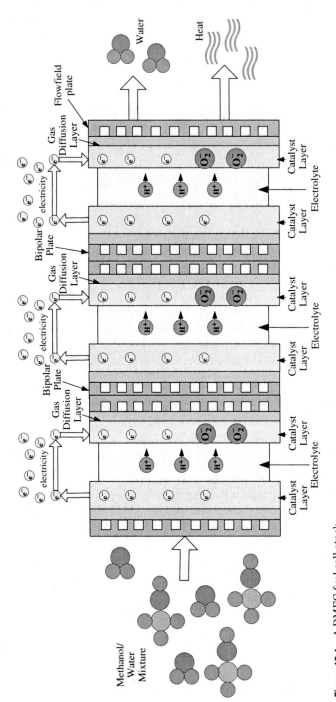

Figure 13-1 A DMFC fuel cell stack.

TABLE 13-1 Fuel Cell Design Parameters

Requirements	• Power Density/Voltage • Efficiency • Start-up • Transient Response • Weight Size/Shape
Operation Conditions	• Temperature • Pressure • Humidity • Flow Rate
Porous Electrode Layer or Gas Distribution Layer	• Thickness • Material properties such as porosity, conductivity, temperature resistance, chemical resistance
Electrolyte Layer	• Thickness • Material properties such as porosity, conductivity, temperature resistance, chemical resistance
Catalyst Layer	• Thickness/Loading • Composition
Fuel Cell Design	• Bipolar Plate Material • Channel Design/Layout • Channel Size • Rib Size • End Plate Material • Cooling Channels • Cell Interconnects

Appendix I shows a more detailed list than Table 13-1, and will help the designer to begin thinking about the important details of stack design.

Fuel cell design was discussed in Chapter 11, and bipolar plate, end plate, and gasket designs and materials were discussed in Chapter 12. This chapter explains how these components are integrated into a fuel cell stack and the potential issues that may be encountered.

13.1 Fuel Cell Stack Sizing

In order to design a fuel cell stack, two independent variables must be considered. The known requirements are the maximum power, voltage, and/or current. Recall that power output is a product of stack voltage and current:

$$W_{FC} = V_{st} \cdot I \tag{13-1}$$

TABLE 13-2 Common Power and Voltage Requirements for Several Fuel Cell Applications

Application	Maximum power requirement	Voltage requirement	Average power supply
Cell Phone	1.2 to 3 W	1.2 to 4.2 V	1.7 W
Laptop	40 W	8 to 12.6 V	15 W
Fuel Cell Powered Scooter	5.9 to 7.7 kW	38 to 51 V	1.2 kW
Automobile	120 kW	255 to 284 V	50 to 85 kW
Backup Power Source	5 kW	24 to 30 V	1.26 kW
Stationary Power	10 to 500 kW	12 to 480 VAC	5 kW

The maximum power and voltage required is dependent upon the application. Therefore, the first step in designing the fuel cell stack is to determine the power requirement for the desired application. Table 13-2 shows the maximum, average power, and voltage requirements for some example applications.

The specific power/voltage requirements will depend on the particular application, as well as the industry standards that exist. It is usually safest to go with the highest possible power and voltage spike that may occur during device operation.

Example 13-1 Design a fuel cell for the laptop specified in Table 13-2. The fuel cell should be capable of providing a maximum voltage of about 13 volts with a power of 40 watts. Using equation 13-1, the stack current can be calculated as follows:

$$W = V_{st} \cdot I$$

$$40 \text{ watts} = (13 \text{ V}) * I$$

$$I = 3.08 \text{ A}$$

For the average power of 15 watts and 12 volts, equation 13-1 should be used again to determine the stack current:

$$15 \text{ watts} = (12 \text{ V}) * I$$

$$I = 1.25 \text{ A}$$

If a current density of 0.6 A/cm^2 is assumed, then using equation 13-2, the total active cell area required is 5.13 cm^2. If each cell area is assumed to be 2 cm^2, then about 2.6 or 3 cells are needed in order for the fuel cell to provide maximum power.

Other initial considerations that are helpful when designing a fuel cell stack are the current and power density. These are often unavailable initially, and can be calculated from the desired power output, stack voltage, efficiency, and volume and weight limitations. The current is a product of the current density and the cell active area:

$$I = i * A_{cell} \qquad (13\text{-}2)$$

The cell potential and the current density are related by the polarization curve:

$$V_{cell} = f(i) \qquad (13\text{-}3)$$

The fuel cell polarization curve is determined experimentally from fuel cell output, and helps the designer size the stack to obtain the appropriate power. Figures 13-2 through 13-8 show example polarization curves for single PEMFC, DMFC, SOFC, PAFC, MCFC, AFC, and PCFC cells. The designer selects the nominal operating point on fuel cell polarization curves as the cell voltage and corresponding cell current density. Most fuel cell developers use a nominal voltage of 0.6 to 0.7 V at nominal power. Fuel cell systems can be designed at nominal voltages of 0.8 V per cell or higher if the correct design, materials, operating conditions, balance-of-plant, and electronics are selected. Plant components, electronics, fuel options, and operating conditions will be discussed in Chapters 14 through 16.

As Figures 13-2 through 13-8 demonstrate, the fuel cell performance is determined by the pressure, temperature, and humidity based upon the application requirements, and can often be improved (depending upon fuel cell type) by increasing the temperature, pressure, and humidity, and optimizing other important fuel cell variables. The ability to increase these variables is application-dependent because system issues, weight, and cost play important factors when optimizing certain parameters. Figures 13-2 through 13-8 present actual data from the literature, but

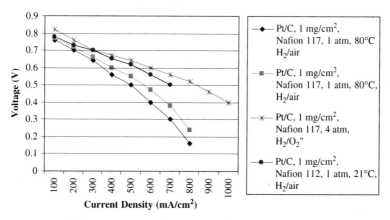

Figure 13-2 Polarization curves for PEMFC single cells [33, 34].

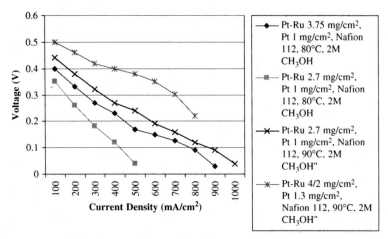

Figure 13-3 Polarization curves for DMFC single cells [29, 30].

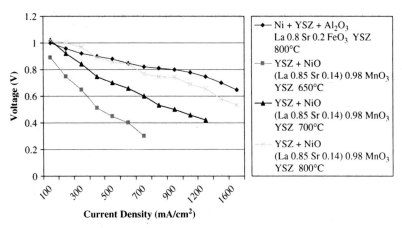

Figure 13-4 Polarization curves for SOFC single cells [13, 14].

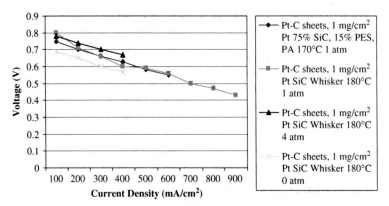

Figure 13-5 Polarization curves for PAFC single cells [16, 17, 18].

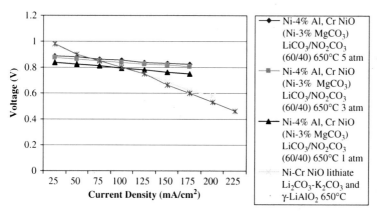

Figure 13-6 Polarization curves for MCFC single cells [26, 27, 28].

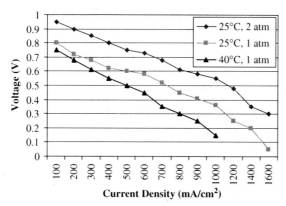

Figure 13-7 Polarization curves for AFC single cells—anode: PTFE 8 percent; Raney Ni 77 percent, Copper 15 percent, cathode: PTFE powder 10 percent, Ag 90 percent, Electrolyte: 30 percent wt KOH embedded in 3-mm thick abestos [22, 24].

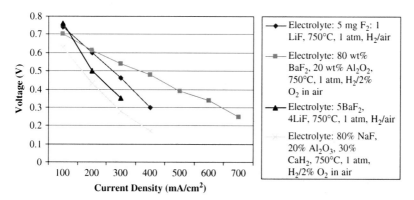

Figure 13-8 Polarization curves for PCFC single cells [15].

many commercial companies have better performance than the data shown. The results achieved at companies are usually not published. Chapter 16 covers fuel cell operating conditions in more detail.

When designing a fuel cell, the designer needs to determine the power density the fuel cell needs to be operated at under maximum power. If the maximum output will be at low power density, the left portion of the polarization curve is used, and the fuel cell area and the stack size will be larger [4]. For a small, portable fuel cell, the efficiency must be maximized, and the power density per area must be high. A high-power density results in lower voltages, and a smaller and cheaper stack.

13.2 Number of Cells

The number of cells in the stack is often determined by the maximum voltage requirement and the desired operating voltage. The total stack potential is a sum of the stack voltages or the product of the average cell potential and number of cells in the stack:

$$V_{st} = \sum_{i=1}^{N_{cell}} V_i = \overline{V}_{cell} * N_{cell} \qquad (13\text{-}4)$$

The cell area must be designed to obtain the required current for the stack. When this is multiplied by the total stack voltage, the maximum power requirement for the stack must be obtained. Most fuel cell stacks have the cells connected in series, but stacks can be designed in parallel to increase the total output current. It is preferable to not have a large amount of cells with a small or very large active area because the cells can be hard to assemble, and there can be large resistive losses. With fuel cells that have large active areas, it can be difficult to achieve uniform temperature, and if applicable, humidity, and water management conditions. Table 13-3 shows the average cell voltages, current density, and power density for various fuel cell types. The current densities displayed in Table 13-3 are the conservative values found in the literature,

TABLE 13-3 Cell Voltages and Current Density for Various Fuel Cell Types

Fuel cell type	Average voltage (full load voltage) (V)	Current density (mA/cm^2)
PEMFC	0.60 to 0.70	300 to 1000
DMFC	0.35 to 0.40	200 to 600
AFC	0.70 to 0.80	300 to 800
MCFC	0.60 to 0.80	200 to 600
PAFC	0.60 to 0.70	300 to 600
SOFC	0.70 to 0.80	300 to 1200
PCFC	0.50 to 0.60	200 to 600

and often can be much higher (up to 1500 mA/cm²) depending upon the fuel cell type and system.

The cell voltage and current density is the operating point at nominal power output, and can be selected at any point on the polarization curve. The average voltage and corresponding current density selected can have a large impact upon stack size and efficiency. A higher cell voltage means better cell efficiency, which could be based upon the fuel cell materials, flow channel design, and optimization of system temperature, heat, humidity, pressure, and reactant flow rates. The fuel cell stack efficiency can be approximated with the following equation:

$$\eta_{stack} = \frac{V_{cell}}{1.482} \qquad (13.5)$$

Example 13.2 Design the fuel cell voltage, current, cell area, and number of cells to power a scooter with a power requirement of 5.9 kW.

According to the polarization curves presented in Figure 13-2, 0.6 V can be obtained at about 500 mA·cm^{-2}. To be conservative and leave room for unusual bursts of speed, a point below this power peak is often chosen. Since the voltage and current density in Figure 13-2 are already conservative, we will use 0.6 V and 500 mA/cm² as the operating voltage and current density.

The number of cells depends upon the required operating voltage. The electric scooter industry in Taiwan is standardizing on 48-V electric motors, so the number of cells is chosen to have the stack operate in the vicinity of 48 V at the most common power demand. To meet this requirement, the total current of the fuel cell is: 5900 W/0.6 V = 9833.3 A. The number of cells is a function of desired operating voltage which is 48 V/0.6 V = 80 cells. The total area can be calculated by 9833.3 A/0.5 A per cm² = 19,666.6 cm². The total area per cell is 19,666.6 cm²/80 cells ~ 246 cm² per cell.

13.3 Stack Configuration

In the traditional bipolar stack design, the fuel cell stack has many cells stacked together in series, and the cathode of one cell is connected to the anode of the next cell. The main components of the fuel cell stack are the fuel cell (MEAs), gaskets, bipolar plates with electrical connections, and end plates. The cells are connected together by bolts, rods, clamps or fused together. When contemplating the appropriate fuel cell design, the following should be considered:

- Fuel and oxidant should be uniformly distributed through each cell, and across their surface area.
- The temperature must be uniform throughout the stack.

- If designing a fuel cell with a polymer electrolyte, the membrane must not dry out or become flooded with water.
- The resistive losses should be kept to a minimum.
- The stack must be properly sealed to insure no gas leakage.
- The stack must be sturdy and able to withstand the necessary environments it will be used in.

The most common fuel cell configuration is shown in Figure 13-9, and has been shown throughout the book. Each fuel cell (MEA) is separated by a plate with flow fields on both sides to distribute the fuel and oxidant. The fuel cell stack end plates have only a single-sided flow field. The majority of fuel cell stacks, regardless of fuel cell type, size, and fuels used are of this configuration.

Another type of fuel cell configuration that is common in high-temperature fuel cells is the tubular configuration, which is joined electrically in series through a full length contact interconnection. The cells are typically 30 cm or greater in length, and the inner porous support tube is to maintain the tubular structure [38]. The ohmic losses are typically small in this fuel cell type since the current travels from one cell to the next through a short path. Connecting cells in parallel is easy when the tubes are densely packed. An example of the tubular configuration is shown in Figure 13-10.

Many other designs have been used for fuel cell stacks. Recently, Lynnetech patented (U.S. Patent No. 7,125,625) an interesting design

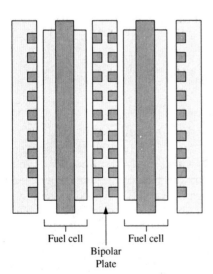

Figure 13-9 Typical fuel cell stack configuration (a two-cell stack).

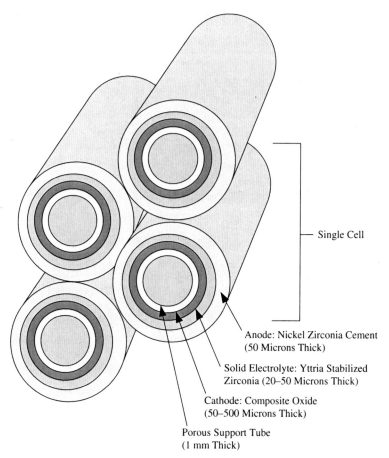

Figure 13-10 Alternate SOFC fuel cell stack configuration (four-cell stack) [38].

that uses a post-type flow field instead of the traditional bipolar plates, as shown in Figure 13-11. The electronically conductive posts form the reactant flow fields on either side of the gas barrier.

There is much more variability in fuel cell design and configuration with MEMs (micro-electro-mechanical) fuel cells that are 1 cm² or less in area than with the larger fuel cell stacks. A noteworthy design is shown in Figure 13-12. The flow fields are made of silicon, and the proton exchange membrane is wedged between the two sets of flow fields. Platinum is deposited on both microcolumns to act as an electrocatalyst and current collector. The flow-field pore diameter was carefully controlled to use capillary pressure in order to distribute

258 Chapter Thirteen

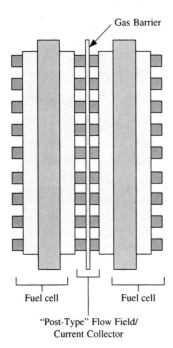

Figure 13-11 Cross-sectional view of a fuel cell stack with post-type flow field/current collectors [43].

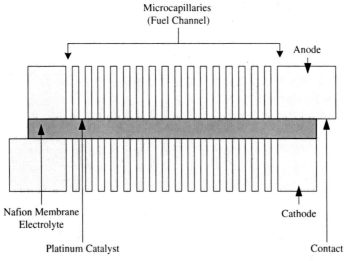

Figure 13-12 Cross-sectional view of the porous silicon-based Micro-DEFC Stack [39].

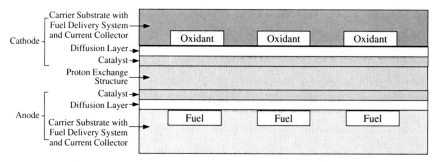

Figure 13-13 Cross-sectional view of a traditional microfuel cell stack [40].

the fuel correctly and minimize crossover. The area of each electrode is 1 cm² [39].

In microfuel cells, the two basic designs are the traditional bipolar design (as shown in Figure 13-13) and the planar design (as shown in Figure 13-14). The planar design is two dimensional and requires a large surface area to deliver similar performance to the bipolar configuration. The fuel and oxidant are delivered through channel networks at a single side of the fuel cell. The fuel and oxidant channels are interdigitated so that the reaction can occur between them [40].

An interesting microfuel cell design is shown in Figure 13-15. This fuel cell structure is usually made of silicon and occurs at small enough dimensions to allow the fuel and oxidant to flow in the laminar flow regime without mixing [41]. The fuel and oxidant have separate entrances. The electrode materials are deposited on to the silicon structure by sputtering or evaporation.

The micro fuel cell design in Figure 13-16 is from U.S. Patent No. 7,029,781 and has dividers separating the anodic and cathodic channels that carry the fuel and the oxidant, respectively. This design allows a good catalyst surface-to-volume ratio and permits the reactants to be carried in counterflow to provide better fuel cell cooling.

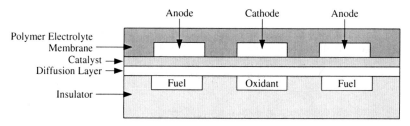

Figure 13-14 Cross-sectional view of planar microfuel cell stack [40].

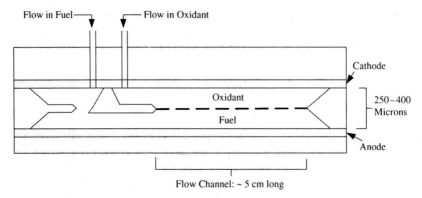

Figure 13-15 Cross-sectional view of a membraneless laminar flow microfuel cell stack [41].

13.4 Distribution of Fuel and Oxidants to the Cells

Fuel cell performance is dependent upon the flow rate of the reactants. Uneven flow distribution can result in uneven performance between cells. Even reactant flow through the cell is normally accomplished by feeding each cell in the stack in parallel. Reactant gases should be supplied in parallel to all cells in the same stack through common manifolds. Some stacks rely on external manifolds, while others use an internal manifolding system. One advantage of external manifolding is its simplicity, which allows a low pressure drop in the manifold, and permits good flow distribution between cells. One of the disadvantages is the gases are flowing in crossflow, which can cause an uneven temperature distribution at the electrodes. Each external manifold must have an insulating gasket to form a seal with the edges of the stack to prevent gas leakage. Internal manifolding distributes gases through channels in the fuel cell itself. An advantage of internal manifolding

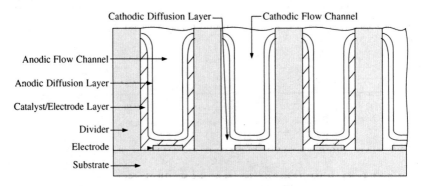

Figure 13-16 Cross-sectional view of a microfuel cell stack [42].

is more flexibility in the direction of flow of the gases. One of the most common methods is ducts formed by the holes in the separator plates that are aligned once the stack is assembled. Internal manifolding allows a great deal of flexibility in the stack design. The main disadvantage is that the bipolar plate design may get complex, depending on the fuel flow channel distribution design. The manifolds that feed gases to the cells and collect gases have to be properly sized. The pressure drop through the manifolds should be an order of magnitude lower than the pressure drop through each cell in order to ensure uniform flow distribution. When analyzing the flow for the cells:

1. The flow into each junction should equal the flow out of it.
2. The flow in each segment has a pressure drop that is a function of the flow rate and length through it.
3. The sum of the pressure drops around a closed loop must be zero.

Some of the factors that need to be considered when designing manifold stacks include manifold structure, size, number of manifolds, overall gas-flow pattern, gas channel depth, and the active area for electrode reactions. The manifold holes can vary in shape from rectangular to circular. The area of the holes is important because it determines the velocity and type of flow. The flow pattern is typically a U-shape (reverse flow) where the outlet gas flows in the opposite direction to the inlet gas, or a Z-shape (parallel flow) where the directions of the inlet and outlet gas-flows are the same as shown in Figures 13-17 and 13-18. The pressure change in the manifolds is much lower than in the flowfield channels in order to insure a uniform flow distribution among cells piled in a stack [58].

The pressure drop can be calculated from the Bernoulli equation:

$$\Delta P(i) = -\rho \frac{[u(i)]^2 - [u(i-1)]^2}{2} + f\rho \frac{L_s}{D_H} \frac{[u(i)]^2}{2} + K_f \rho \frac{[u(i-1)]^2}{2} \quad (13\text{-}6)$$

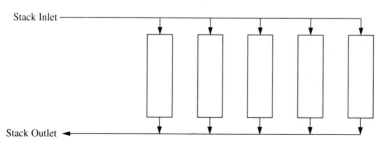

Figure 13-17 A U-type manifold.

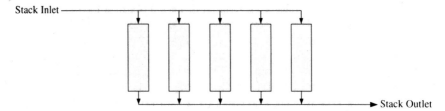

Figure 13-18 A Z-type manifold.

where ρ is the density of the gas (kg/m³), v is the velocity (m/s), f is the friction coefficient, L_s is the length of the segment (m), D_H is the hydraulic diameter of the manifold segment (m), and K_f is the local pressure loss coefficient.

For laminar flow (Re <2300), the friction coefficient f for a circular conduit is:

$$f = \frac{64}{Re} \qquad (13\text{-}7)$$

The walls of the fuel cell manifolds are considered "rough" when the stack has bipolar plates clamped together. The friction coefficient for turbulent flow is a function of wall roughness. The friction coefficient is [3]:

$$f = \frac{1}{\left(1.14 - 2\log\frac{\varepsilon}{D}\right)^2} \qquad (13\text{-}8)$$

where $\frac{\varepsilon}{D}$ is the relative roughness, which can be as high as 0.1.

13.5 Cell Interconnection

As described previously, the voltage per cell is small at approximately 0.6 to 0.7 V. In order to generate a useful voltage, many cells have to be connected in series. The simplest way of accomplishing this is to connect the anode to each cathode in a straight line. The electrons must be able to flow across the face of the electrode to the current collector. The electrons must be able to flow to the current collector without a voltage loss. This method would only be used if the electrode was a good conductor, or the current flows were very low.

The typical method of cell interconnection has been described many times throughout this book—the use of bipolar plates as cell interconnects (see Chapter 12). The bipolar plate electrically connects one cathode to the anode of the next cell. At the same time, these plates feed fuel to the anode, and oxygen to the cathode. Although many fuel cell types use bipolar plates, not all fuel cell types do. Sections 13.5.1–13.5.2 describe alternate cell interconnection types.

13.5.1 SOFCs

The interconnect material used for SOFCs are metals that can withstand temperatures between 800 and 1000°C. Finding materials for the SOFC interconnects is a challenge because many of these materials have a thermal expansion coefficient mismatch with the YSZ electrolyte. New alloy types have been developed to overcome these barriers, such as the $Cr_5Fe-1Y_2O_3$ alloy developed by Siemens [61]. Unfortunately, this type of material can poison the cathode with chromium. Chromium can evaporate from the interconnect and deposit at the $(La,Sr)MnO_3$ (LSM) /YSZ/gas three-phase boundary, which results in rapid deactivation of the catalyst [61]. For lower-temperature SOFCs, cheaper materials are available for the interconnects that do not contain chromium (see Chapter 12). Some material types form oxide coatings, which can limit electrical conductivity and act as a mass transport barrier.

For tubular SOFC designs, the interconnect material most frequently used is lanthanum chromite, which is enhanced with magnesium or other alkaline materials to promote conductivity [61]. An example of this type of interconnect is shown in Figure 13-19. The material has to be

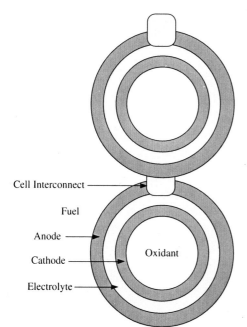

Figure 13-19 Cell interconnect in a tubular SOFC.

sintered at a very high temperature (1625°C) to produce a dense phase. As mentioned previously, the fabrication of SOFCs is a challenge because all of the materials have to be mechanically and chemically compatible and must have similar thermal expansion coefficients. The layers also have to be deposited so that good adherence is achieved between materials.

13.5.2 AFCs

Most of the current AFC systems use electrodes that have a PTFE layer over the gas side. PTFE is an insulator; therefore, it is difficult to make an electrical connection to the face of the electrode. The normal bipolar plate configuration cannot be used in this fuel cell type. The connections are normally made to the edge of the electrode. Often, wires are simply used to connect the positive of one cell to the negative of another. Due to this, complex series/parallel connections can be made. The cathodes in AFCs often face each other so one channel can feed two cathodes. Complex connection patterns are often used to reduce the problem of internal currents within the electrolyte of the fuel cell stack. These arise because the ionically conductive electrolyte is in contact with all cells within the stack. Cells in parallel can help to reduce the voltage while increasing the current, and an alternating pattern can sometimes helpfully balance electric fields in the electrolyte [61].

13.6 Stack Clamping

The stacking design and cell assembly can significantly affect the performance of fuel cells. Adequate contact pressure is needed to hold the fuel cell stack components together to prevent leaking of the reactants between the layers and minimize the contact resistance between layers. The clamping force is equal to the force required to compress the gasket, fuel cell layers, and internal force. The assembly pressure affects the characteristics of the contact interfaces between components. Due to thin dimensions and the low mechanical strength of the fuel cell layers versus the gaskets, bipolar plates, and end plates, the most important goal in the stack design and assembly is to achieve a proper and uniform pressure distribution. If inadequate or nonuniform assembly pressure is used, there will be stack sealing problems, such as fuel leakage, internal combustion, and unacceptable contact resistance. Too much pressure may damage the fuel cell layers, resulting in a broken porous structure and a blockage of the gas diffusion passage. In both cases, it will decrease the cell performance. Every stack has a unique assembly pressure due to differences in fuel cell materials and stack design. The torque on the bolts can be calculated from the following equation [3]:

$$T_t = \frac{F_{clamp} K_b D_b}{N_b} \tag{13-9}$$

where T_t is the tightening torque, Nm, F_{clamp} is the clamping force, N, K_b is the friction coefficient, D_b is the bolt nominal diameter, m, and N_b is the number of bolts.

13.7 Water Management for PEMFCs

As mentioned previously, one of the primary challenges in obtaining optimal performance from PEMFCs is maintaining the MEA water balance. The water content of the polymer membrane effects several overpotentials, and both drying out and flooding can adversely affect the cell performance. Proton conductivity of the polymer membrane is dependent on water content. If the MEA is not adequately humidified, protonic conductivity decreases, which increases the cell resistance. Therefore, the performance of the drying out of the MEA also increases the activation overpotentials.

Excess water can also present a problem. It can prevent reactant diffusion to the catalyst sites by flooding of the electrodes, gas diffusion backings, or gas channels if the water removal is not efficient, which increases diffusion overpotentials. In order for the membrane to remain adequately hydrated, many parameters, such as current density, temperature, reactant flow rates, pressures, humidification, cell design, and component materials need to be optimized [6].

Water in a PEM fuel cell can come from the humidification of the reactant gases or the cathode reaction. The humidity of one or both of the reactant gases depends upon the gas temperature, pressure, flow rate, channel design, and membrane thickness. The water production in the cathode is proportional the current density. The water production can be controlled through proper flow channel and/or heater design (see Chapters 12 and 9).

Water is transported mainly through the gas channels, but to a lesser extent, through the membrane and the electrodes. As described in Chapter 7, two transport processes occur in the polymer membrane as shown in Figure 13-20.

The first transport process occurs when a load is put on a fuel cell, and hydrogen protons travel through the polymer membrane from the anode to the cathode and carry water molecules with them. This is termed the electro-osmotic drag (as mentioned in Chapter 7). The average number of water molecules "dragged" by one proton is called the electro-osmotic

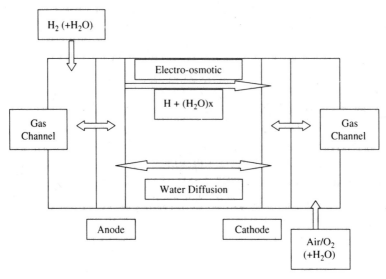

Figure 13-20 Water transport processes in a PEM fuel cell.

drag coefficient. The second transport process occurs when the concentration gradient in the cathode drives the diffusion of water through the membrane. This process is called back diffusion.

The accumulation of water at the cathode occurs from both the electro-osmotic drag and the production of water at the cathode. The water at the cathode can be either transported to the flow channel through the gas diffusion backing, evaporated through heaters, or diffused through the membrane toward the anode [6].

One of the methods of removing water from a PEM fuel cell is through the reactant streams. Undersaturated reactant gases pick up the liquid water in the flow channels. The water removal rate and the ratio of liquid and gaseous water depends upon the operating conditions and cell design. The operating temperature, pressure, and flow rates have a substantial effect on water vapor content of the gases, water evaporation rate, and vapor pressure. Therefore, the efficiency of water removal depends upon these parameters. Chapter 16 discusses fuel cell operating parameters in more detail.

13.7.1 Water management methods

When the reaction product water produced at the cathode is not sufficient to keep the membrane hydrated, the use of humidified reactant gases is necessary. This need arises when the membrane becomes dehydrated due to the high current density requirement at the anode. This phenomenon occurs when the electro-osmotic drag is stronger than the water back diffusion, which leads to a greater flux of water from the anode to the cathode side. This problem is exacerbated by thick polymer membranes since they require more water to remain hydrated [6].

The humidification of the anode can be easily controlled through the hydrogen humidification temperature. Controlling the water balance at the cathode is much more difficult because water accumulates easily due to the reaction and the electro-osmotic drag. The gas diffusion layers have an important role in removing water from the electrodes. The water concentration gradient depends upon the properties (such as porosity and hydrophobic and hydrophilic chains) of the gas diffusion backings. The flow channel geometry also has an effect on the water content in the fuel cell. The width, length, depth, and space between fields affect the flow velocity and the rate of diffusion of the reactants to the electrode layers. It also determines how water droplets flow through the cell [6].

Depending upon the fuel cell application, the use of gas humidifier subsystems may not be desirable due to additional weight, size, system complexity, and parasitic power required. Several methods of controlling water in the fuel cell are commonly used when humidifying reactant gases is not possible [3,6]:

- Direct water humidification of the MEA
- Thinner membranes to enhance water back diffusion
- Internal membrane humidification by embedded catalyst particles
- Structural improvements of the gas diffusion backings

For small, portable applications, the dead-ended bipolar design works well to prevent water accumulation. The dead-ended design reduces the accumulation of water on the anode. The electro-osmotic drag and water back diffusion are balanced, and accumulation on the anode does not occur.

Water management at the cathode is mainly affected by the temperature, cathode flow, channel structure, air flow rate, and the gas diffusion backing layer. If air is used in the fuel cell, it can either be controlled by natural convection or forced into the stack by a small fan or blower. Natural convection is the simplest solution for small, portable fuel cells, but forced convection can make the fuel cell more efficient. However, water management is more difficult with forced convection.

13.8 Putting the fuel cell stack together

Continuing with the fuel cell design from Chapters 11 and 12, the stack will now be put together. After the MEAs, bipolar plates, end plates, and gaskets have all been designed or selected, the next step is simply to put the stack together. The selection of clamping bolts or other methods used for clamping the stack is easy. Make sure the bolts or clamps are

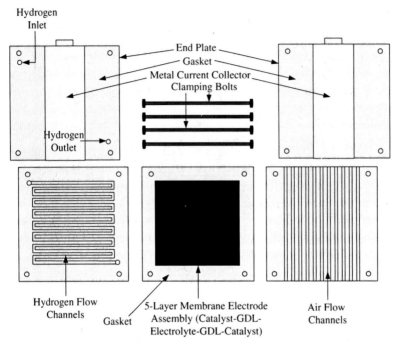

Figure 13-21 Single fuel cell stack parts.

not metallic so they are isolated from the electric current in the bipolar plates. For small stacks, nylon screws or bolts work well. There is a variety of materials and methods that can be used for clamping the stack together. The pieces of a very basic single cell air-breathing PEM stack are shown in Figure 13-21.

Chapter Summary

Many parameters must be considered when designing a fuel cell. Some of the most basic design considerations include: power required, size, weight, volume, cost, transient response, and operating conditions. From these initial requirements, the more detailed design requirements (such as the number of cells, material and component selections, flow field design, etc) can be chosen. The most commonly used stack configuration is the bipolar configuration, which has been described in previous chapters. This chapter also presented many alternative stack configurations, such as the tubular, post-type flow field, planar, and membraneless stack configurations. Common manifold types for even reactant flow through the cell and alternate cell interconnections for electron flow through the stack are also presented. The last topics discussed were the importance of stack clamping and water management in the PEM fuel cell.

Problems

1. Design a fuel cell stack that has to operate at 80°C with air and hydrogen pressures of 1 atm. The Pt/C loading is 1 mg/cm^2 and the cells use the Nafion 117 electrolyte. The total power should be 250 Watts. Use the appropriate polarization curve in Figure 13-2 to help design the stack.

2. Design a fuel cell stack that has to operate at 90°C with 2 M CH_3OH. The Pt-Ru loading is 2.7 mg/cm^2, the Pt/C loading is 1 mg/cm^2, and the cells use the Nafion 112 electrolyte. The total power should be 50 Watts. Use the appropriate polarization curve in Figure 13-3 to help design the stack.

3. Design an SOFC stack that has to operate at 700°C. The total power should be 250 Watts. Use the appropriate polarization curve in Figure 13-4 to help design the stack.

4. Design a PAFC stack that has to operate at 180°C and 1 atm. The total power should be 100 Watts. Use the appropriate polarization curve in Figure 13-5 to help design the stack.

5. Design an AFC stack that has to operate at 25°C and 1 atm. The total power should be 50 Watts. Use the appropriate polarization curve in Figure 13-6 to help design the stack.

Bibliography

[1] Sousa, Ruy Jr., and Ernesto Gonzalez. "Mathematical Modeling of Polymer Electrolyte Fuel Cells." *Journal of Power Sources*. Vol. 147, 2005, pp. 32–45.
[2] O'Hayre, Ryan, Suk-Won Cha, Whitney Colella, and Fritz B. Prinz. *Fuel Cell Fundamentals*. 2006. New York: John Wiley & Sons.
[3] Barbir, Frano. *PEM Fuel Cells: Theory and Practice*. 2005. Burlington, MA: Elsevier Academic Press.
[4] Lin, Bruce. "Conceptual Design and Modeling of a Fuel Cell Scooter for Urban Asia." Princeton University, Masters Thesis. 1999.
[5] Mench, M.M., Z.H. Wang, K. Bhatia, and C.Y. Wang. "Design of a Micro-direct Methanol Fuel Cell." Electrochemical Engine Center, Department of Mechanical and Nuclear Engineering, Pennsylvania State University. 2001.
[6] Mench, Matthew M., Cao-Yang Wang, and Stephan T. Tynell. "An Introduction to Fuel Cells and Related Transport Phenomena." Department of Mechanical and Nuclear Engineering, Pennsylvania State University. Draft. http://mtrl1.mne.psu.edu/Document/jtpoverview.pdf Last Accessed March 4, 2007.
[7] You, Lixin, and Hongtan Liu. "A Two-Phase Flow and Transport Model for PEM Fuel Cells." *Journal of Power Sources*. Vol. 155, 2006, pp. 219–230.
[8] Springer et al. "Polymer Electrolyte Fuel Cell Model." *J. Electrochem. Soc.* Vol. 138, No. 8, 1991, pp. 2334–2342.
[9] Li, Xianguo. *Principles of Fuel Cells*. 2006. New York: Taylor & Francis Group.
[10] Springer et al. "Polymer Electrolyte Fuel Cell Model."
[11] Rowe, A. and X. Li. "Mathematical Modeling of Proton Exchange Membrane Fuel Cells." *Journal of Power Sources*. Vol. 102, 2001, pp. 82–96.
[12] Chen, R. and T.S. Zhao. "Mathematical Modeling of a Passive Feed DMFC with Heat Transfer Effect." *Journal of Power Sources*. Vol. 152, 2005, pp. 122–130.
[13] Subhash, Singhall C. "*High Temperature Solid Oxide Fuel Cells Fundamentals, Design and Applications.*" Pan American Advanced Studies Institute. Rio de Janeiro,

Brazil. Pacific Northwest National Laboratory, U.S. Department of Energy. July, 2003.
[14] Chung, Brandon W., Christopher N. Chervin, Jeffrey J. Haslam, Ai-Quoc Pham, and Robert S. Glass. "Development and Characterization of a High Performance Thin-Film Planar SOFC Stack." *Journal of the Electrochemical Society*. Vol. 152, No. 2, 2005, pp. A265–A269.
[15] Zhu, Bin. "Proton and Oxygen Ion-Mixed Conducting Ceramic Composites and Fuel Cells." *Solid State Ionics*. Vol. 145, 2001, pp. 371–380.
[16] Yoon, K.H., J.Y. Choi, J.H. Jang, Y.S. Cho, and K.H. Jo. "Electrode/Matrix Interfacial Characteristics in a Phosphoric Acid Fuel Cell." *Journal of Applied Electrochemistry*. Vol. 30, 2000, pp. 121–124.
[17] Yoon, K.H., J.H. Jang, and Y.S. Cho. "Impedance Characteristics of a Phosphoric Acid Fuel Cell." *Journal of Materials Science Letters*. Vol. 17, 1998, pp. 1755–1758.
[18] Song, Rak-Hyun and Dong Ryul Shin. "Influence of CO Concentration and Reactant Gas Pressure on Cell Performance in PAFC." *International Journal of Hydrogen Energy*. Vol. 26, 2001, pp. 1259–1262.
[19] Soler, J., T. Gonzalez, M.J. Escudero, T. Rodrigo, and L. Daza. "Endurance test on a single cell of a novel cathode material for MCFC." *Journal of Power Sources*. Vol. 106, 2002, pp. 189–195.
[20] Silva, V.S., J. Schirmer, R. Reissner, B. Ruffmann, H. Silva, A. Mendes, L.M. Madeira, and S.P. Nunes. "Proton Electrolyte Membrane Properties and Direct Methanol Fuel Cell Performance." *Journal of Power Sources*. Vol. 140, 2005, pp. 41–49.
[21] Yamada, Masanori and Itaru Honma. "Biomembranes for Fuel Cell Electrodes Employing Anhydrous Proton Conducting Uracil Composites." *Biosensors and Bioelectronics*. Vol. 21, 2006, pp. 2064–2069.
[22] Rowshanzamir, S. and M. Kazemeini. "A New Immobilized-Alkali H_2/O_2 Fuel Cell." *Journal of Power Sources*. Vol. 88, 2000, pp. 262–268.
[23] Coutanceau, C., L. Demarconnay, C. Lamy, and J.M. Leger. "Development of Electrocatalysts for Solid Alkaline Fuel Cell (SAFC)." *Journal of Power Sources*. Vol. 156, 2006, pp. 14–19.
[24] Verma, A and S. Basu. "Direct Use of Alcohols and Sodium Borohydride as Fuel in an Alkaline Fuel Cell." *Journal of Power Sources*. Vol. 145, 2005, pp. 282–285.
[25] Gulzow, E., M. Schulze, and U.Gerke. "Bipolar Concept for Alkaline Fuel Cells." *Journal of Power Sources*. Vol. 156, 2006, pp. 1–7.
[26] Morita, H., M. Komoda, Y. Mugikura, Y. Izaki, T. Watanabe, Y. Masuda, and T. Matsuyama. "Performance Analysis of Molten Carbonate Fuel Cell Using a Li/Na Electrolyte." *Journal of Power Sources*. Vol. 112, 2002, pp. 509–518.
[27] Wee, Jung-Ho. "Performance of a Unit Cell Equipped with a Modified Catalytic Reformer in Direct Internal Reforming Molten Carbonate Fuel Cell." *Journal of Power Sources*. Vol. 156, 2006, pp. 288–293.
[28] Bove, Roberto and Piero Lumghi. "Experimental Comparison of MCFC Performance Using Three Different Biogas Types and Methane." *Journal of Power Sources*. Vol. 145, 2005, pp. 588–593.
[29] Nakagawa, Nobuyoshi and Yikun Xiu. "Performance of a Direct Methanol Fuel Cell Operated." *Journal of Power Sources*. Vol. 118, 2003, pp. 248–255.
[30] Lim, C. and C. Y. Wang. "Development of High Power Electrodes for a Liquid Feed Direct Methanol Fuel Cell." *Journal of Power Sources*. Vol. 113, 2003, pp. 145–150.
[31] Wong, C.W., T.S. Zhao, Q. Ye, and J.G. Liu. "Experimental Investigations of the Anode Flow Field of a Micro Direct Methanol Fuel Cell." *Journal of Power Sources*. Vol. 155, 2006, pp. 291–296.
[32] Liu, J.G., T.S. Zhao, Z.X. Liang, and R. Chen. "Effect of Membrane Thickness on the Performance and Efficiency of Passive Direct Methanol Fuel Cells." *Journal of Power Sources*. Vol. 153, 2006, pp. 61–67.
[33] Yan, Qiangu, Hossein Toghiani, and Junxiao Wu. "Investigation of Water Transport Through Membrane in a PEM Fuel Cell by Water Balance Experiments." *Journal of Power Sources*. Vol. 158, 2006, pp. 316–325.
[34] Fabian, Tibor, Jonathan D. Posner, Ryan O'Hayre, Suk-Won Cha, John K. Eaton, Fritz B. Prinz, and Juan G. Santiago. "The Role of Ambient Conditions on the

Performance of a Planar, Air-Breathing Hydrogen PEM Fuel Cell." *Journal of Power Sources.* Vol. 161, 2006, pp. 168–182.
[35] Kim, Jun-Yeop, Oh Joong Kwon, Sun-Mi Hwang, Moo Seong Kang, and Jae Jeong Kim. "Development of a Miniaturized Polymer Electrolyte Membrane Fuel Cell with Silicon Separators." *Journal of Power Sources.* Vol. 161, 2006, pp. 432–436.
[36] Wells, Brian and Henry Voss. "DMFC Power Supply for All-Day True Wireless Mobile Computing." *Polyfuel, Inc.* May 17, 2006. http://www.hydrogen.energy.gov/pdfs/review06/fcp_39_wells.pdf. Last accessed November 4, 2006.
[37] Bosco, Andrew. "General Motors Fuel Cell Research. Hy-Wire." http://www.ansoft.com/workshops/altpoweree/Andy_Bosco_GM.pdf. Last accessed November 4, 2006.
[38] U.S. Patent No. 4,490,444. "High Temperature Solid Oxide Fuel Cell Configurations and Interconnections." Arnold O. Isenberg, Forest Hills, PA. Westinghouse Electric Corporation. December 25, 1984.
[39] Aravamudhan, Shyam, Abdur Rub Abdur Rahman, and Shekhar Bhansali. "Porous Silicon-Based Orientation Independent, Self-Priming Micro-Direct Ethanol Fuel Cell." *Sensors and Actuators A.* Vol. 123–124, 2005, pp. 497–504.
[40] Nguyen, Nam-Trung and Siew Hwa Chan. "Micromachined Polymer Electrolyte Membrane and Direct Methanol Fuel Cells – A Review." *J. Micromech. Microeng.* Vol. 16, 2006, pp. R1–R12.
[41] U.S. Patent Application Publication No. US 2006/0003217 A1. Planar Membraneless Microchannel Fuel Cell. Cohen, Jamie L., David James Volpe, Daron A. Westly, Alexander Pechenik, and Hector D. Abruna. January 5, 2006.
[42] U.S. Patent No. 7,029,781 B2. Microfuel Cell Having Anodic and Cathodic Microfluidic Channels and Related Methods. Lo Priore, Stefano, Michele Palmieri and Ubaldo Mastromatteo. STMicroelectronics, Inc. Carollton, TX. April 18, 2006.
[43] U.S. Patent No. 7,125,625 B2. Electrochemical Cell and Bipolar Assembly for an Electrochemical Cell. Cisar, Alan j., Craig C. Andrews, Charles J. Greenwald, Oliver J. Murphy, Chris Boyer, Rattaya C. Yalamanchili, and Carlos E. Salinas. Lynntech, Inc. College Station, TX. October 24, 2006.
[44] Choban, Eric R., Larry J. Markoski, Andrzej Wieckowski, and Paul J. A. Kenis. "Microfluidic Fuel Cell Based on Laminar Flow." *Journal of Power Sources.* Vol. 128, 2004, pp. 54–60.
[45] Aravamudhan, Shyam et al. "Porous Silicon-Based Orientation Independent, Self-Priming Micro-Direct Ethanol Fuel Cell."
[46] Nguyen, Nam-Trung and Siew Hwa Chan. "Micromachined Polymer Electrolyte Membrane and Direct Methanol Fuel Cells – A Review."
[47] U.S. Patent Application Publication No. US 2006/0003217 A1. Planar Membraneless Microchannel Fuel Cell.
[48] U.S. Patent No. 7,029,781 B2. Microfuel Cell Having Anodic and Cathodic Microfluidic Channels and Related Methods.
[49] Cha, S.W., R. O'Hayre, Y. Saito, and F.B. Prinz. "The Influence of Size Scale on the Performance of Fuel Cells." *Solid State Ionics.* Vol. 175, 2004. pp. 789–795.
[50] Heinzel, A., C. Hebling, M. Muller, M. Zedda, and C. Muller. "Fuel Cells for Low Power Applications." *Journal of Power Sources.* Vol. 105, 2002, pp. 250–255.
[51] Lu, G. Q. and C. Y. Wang. "Development of Micro Direct Methanol Fuel Cells for High Power Applications." *Journal of Power Sources.* Vol. 144, 2005, pp. 141–145.
[52] Blum, A., T. Duvdevani, M. Philosoph, N. Rudoy, and E. Peled. "Water Neutral Micro Direct Methanol Fuel Cell (DMFC) for Portable Applications." *Journal of Power Sources.* Vol. 117, 2003, pp. 22–25.
[53] Dyer, C. K. "Fuel Cells for Portable Applications." *Journal of Power Sources.* Vol. 106, 2002, pp. 31–34.
[54] Liu, Shaorong, Qiaosheng Pu, Lin Gao, Carol Korzeniewski, and Carolyn Matzke. "From Nanochannel Induced Proton Conduction Enhancement to a Nanochannel Based Fuel Cell." *Nano Letters.* Vol. 5, No. 7, 2005, pp. 1389–1393.
[55] Mitrovski, Svetlana M., Lindsay C. E. Elliott, and Ralph G. Nuzzo. "Microfluididc Devices for Energy Conversion: Planar Integration and Performance of a Passive, Fully Immersed H_2-O_2 Fuel Cell." *Langmuir.* Vol. 20, 2004, pp. 6974–6976.

[56] Modroukas, Dean, Vijay Modi, and Luc G. Frechette. "Micromachined Silicon Structures for Free-Convection PEM Fuel Cells." *J. Micromech. Microeng.* Vol. 15, 2005, pp. S193–S201.
[57] Wozniak, Konrad, David Johansson, Martin Bring, Anke Sanz-Velasco, and Peter Enoksson. "A Micro Direct Methanol Fuel Cell Demonstrator." *J. Micromech. Microeng.* Vol. 14, 2004, pp. S59–S63.
[58] Koh, Joon-Ho, Hai-Kyung Seo, Choong Gon Lee, Young-Sung Yoo, and Hee Chun Lim. "Pressure and Flow Distribution in Internal Gs Manifolds of a Fuel Cell Stack." *Journal of Power Sources.* Vol. 115, 2003, pp. 54–65.
[59] Chang, Paul A.C., Jean St-Pierre, Jurgen Stumper, and Brian Wetton. "Flow Distribution in Proton Exchange Membrane Fuel Cell Stacks." *Journal of Power Sources.* Vol. 162, 2006, pp. 340–355.
[60] Lee, Shuo-Jen, Chen-De Hsu, and Ching-Han Huang. "Analyses of the Fuel Cell Stack Assembly Pressure." *Journal of Power Sources.* Vol. 145, 2005, pp. 353–361.
[61] Larminie, James and Andrew Dicks. Fuel Cell Systems Explained. 2000. John Wiley & Sons, Ltd, Chichester, England.

Chapter

14

Fuel Cell System Design

Fuel cell system designs range from very simple to very complex depending upon the fuel cell application and the system efficiency desired. A fuel cell system can be very efficient with just the fuel cell stack and a few other plant components. An example of a very simple fuel cell system is shown in Figure 14-1. Typically, the larger the fuel cell stack(s), the more complex the fuel cell plant subsystem will be. A more complex fuel cell plant subsystem may also be used to make the fuel cell stack(s) more efficient. The designs and optimization of the fuel cell plant and electrical subsystems are endless.

The following is a brief description of the fuel cell system components shown in Figure 14-1:

- **Oxidant Air Flow:** The oxidant air is filtered for particulates as it is being pumped into the fuel cell from the atmosphere. The air pressure transducer keeps track of the air pressure coming into the fuel cell. The oxidant air is filtered again for particulates, and then humidified before it enters the fuel cell stack.
- **Hydrogen Flow:** The pure hydrogen is stored in a compressed gas cylinder. There can be one or more check valves before the hydrogen enters the system. A mass flow controller and a pressure transducer would also be useful for monitoring the pressure and flow rate.
- **Water and Hydrogen Out:** The hydrogen exits the fuel cell stack by going through a particulate filter. The pressure transducer records the pressure of this stream before it is purged. The water is purged through an external product water vent.

Fuel cell systems can easily become complex when the stack is large and the temperature, pressure, water, and heat are monitored and/or

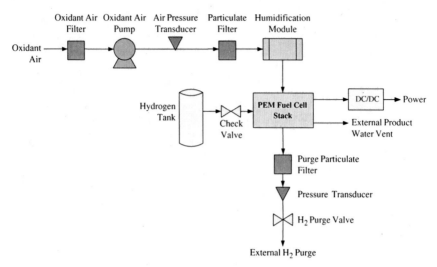

Figure 14-1 Simple PEM fuel cell system.

require fuel processing units. If a carbon-based fuel is converted to hydrogen for electrical power and heat, fuel processing units and gas cleanup units are necessary. Other additional components often found in fuel cell plant and electrical subsystems include: heat exchangers, pumps, fans, blowers, compressors, electrical power inverters, converters and conditioners, water handling devices, and control systems. A slightly more complex fuel cell system is shown in Figure 14-2.

A brief description of the components in Figure 14-2 is as follows:

- **Oxidant Air Flow:** The oxidant air is filtered for particulates as it is being sucked into the fuel cell from the atmosphere. The air pressure transducer keeps track of the air pressure coming into the fuel cell. The oxidant air is filtered again for particulates before it enters the fuel cell stack.

- **Methanol Flow:** The methanol/water mixture is stored in a methanol reservoir. Pumps are required to push the methanol out of its holding tank and into circulation so it can travel to the fuel cell stack. A mixer ensures a uniform solution of methanol and water. The methanol sensor will tell the control system the concentration of methanol (molarity) that will be entering the fuel cell. The cold start heater will heat the methanol to the desired temperature to obtain the correct vapor-to-liquid ratio. Thermocouples will be used to measure the temperature of the methanol input.

- **Water, Methanol, and CO_2 Out:** The water, methanol, and CO_2 solution exiting the fuel cell stack are separated into their liquid and

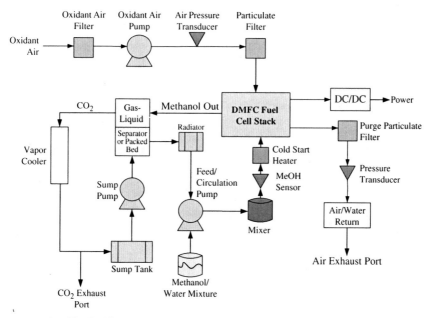

Figure 14-2 The DMFC system.

vapor components via a gas-liquid separator or a packed bed. The gas portion is cooled (to separate the water and excess methanol from the CO_2), and then sent to an exhaust. The water/methanol solution is held in a sump tank. Along with the liquid from the gas–fluid separator (water/methanol), water/methanol is also pumped from the sump tank. These two flows are combined and fed into a radiator, which cools the stream before it enters the methanol flow stream.

- **Air/Water Out:** An air/water purge stream exits the fuel cell stack by going through a particulate filter. The pressure transducer records the pressure of this stream. The air is vented, and the remaining water goes to the sump tank.

Only a few sensors and pressure transducers are included in Figures 14-1 and 14-2. A fully developed control system may consist of thermocouples, pressure transducers, methanol, hydrogen, oxygen or altitude sensors, and mass flow controllers, which will measure and record data, and use feedback to control temperature, humidity, pressure or flowrates using a data acquisition and control program.

As described in Chapters 11 and 12, the fuel cell catalyst, membranes, and flow-field plates are very important areas for fuel cell improvement, but stack optimization is as equally important. This chapter focuses on selecting and designing the fuel and electrical subsystems.

14.1 Fuel Subsystem

The fuel subsystem is very important because the reactants need to move around the fuel cell plant in order to be pressurized, cooled, heated, humidified and ultimately delivered to the fuel cell. In order to accomplish any of these tasks, plant components such as blowers, compressors, pumps, and humidification systems have to be used. Other plant components, such as turbines are also useful because they can harness energy from the heated exhaust gases leaving the fuel cell. These plant components are mature technologies, and can be found from a large number of companies for a variety of applications. Sections 14.1.1–14.1.5 describe these plant components, and give some of the relevant equations needed for designing the fuel cell plant subsystem.

14.1.1 Humidification systems

In PEMFC's, a hydrogen humidification system may be required to prevent the fuel cell PEM from dehydrating under load. As discussed previously, water management is a challenge in the PEM fuel cell because there is ohmic heating under high current flow, which will dry out the polymer membrane and slow ionic transport. In extreme cases, the membrane can be physically damaged. Small fuel cell stacks, or stacks that are not operating continuously at the maximum power, may not require any humidification, or the stack may be able to humidify itself. In larger fuel cell systems, either the air or the hydrogen or both the air and hydrogen must be humidified at the fuel inlets. The following are ways to humidify the gases:

- Bubbling the gas through water (dewpoint humidification)
- Water or steam injection
- Flash evaporation
- Exchange of water and heat through a water permeable medium
- Exchange of water and heat on an absorbent surface

Examples of these humidification methods are shown in Figure 14-3. In laboratory systems, bubbling or dewpoint humidification is the most common method. The desired level of humidification is achieved by controlling the temperature of water with heaters. The device should be designed so the fuel does not carry water droplets. Many commercial humidification devices are available, but many fuel cell developers choose to create their own for their particular fuel cell system. A commercial fuel cell humidification system is illustrated in Figure 14-4.

Fuel Cell System Design   277

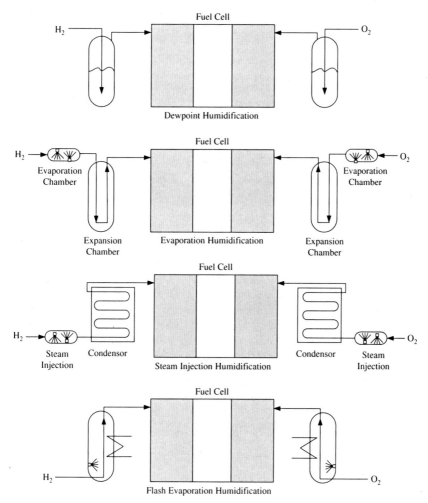

Figure 14-3 Conventional humidification methods: (a) dewpoint humidification, (b) evaporation humidification, (c) steam injection humidification, and (d) flash evaporation humidification.

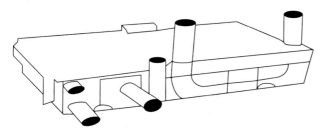

Figure 14-4 Example humidification system.

The following are a few pointers for designing the humidification system [13]:

- Minimize the heat loss in the system.
- The water should be delivered in the vapor phase at or near the inlet gas temperature required by the fuel cell. This may mean heating the piping after the humidification unit to the fuel cell.
- The level of humidification may need to be controlled depending upon the fuel cell system design.

When the total pressure is constant, the humidity depends upon the partial pressure of vapor in the mixture. For a vapor-gas system where the vapor is component A and the fixed phase is component B [24]:

$$\Phi = \frac{M_A p_A}{M_B(p_{tot} - p_A)} \tag{14-1}$$

For an air-water system:

$$\Phi = \frac{M_{H_2O} p_{H_2O}}{M_{air}(p_{tot} - p_{H_2O})} = \frac{18 * p_{H_2O}}{29 * (p_{tot} - p_{H_2O})}$$

The mole fraction of the vapor is:

$$x_{vap} = \frac{\dfrac{\Phi}{M_{H_2O}}}{\dfrac{1}{M_{air}} + \dfrac{\Phi}{M_{H_2O}}} \tag{14-2}$$

The saturation humidity is where the gas has vapor in equilibrium with the liquid at gas temperature:

$$\Phi_s = \frac{M_{H_2O} p^0_{H_2O}}{M_{air}(p_{tot} - p^0_{H_2O})} = \frac{18 * p^0_{H_2O}}{29 * (p_{tot} - p^0_{H_2O})} \tag{14-3}$$

The percent humidity is the ratio of the actual humidity to the saturation humidity:

$$\Phi_{water} = 100 * \frac{\Phi}{\Phi_s} = \Phi_R \frac{p_{tot} - p^0_{H_2O}}{p_{tot} - p_{H_2O}} \tag{14-4}$$

The heat that is required to increase the temperature of one pound of gas and the vapor it contains by 1°F for the air-water system is:

$$c_s = (c_p)_{H_2O} + \Phi(c_p)_{air} \approx 0.24 + 0.45*\Phi \qquad (14\text{-}5)$$

The wet bulb temperature is the dynamic equilibrium temperature attained by the liquid surface when the rate of heat transfer to the surface by convection equals the rate of heat required for evaporation away from the surface [24]. The partial pressure and the vapor pressure are usually small relative to the total pressure, therefore, the wet bulb equation can be expressed in terms of humidity conditions [24]:

$$\Phi_s - \Phi = h_c \frac{T - T_w}{\lambda k_1} \qquad (14\text{-}6)$$

For the air-water system, this equation becomes:

$$\Phi_s - \Phi = 18 h_c \frac{T - T_w}{29 \lambda k_1}$$

The adiabatic saturation temperature is reached when a stream of air is mixed with water at a temperature, T_s, in an adiabatic system. This can be expressed as:

$$\Phi_s - \Phi = c_s \frac{T - T_s}{\lambda} \qquad (14\text{-}7)$$

A humidity (psychrometric) chart provides a way to determine the properties of a gas-vapor mixture. Figure 14-5 shows an example of a psychrometric chart for a mixture of air and water. Any point on the chart represents a specific mixture of air and water. Points above and below the saturation lines represent a mixture of saturated air as a function of air temperature. The curved lines between the saturation line and the temperature axis represent mixtures of air and water at specific percentage humidities.

An easy method of humidifying the fuel is by direct water/steam injection, and is illustrated in Figure 14-6. The amount of water that needs to be injected can be calculated using the following equation [13]:

$$\dot{m}_{H_2O} = \dot{m}_{air} \frac{M_{H_2O}}{M_{air}} \left[\frac{\varphi P_{sat}(T)}{P - \varphi P_{sat}(T)} - \frac{\varphi_{amb} P_{sat}(T_{amb})}{P_{amb} - \varphi_{amb} P_{sat}(T_{amb})} \right] \qquad (14\text{-}8)$$

where φ and φ_{amb} are the relative humidity, T and T_{amb}, are the temperatures, and P_{amb} and P are the pressures at the fuel cell inlet and of the ambient air, respectively.

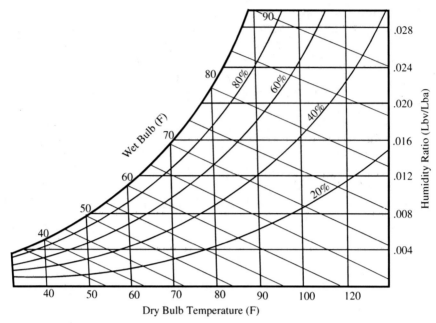

Figure 14-5 Example psychrometric chart [25].

A metering pump can be used to dose the exact amount of water needed. The water should be preferably injected in the form of a fine mist to maximize the contact area between the water and the air. In certain systems, a simple injection of water may not be sufficient to humidify the stream because humidification also requires heat. Therefore, sometimes additional heat must be provided to the water through a heat source in the fuel cell system or through a separate heat exchanger. In addition, the length of the pipe required for sufficient mixing after the steam injection needs to be calculated to insure a uniformly humidified gas is entering the fuel cell.

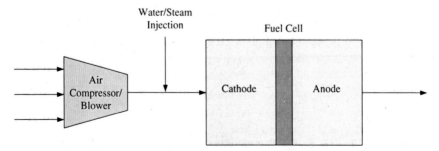

Figure 14-6 The water/steam injection humidification system.

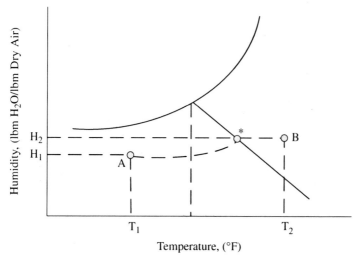

Figure 14-7 Example humidification process.

Commercial humidifiers typically use heating coils and a warm water spray to bring the gas to the desired temperature and humidity. Figure 14-7 shows a process that can occur in a humidifying device. Point A represents the entering air with the dry bulb temperature, T_1 and humidity H_1. A dry bulb temperature of T_2 and humidity of H_2 is desired (point B). The method of reaching point B is by going from point A to point * by water spray, and then heating to reach point B.

14.1.2 Fans and Blowers

An economical way to cool a fuel cell or to provide air to a fuel cell is to use fans or blowers. When using a fan or blower to provide air to the fuel cell, the exhaust is open to the environment. The fan or blower is driven by an electric motor, which requires power from the fuel cell or other source to run. One of the most commonly used fans is the axial fan, which is effective in moving air over parts, but not effective across large pressure differentials. The back pressure of this fan type is very low at 0.5 cm of water [18]. These fans are well suited for many hydrogen–air PEM fuel cell designs. The actual fan power is given by the following equation:

$$W_{act} = \frac{W_{ideal}}{\eta_s} \qquad (14\text{-}9)$$

The ideal power can be calculated by:

$$W_{ideal} = mc_{p,avg}(T_2 - T_1) + \frac{v_2^2}{2} - \frac{v_1^2}{2} \tag{14-10}$$

where $c_{p,avg}$ is the specific heat at the average temperature of the inlet and outlet. The ideal exit temperature can be calculated from equation:

$$\frac{T_2}{T_1} = \left(\frac{P_2}{P_1}\right)^{(\gamma-1)/\gamma} \tag{14-11}$$

where T_2 is the isentropic temperature, and γ is the ratio of the specific heat capacities of the gas, C_p/C_v. The actual exit temperature is then:

$$T_2 = T_1 + \frac{1}{c_{p,avg}}\left\{\frac{W_{act}}{m} - \left(\frac{v_2^2}{2} - \frac{v_1^2}{2}\right)\right\} \tag{14-12}$$

The operating information for an actual fan can be obtained from manufacturer's data. The actual speed and power required can be found from the manufacturer's table once the inlet volume rate and pressure boost are specified. Fan data can sometimes be represented in terms of dimensionless parameters. These can be defined as:

Discharge Coefficient: $\quad C_Q = \dfrac{\dot{V}}{ND^3}$ $\qquad$ (14-13)

Pressure Coefficient: $\quad C_H = \dfrac{\Delta P}{\rho N^2 D^2}$ $\qquad$ (14-14)

Isentropic Efficiency: $\quad \eta_{fan} = \dfrac{W_{ideal}}{W_{actual}}$ $\qquad$ (14-15)

Specific Speed: $\quad N_s = \dfrac{N\dot{V}^{1/2}\rho^{3/4}}{(\Delta P)^{3/4}}$ $\qquad$ (14-16)

where $\dot{V}$ is the volumetric flow rate, ρ is the density of the fluid, D is the wheel blade diameter, N is the fan speed, ΔP is the fan pressure boost and W is the fan power. The ideal fan power is:

$$W_{ideal} = m\left[c_{p,avg}T_1\left(\left(\frac{P_2}{P_1}\right)^{(\gamma-1)/\gamma} - 1\right) + \left(\frac{v_2^2}{2} - \frac{v_1^2}{2}\right)\right] \tag{14-17}$$

The fan power can be calculated using the following equation:

$$W_{act} = V\left[\Delta P + \frac{\rho v_2^2 - v_1^2}{2}\right] \quad (14\text{-}18)$$

Since the total pressure can be expressed as: $P_0 = P + \frac{\rho v_2^2}{2}$, then:

$$W_{act} = V * \Delta P_0 \quad (14\text{-}19)$$

Larger pressure differences can be obtained by using centrifugal fans. Centrifugal fans have air or gases entering in the axial direction, and discharge air or gases in the radial direction [21]. These are used for circulating cooling air through small- to medium-sized fuel cells. The pressure created by these fans is from 3 to 10 cm of water [18].

Blowers are also used in atmospheric systems to draw air into the fuel cell. An example of a radial blower shown in Figure 14-8. The blower is typically powered by a battery for startup, and then some of the power output of the fuel cell is used to keep the blower running (like other plant components). The blower power required is

$$W = (\Delta PV)/\eta_{blower} \quad (14\text{-}20)$$

where η_{blower} is the blower efficiency.

Many fans and blowers are manufactured commercially, and when designing the fuel cell system, the important factors to consider are the air flow rate required, the robustness, efficiency, noise, cost, and, if applicable, the power requirements and weight. Fan and blower performance also vary with the temperature, speed, and density of the gas handled; therefore, it is important to consider these factors when selecting a fan or blower since most performance parameters are tested at standard conditions. When selecting a blower, a spark-free brushless motor may also offer a number of advantages for the fuel cell system (especially if hydrogen is being used as the fuel), such as high reliability,

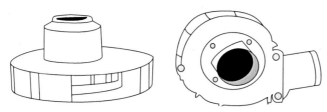

Figure 14-8 Example radial blower.

a long service life, and microprocessor-based electronics that allow rapid control and response time.

Example 14-1 Calculate the power consumption of a blower with a pressure drop of 0.5, and a flow rate of 0.013067 ft^3/min. The blower has a 50-percent efficiency.

$$W = (\Delta PV)/\eta_{blower} = (0.5 * 0.013067 \text{ ft}^3/\text{min})/0.50 = 0.013 \text{ W}$$

The power consumption is 0.013 W.

Example 14-2 A fan is used to move air into a fuel cell system at 20°C and 100 kPa. The fan moves 0.05 m^3/s of air with a pressure boost 0.6 kPa. The actual power is 50 W, and the outlet velocity is 1 m/s. Determine the efficiency of the fan.

With the pressure boost, P_2 = 100.6 kPa. The ideal work must first be calculated to determine the efficiency:

$$w_{ideal} = c_p(T_2 - T_1)$$

where T_2 comes from our isentropic ratio, and γ is 1.38:

$$\frac{T_2}{T_1} = \left(\frac{P_2}{P_1}\right)^{(\gamma-1)/\gamma} = (293.15) * \left(\frac{100.6}{100}\right)^{(1.38-1)/1.38} = 293.63 \text{ K}$$

$$w_{ideal} = (1.005) * (293.63 - 293.15) = 0.4858 \text{ kJ/kg}$$

Then:

$$\eta_s = \frac{W_{ideal}}{W_{act}} = \frac{W_{ideal}}{\dot{w}_{act}/\dot{m}}$$

where $\dot{m} = \dot{v}/V = \dot{v}/(RT/P) = (0.05*100)/(0.286*293.15) = 0.0596$ kg/s

Therefore,

$$\eta_s = \frac{W_{ideal}}{W_{act}} = \frac{W_{ideal}}{\dot{w}_{act}/\dot{m}} = \frac{0.4858 \text{ kJ/kg}}{(0.05 \text{ kW}/0.0596 \text{ kg/s})} = 0.579$$

14.1.3 Compressors

Compressing the air input increases the concentration of oxygen per volume per time, and thus increases the fuel cell efficiency. This means that a smaller and lighter fuel cell stack can be used. In addition, the drop-off in voltage caused by mass transport problems is delayed until higher current densities. If the pressure is higher, a lower volumetric flow rate can be used for the same molar flow rate. Humidification is also easier since less water is required for saturation (per mole of air). The flow-field design is less restrictive because larger pressure drops can be tolerated in the flow field. A compressor (like a fan or blower) is driven by an electric motor, which requires power from the fuel cell or other source to run. The compression can be isothermal or adiabatic. Isothermal compression allows temperature equilibration with the environment, and adiabatic uses compression without any heat exchange with the environment.

Two types of compressors can be utilized in fuel cell systems: positive displacement compressors, such as the Roots, Screw, diaphragm, or piston compressors; and continuous flow compressors, such as the centrifugal/radial compressor or the axial flow compressor. The regulation of flow is different for these two types of compressors.

Positive displacement compressors, such as the Roots compressor, are less expensive and can be used over a wide range of flow rates. This compressor type only gives useful efficiencies over small pressure differences. The screw or Lysholm compressor has two screws that counter rotate, driving the gas between the two screws forward, and compressing it. The screw compressor has two configurations: an external motor drives one rotor, and the second rotor is turned by the first. The second configuration uses a synchronizing gear to connect to two screws. The advantage of the Lysholm compressor is that it can provide a wide range of compression ratios. It has the capability to increase the pressure up to eight times the input pressure, but is expensive to manufacture [18].

The most common type of compressor is the centrifugal compressor. It uses kinetic energy to create a pressure increase. It is low in cost and can provide a wide range of flow rates. This type of compressor cannot be operated in the low flow region, which means that the pressure must be regulated. The centrifugal compressor can be operated with high efficiencies through a high range of flow rates by changing both the flow rate and the pressure. This compressor type is commonly found on engine turbocharging systems. Figure 14-9 shows an example of a motor-driven turbocompressor for PEM fuel cells.

The axial flow compressor uses large blades to push air through the device with a decreased cross-sectional area. These compressors are expensive to manufacture, yet the efficiency is high over a large range of flow rates.

The efficiency of the compressor is important for the overall efficiency of the fuel cell system. The efficiency and ideal exit temperature can be found by using the ratio of actual work done to raise the pressure from P_1 to P_2 (like in Section 14.1.2):

$$\frac{T_2}{T_1} = \left(\frac{P_2}{P_1}\right)^{(\gamma-1)/\gamma} \tag{14-21}$$

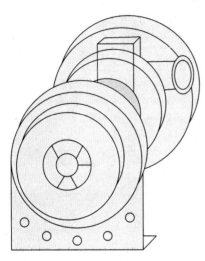

Figure 14-9 Example of a motor-driven turbocompressor for PEM fuel cells.

where T_2 is the isentropic temperature, and γ is the ratio of the specific heat capacities of the gas, C_p/C_v. To use this equation, some general assumptions are made [19]:

- The heat flow from the compressor is negligible.
- The kinetic energy of the gas as it flows through the compressor is negligible.
- The gas is ideal; therefore, the specific heat at constant pressure, C_p, is constant.

The actual work done by the system is

$$W = c_p (T_2 - T_1) m \qquad (14\text{-}22)$$

where m is the mass of the gas compressed (air flow rate, g/s), T_1 and T_2 are the inlet and exit temperatures, respectively, and c_p is the specific heat at constant pressure (J/gK). The efficiency is the ratio of these two quantities of work:

$$\eta = \text{isentropic work/real work} = \frac{c_p(T_2 - T_1)m}{c_p(T_2 - T_1)m} \text{ and } \eta_c = \frac{(T_2 - T_1)}{(T_2 - T_1)} \qquad (14\text{-}23)$$

The change in temperature at the end of compression can be found from the equation:

$$\Delta T = T_2 - T_1 = \frac{T_1}{\eta_c}\left(\left(\frac{P_2}{P_1}\right)^{\frac{\gamma-1}{\gamma}} - 1\right) \qquad (14\text{-}24)$$

where γ is the ratio of specific heats (for diatomic gases, k = 1.4). The total efficiency is the compressor efficiency multiplied by the mechanical efficiency of the shaft:

$$\eta_T = \eta_m \times \eta_c \qquad (14\text{-}25)$$

The power required to increase the temperature of the gas is defined as:

$$W = c_p \Delta T \dot{m} \qquad (14\text{-}26)$$

Taking into account the inefficiencies with the compression process, and substituting this into the previous equation, then:

$$P_{compressor} = c_p \frac{T_1}{\eta_c}\left(\left(\frac{P_2}{P_1}\right)^{\frac{\gamma-1}{\gamma}} - 1\right)\dot{m} \qquad (14\text{-}27)$$

Many compressors are manufactured commercially, and when designing the fuel cell system, the important factors to consider are the temperature, pressure, type of gas handled, reliability, efficiency, and corrosion-free materials.

Example 14-3 Air at 100 kPa and 298 K enter a compressor at 2 kg/s and receives a pressure boost of 100 kPa. Determine the power required, adiabatic efficiency and exit temperature for an (a) ideal compressor, and (b) adiabatic compressor with 75% efficiency.

Using the ideal gas law:

$$v_1 = \frac{RT_1}{P_1} = \frac{0.287 * 298}{100} = 0.8553$$

The volumetric flow rate can be calculated by:

$$V_1 = v_1 * m = 0.8553 * 2 = 1.710 \frac{m^3}{s}$$

(a) The exit pressure can be estimated by:

$$P_2 = P_1 + \Delta P = 100 + 100 = 200 \; kPa$$

The ideal calculation of T_2 can be estimated by:

$$\frac{T_2}{T_1} = \left(\frac{P_2}{P_1}\right)^{(\gamma-1)/\gamma} = (298) * \left(\frac{200}{100}\right)^{(1.38-1)/1.38} = 360.68 \; K$$

The ideal work can be calculated by:

$$W = c_{p,\;avg}(T_2 - T_1)m = 1.005 * (360.68 - 298) * (2) = 125.98 \; kW$$

The efficiency of an ideal compressor is 100%.

(b) For an adiabatic compressor with an efficiency of 75%:

$$W_{act} = \frac{W_{ideal}}{\eta_s} = \frac{125.98}{0.75} = 167.98 \ kW$$

The actual exit temperature can be calculated by:

$$T_2 = T_1 + \frac{W_{act}}{m * c_{p,avg}} = 298 + \frac{167.98}{2 * 1.005} = 381.57 \ K$$

Example 14-4 A fuel cell stack with an output power of 50 kW operates with a pressure of 2 bar. Air is fed to the stack using a screw compressor at 1.0 bar and 22°C with a stoichiometry of 1.5. The average cell voltage is 0.7 V. The rotor speed factor is 300 rev/minK$^{1/2}$, and the efficiency is 0.6. Find the required rotational speed of the air compressor, the temperature of the air as it leaves the compressor, and the power of the electric motor needed to drive the compressor.

The mass flow rate of air should be found first:

$$m = 3.57 \times 10^{-7} \times \lambda \times \frac{P_e}{V_c} \ kg/s, \text{ where } \lambda \text{ is the stoichiometry}$$

$$m = \frac{3.57 \times 10^{-7} \times 1.5 \times 50000}{0.70} = 0.03825 \ kg/s$$

This should then be converted to the mass flow factor:

$$m_{ff} = \frac{0.03825 \times \sqrt{295.15}}{1.0} = \frac{0.657 \ kg}{s\sqrt{K}bar}$$

The mass flow factor helps to find the rotor speed factor and efficiency from many standard compressor performance charts. In this example, the rotor speed factor was given as 300 rev/minK$^{1/2}$, and the efficiency is 0.6. The rotor speed calculation is

$$300 \times \sqrt{295.15} = 5153.98 \ rpm$$

The temperature rise is calculated:

$$\Delta T = \frac{295.15}{0.6}\left(\left(\frac{2.0}{1.0}\right)^{0.286} - 1\right) = 107.86 \ K$$

Since the inlet air temperature is 22°C, the exit temperature will be 129.86°C. If this was a PEMFC, the stack would need cooling, but if this was a PAFC, this would be a good preheating step. The compressor power is

$$W_{compressor} = 1004 \times \frac{295.15}{0.6}\left(\left(\frac{2.0}{1.0}\right)^{0.286} - 1\right) * 0.03825 = 4142.15 \ kW$$

This power ignores the mechanical losses in the bearings and drive shafts. Many PEMFCs also require that the inlet air be humidified, which will alter the specific heat capacity and the ratio of specific heat capacities, and thus alter the performance of the compressor. Sometimes the air will also be humidified after the compression when the air is hotter.

14.1.4 Turbines

In pressurized fuel cell systems, the outlet gas is typically warm and pressurized (though lower than the inlet pressure). This hot gas from fuel cells can be turned into mechanical work through the use of turbines. This energy can be used to generate work that may offset the work needed to compress the air. The efficiency of the turbine determines whether it should be incorporated into the fuel cell system [18]. Figures 14-10 and 14-11 show examples of a SOFC and MCFC fuel cell system with turbines.

Like compressors and fans, the ideal efficiency and exit temperatures of the turbine can be determined using the following equation:

$$\frac{T_2}{T_1} = \left(\frac{P_2}{P_1}\right)^{(\gamma-1)/\gamma} \tag{14-28}$$

By substituting the proper equations, the efficiency becomes

$$\eta_{turbine} = \frac{T_1 - T_2}{T_1 - T_2} = \frac{T_1 - T_2}{T_1}\left(1 - \left(\frac{P_2}{P_1}\right)^{\frac{\gamma-1}{\gamma}} - 1\right) \tag{14-29}$$

where η_c is the ratio between the actual work and the ideal isentropic work between P_1 and P_2. The temperature at the end of expansion is

$$\Delta T = T_2 - T_1 = \eta_c T_1\left(\left(\frac{P_2}{P_1}\right)^{\frac{\gamma-1}{\gamma}} - 1\right) \tag{14-30}$$

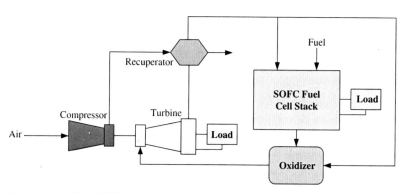

Figure 14-10 The SOFC fuel cell system with turbine/compressor.

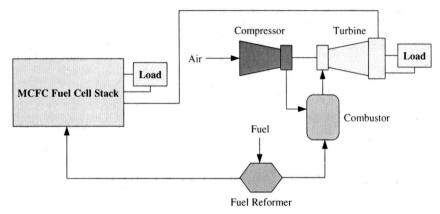

Figure 14-11 The MCFC fuel cell system with turbine/compressor.

The power of the turbine can be found from the same equation as the compressor:

$$W = c_p \Delta T \dot{m} \qquad (14\text{-}31)$$

$$P_{turbine} = c_p \eta_c T_1 \left(\left(\frac{P_2}{P_1} \right)^{\frac{\gamma-1}{\gamma}} - 1 \right) \dot{m} \qquad (14\text{-}32)$$

Due to the inefficiency of the compression and expansion process, a turbine may only recover a portion of work that is required of other system components. If the temperature of the exhaust is high, the turbine may generate all of the required power need for the other fuel cell subsystems.

Example 14-5 For the fuel cell analyzed in Example 14-4, the exit temperature is 100°C (383.15 K) and the pressure is 1.8 bar. The efficiency and rotor speed of the turbine is 0.55 and 4000 rev/(minK$^{1/2}$). Use cp = 1100 J/kgK and γ = 1.38. What power will be available from the exit gases?

The cathode exit gas mass would have been increased by the water present in the fuel cell, but since the mass change will be very small, we will consider it negligible for this problem, so the value of 0.03825 kg/s will be used. First, calculate the mass flow factor:

$$m_{ff} = \frac{0.03825 \times \sqrt{383.15}}{1.8} = \frac{0.416 \; kg}{s\sqrt{K}\,bar}$$

The rotor speed is

$$Sp_{rotor} = 5000 \times \sqrt{383.15} = 97{,}871.09 \; RPM$$

Since the efficiency is 0.55, the available power is

$$P_{turbine} = 1100 \times 0.55 \times 383.15\left(\left(\frac{1.0}{1.8}\right)^{0.275} - 1\right) * 0.03825$$

$$\approx -1323.37 \; kW$$

This is the amount of power given out, which is a useful addition to the 50 kW generated by the fuel cell, but it is not nearly enough to drive the compressor.

14.1.5 Fuel cell pumps

Pumps, like blowers, compressors and fans, are among the most important components in the fuel cell plant system. These components are required to move fuels, gases, and condensate through the system, and are important factors in the fuel cell system efficiency. Small- to medium-sized PEMFCs for portable applications have a back pressure of about 10 kPa or 1 m of water [19]. This is too high for most axial or centrifugal fans, as discussed earlier. Figure 14-12 shows examples of diaphragm pumps. Larger diaphragm pumps can also be used against a back pressure of 1 to 2 meters [18].

The features of these pumps include

- Low cost
- Reliable for long periods of time
- Can operate against pressures of 10 kPa
- Available in a wide range of sizes

Choosing the correct pump for the fuel cell application is important. Like in fans, blowers and compressors, factors to consider are efficiency, reliability, corrosion-free materials, and the ability to work with the required temperatures, pressures and flow rates for the specific fuel

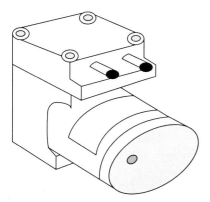

Figure 14-12 Example micro diaphragm pump.

cell system. The appropriate matching of a high-efficiency pump with the appropriate motor speed/torque curve may allow for a more efficient fuel cell stack and system. A spark-free brushless motor may also offer a number of advantages for the fuel cell system (especially if hydrogen is being used as the fuel), such as high reliability, a long service life, and microprocessor-based electronics that allow rapid control and response time. Many types of low-cost pumps that can be used for fuel cell stacks are made for various other applications, such as fish tanks, gas sampling equipment, and small-scale chemical processing plants. Many of these pumps are silent and can be used for several years. The equations that describe pump performance characteristics are the same as the fan performance characteristics equations 14-13–14-16.

Example 14-6 A pump needs to be selected to move 10 kg/s of water with a pressure boost of 100 kPa. Calculate the work required, exit temperature and if applicable, the pump speed and diameter for (a) the ideal pump, (b) an axial flow pump at maximum efficiency, and (c) a centrifugal pump at maximum efficiency. The maximum efficiency for an axial flow pump has the following parameters:

$$(C_Q)_{max} = 0.6398 \quad (C_H)_{max} = 1.4922 \quad \eta_{max} = 0.8488$$

The maximum efficiency for a centrifugal flow pump has the following parameters:

$$(C_Q)_{max} = 0.11509 \quad (C_H)_{max} = 5.3317 \quad \eta_{max} = 0.93508$$

(a) The ideal work can be calculated by:

$$W_{ideal} = \dot{m}v\Delta P = (10)(1.002*10^{-3})*100 = 1.002 \; kW$$

The exit temperature is equal to the inlet temperature.

(b) The actual work can be calculated by:

$$W_{act} = \frac{W_{ideal}}{\eta_{max}} = \frac{1.002}{0.8488} = 1.18 \; kW$$

The temperature change is then given by:

$$\Delta T = \frac{W_{act} - W_{ideal}}{\dot{m}c_p} = \frac{1.18 - 1.002}{(10)(4.181)} = 0.00427 \; K$$

At maximum efficiency, the following equations apply for the pump diameter and speed:

$$D = \left[\frac{(C_H)_{max} \dot{m}^2}{\rho \Delta P (C_Q)_{max}^2}\right]^{1/4} \quad \text{and} \quad N = \frac{\dot{m}}{\rho D^3 (C_Q)_{max}}$$

$$D = \left[\frac{(C_H)_{max} m^2}{\rho \Delta P (C_Q)_{max}^2}\right]^{1/4} = \left[\frac{1.4922*(10)^2}{(998)*(100)*0.6398^2}\right]^{0.25} = 0.246 \ m$$

$$N = \frac{m}{\rho D^3 (C_Q)_{max}} = \frac{10}{998*0.246^3*0.6398} = 1.0546 \ rps = 63.28 \ rpm$$

(c) The actual work can be calculated by:

$$W_{act} = \frac{W_{ideal}}{\eta_{max}} = \frac{1.002}{0.93508} = 1.072 \ kW$$

The temperature change is then given by:

$$\Delta T = \frac{W_{act} - W_{ideal}}{m c_p} = \frac{1.072 - 1.002}{(10)(4.181)} = 0.00166 \ K$$

At maximum efficiency:

$$D = \left[\frac{(C_H)_{max} m^2}{\rho \Delta P (C_Q)_{max}^2}\right]^{1/4} = \left[\frac{5.3317*(10)^2}{(998)*(100)*0.11509^2}\right]^{0.25} = 0.403 \ m$$

$$N = \frac{m}{\rho D^3 (C_Q)_{max}} = \frac{10}{998*0.403^3*0.11509} = 1.330 \ rps = 79.81 \ rpm$$

Example 14-7 A 1 kW PEMFC operates at a temperature of 80°C, and 3 atm with a cell voltage of 0.65 V as shown in Figure 14-13. The air stoichiometry is 2.5. The compressor and turbine efficiency is 0.6. The temperature and pressure of the air entering the compressor is 25°C and 1 atm. (a) Find the amount of power required for the compressor, and the associated temperature change. (b) Is a humidifier needed for this fuel cell system? (c) Would putting a turbine into the fuel cell system be useful? If so, calculate the associated temperature change with the turbine. (d) Calculate the amount of heat generated by the stack.

(a) With the molar mass of air being 28.97×10^{-3} kg/mol and the mole fraction of air that is oxygen is 0.21, the inlet air flow rate can be estimated by:

$$m_{air,in} = \frac{S_{O_2} \times M_{air} \times P}{4 \times x_{O_2} \times V_{cell} \times F}$$

$$m_{air,in} = \frac{2.5 \times 28.97 \times 10^{-3} \times 1000}{4 \times 0.21 \times 0.65 \times 96{,}485} = 0.001375 \ kg/s$$

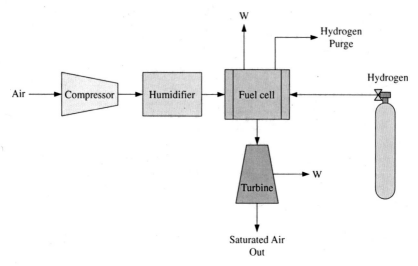

Figure 14-13 Fuel Cell System for Example 14-7.

Using air at 25°C, $\gamma = 1.4$, the temperature rise associated with the compression of air can be calculated by:

$$\Delta T = T_2 - T_1 = \frac{T_1}{\eta_c}\left(\left(\frac{P_2}{P_1}\right)^{\frac{\gamma-1}{\gamma}} - 1\right)$$

$$\Delta T = \frac{298}{0.6}\left(\left(\frac{303{,}975.03}{101{,}325.01}\right)^{\frac{1.4-1}{1.4}} - 1\right) = 183.35\ K$$

With $c_p = 1004$ J/kg $*$ K, the power needed to compress the air is:

$$P_{compressor} = c_p \frac{T_1}{\eta_c}\left(\left(\frac{P_2}{P_1}\right)^{\frac{\gamma-1}{\gamma}} - 1\right)\dot{m}$$

$$P_{compressor} = 1004 * \frac{298}{0.6}\left(\left(\frac{303{,}975.03}{101{,}325.01}\right)^{\frac{1.4-1}{1.4}} - 1\right) * 0.001375\ kg/s$$

$$= 253.12\ W$$

There is a large temperature rise, which will need to be compensated for by cooling the air before it enters the cell.

(b) The exit air flow rate can be estimated by:

$$m_{air,out} = \frac{S_{O_2} \times M_{air} \times P}{4 \times x_{O_2} \times V_{cell} \times F} - \frac{32 \times 10^{-3} \times P}{4 \times V_{cell} \times F}$$

$$m_{air,out} = \frac{2.5 \times 28.97 \times 10^{-3} \times 1000}{4 \times 0.21 \times 0.65 \times 96{,}485}$$

$$- \frac{32 \times 10^{-3} \times 1000}{4 \times 0.65 \times 96{,}485} = 0.001246\ kg/s$$

If the exit air is at 100% humidity, and the saturated vapor pressure of water at 80°C is 47.39 kPa, then the pressure of the dry air is the total pressure of the exit air minus the saturated vapor pressure. If we estimate the exit pressure to be less than the entry pressure due to pressure drop through the flowfields, we estimate a 20 kPa pressure drop, then 303,975.03 − 20,000 = 283,975.03 Pa, and the pressure of dry air is 283,975.03 Pa − 47,390 Pa = 236,585.03 Pa.

With the molecular mass of water being 18 and the molecular mass of air being 28.97, then humidity ratio is:

$$\omega = \frac{m_w}{m_a} = \frac{18 \times P_W}{28.97 \times P_a} = \frac{18 \times 47{,}390}{28.97 \times 236{,}585.03} = 0.1245$$

Therefore, the mass flow rate of water leaving the cell is:

$$m_w = \omega m_a = 0.1245 \times 0.001246 \ kg/s = 0.000155 \ kg/s$$

The rate of water production:

$$m_{H_2O} = \frac{M_{air} \times P}{2 \times V_{cell} \times F} = \frac{28.97 \times 10^{-3} \times 1000}{2 \times 0.65 \times 96{,}485} = 0.000231 \ kg/s$$

The rate that water should enter the cell can be estimated by 0.000155 − 0.000231 = −0.000076, which implies that no water needs to enter the cell. Therefore, a humidifier is not needed. The total exit flow rate is the dry air flow rate plus the water flow rate, which is 0.001246 kg/s + 0.000155 kg/s = 0.001401 kg/s.

(c) If cp = 1100 J/kg ∗ K, and γ = 1.33, and the turbine exit pressure is still above atmospheric pressure (~283,975.03 − 100,000 Pa = 183,975.03), then the turbine power can be calculated by:

$$P_{turbine} = c_p \eta_c T_1 \left(\left(\frac{P_2}{P_1} \right)^{\frac{\gamma-1}{\gamma}} - 1 \right) \dot{m}$$

$$P_{turbine} = 1100 * 0.6 * 353.15 \left(\left(\frac{183{,}975.03}{283{,}975.03} \right)^{\frac{1.33-1}{1.33}} - 1 \right) * 0.001401$$

$$= -33.33 \ W$$

About 33 Watts would make a useful contribution to the 253 Watts needed. The temperature change through the turbine can be calculated by:

$$\Delta T = \eta_c T_1 \left(\left(\frac{P_2}{P_1} \right)^{\frac{\gamma-1}{\gamma}} - 1 \right)$$

$$\Delta T = 0.6 * 353.15 \left(\left(\frac{183{,}975.03}{283{,}975.03} \right)^{\frac{1.33-1}{1.33}} - 1 \right) = -21.62 \ K$$

This would bring the exit gas temperature down to 58.38 K.

(d) The heating rate is:

$$q = P*\left(\frac{1.25}{V_{cell}} - 1\right) = 1000*\left(\frac{1.25}{0.65} - 1\right) = 923.077 \; W$$

14.2 Electrical Subsystem

The electrical output of a fuel cell is not an ideal power source. The output of all fuel cells is a DC voltage that varies widely and has a limited overload capacity. Nevertheless, the output is useful for many applications such as AC grid-connected power generation, and AC- or DC-independent loads. A fuel cell stack is slow to respond to load changes, and may have a slow startup. The fuel cell terminal voltage can alternate as much as ± 50 percent, depending on the load and fuel cell delivery. In order to make fuel cells behave like batteries, the fuel cell either needs to be designed to compensate for power spikes, the loads need to be redesigned to accommodate the limitations of fuel cells, or sophisticated power electronics are required. Figure 14-14 shows a general fuel cell schematic which illustrates the power electronics component as a key element in the fuel cell system.

The two basic power electronics areas that need to be addressed are power regulation and inverters. The electrical power output of a fuel cell is not constant. Increasing the current decreases the voltage proportionally greater than other electronic devices. The fuel cell voltage is typically controlled by voltage regulators, DC/DC converters, and chopper circuits at a constant value that can be higher or lower than the fuel cell operating voltage. The current generated by fuel cells is direct current (DC), which is advantageous for many smaller fuel cell systems. Larger fuel cell systems that connect to the power grid must be converted to alternating current (AC) using inverters. As shown in Figure 14-15, a DC-DC boost converter is required to boost the voltage level for the inverter. This boost converter, in addition to boosting the fuel cell voltage, also regulates the inverter input voltage and isolates the low and

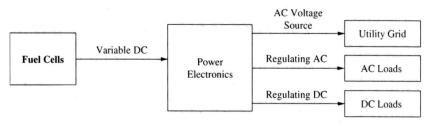

Figure 14-14 Power electronics is a key element [12].

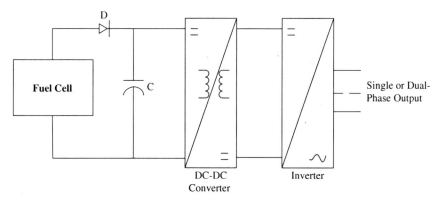

Figure 14-15 A typical fuel cell power electronics interface block diagram [9].

high voltage circuits. The basics of power semiconductor devices can be classified as:

- Diodes
- Thyristors or silicon controlled rectifiers (SCRs)
- Power MOSFET
- Insulated gate bipolar transistor (IGBT)
- Integrated gate-commutated thyristor (IGCT)

These devices will be discussed in more detail in Sections 14.2.1–14.2.2.

14.2.1 Power diodes

Power diodes are used in DC-DC, DC-AC, and AC-DC power conversion. Figure 14-16 shows the diode symbol. Diodes can be considered an ideal switch because they turn on rapidly. At turn-off, the diode current reverses for a recovery time, t_{rr}. Diodes can be classified as a slow-recovery diode, a fast-recovery diode, or a Schottky diode. Slow recovery diodes are designed to have low voltage, and therefore have a larger t_{rr}.

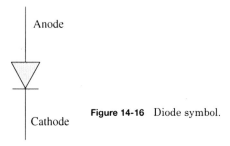

Figure 14-16 Diode symbol.

These are used in line frequency power rectification. Fast-recovery diodes are designed to use in high-frequency circuits, and are used in combination with controllable switches where a small reverse-recovery time is needed. The Schottky diode has a lower forward conduction drop and faster switching time, but a lower blocking voltage. The Schottky diode is used in DC-to-DC converter rectifier stage circuits [22,23].

14.2.2 Switching devices

Fuel cells give a very unregulated voltage, and since most electronic and electrical equipment require a steady voltage, the voltage output must be regulated. The voltage can be regulated by dropping the voltage down to a fixed value under the operating voltage, or boosting the voltage to a fixed voltage. Typically, the voltage is boosted to a higher voltage. The voltage adjustments can be achieved by switching or chopping circuits. The basic electronic switches that will be discussed are the MOSFET, IGBT, and the thyristor (SCR) devices.

Power MOSFET. The metal oxide semiconductor field effect transistor (MOSFET) is a voltage-controlled device that is turned on by applying a voltage to the gate. When the device is "on," the resistance between the drain (d) and source (s) is very low. The current path behaves like a resistor, whose "on" value is R_{DSon}, which can be low as 0.01 ohms for devices that switch low voltages, and 0.1 ohms for devices that can switch higher voltages. Figure 14-17 shows an example MOSFET device. MOSFETs are typically used in systems that have a power of less than 1 kW.

MOSFET current-voltage characteristics have two distinct regions: a contact resistance, which determines the conduction drop, and a constant current region [22]. Common MOSFET applications are switch mode power supplies and battery charging [23].

Insulated gate bipolar transistor (IGBT). The IGBT (insulated gate bipolar transistor) is a device that combines a MOSFET and a conventional bipolar transistor. The IGBT requires a fairly low voltage with negligible current at the gate to turn on. The main current flow is from the collector to

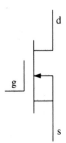

Figure 14-17 Example of a MOSFET device.

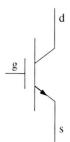

Figure 14-18 Example of an IGBT device.

the emitter, and the voltage does not rise much above 0.6 V at all currents within the device. Figure 14-18 shows an example IGBT device that has a maximum voltage of 1700 and a maximum current of 600 with a switching time of 1 to 4 microseconds [18]. The IGBT is the preferred choice in systems from 1 kW to several hundred kilowatts. The disadvantage of the IGBT is the slower switching times, which makes it a disadvantage in lower power systems. Common IGBT applications are DC-AC inverters for motor control, UPS systems, and low-power lighting [22,23].

Thyristors or silicon controlled rectifiers (SCRs). One of the most common electronic switches is the thyristor, which can only be used as an electronic switch. These devices block the voltage in both the forward and reverse directions. A pulse of current into the gate triggers the transition from the blocking to the conducting state. The device continues to conduct until the current falls to zero. Figure 14-19 shows an example thyristor device.

The energy required to effect the switching is greater in the thyristor than the MOSFET or IGBT. The main advantage of the thyristor for DC switching is that higher currents and voltages can be switched, even though the switching times are much longer. Commercial thyristors are available with very large current and voltage ratings. Thyristors and SCRs have applications in high-power static bypass switches and in UPS systems [22,23].

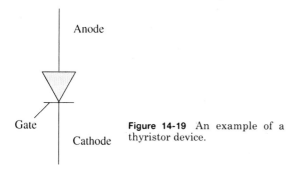

Figure 14-19 An example of a thyristor device.

Integrated gate-commutated thyristor (IGCT). The IGCT is a high-voltage, high-power device, and was recently introduced in 1997 [22]. This device is an asymmetric-blocking gate turn-off thyristor (GTO) that requires a negative turnoff current. This IGCT is commonly used in power distribution systems installations such as medium voltage static transfer switches and industrial drive systems.

14.2.3 Switching regulators

A switching regulator takes small bits of energy from the input voltage source, and moves them to the output. The energy losses are very small, and as a result is that a switching regulator can typically have greater than 85% efficiency. Since their efficiency is less dependent on input voltage, they can power useful loads from high voltage sources. The chopper circuit is shown in Figure 14-20. The components are usually an electronic switch, a diode, and an inductor. Figure 14-20 shows the current path when the current is off and the current flows through the inductor and the load. The stored energy in the inductor keeps the current flowing through the load using the diode.

When the switch is on, as shown in Figure 14-21, the current flows through the inductor and the load.

The output voltage, V_2 can be given by the following equation:

$$V_2 = \frac{t_{ON}}{t_{ON} + t_{OFF}} V_1 \qquad (14\text{-}33)$$

where t_{ON} and t_{OFF} are the "on" and "off" times for the electronic switch, respectively. The main energy losses in the step-down chopper circuit are:

- The switching losses in the electronic switch
- A power loss in the switch while it is turned on
- Diode losses
- Power loss due to the resistance of the inductor

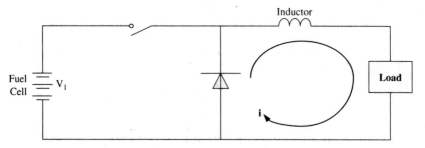

Figure 14-20 Current path when the switch is off.

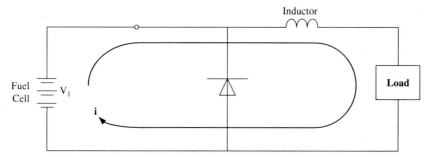

Figure 14-21 Current path when the switch is on.

The efficiency of a step-down chopper is usually over 90 percent, and efficiencies as high as 98 percent are possible. Since fuel cells are low-voltage devices, it is often desirable to boost the voltage. This can be accomplished using switching circuits as shown in Figures 14-22 and 14-23.

If some charge is built in the capacitor, and the switch is on, the electric current is building up in the inductor. The load is supplied by the discharging of the capacitor, and the diode prevents the charge from flowing back through the switch. When the switch is off, the inductor voltage rises. As soon as the voltage rises above the capacitor, the current will flow through the diode, charge the capacitor, and flow through the load.

Losses in the circuit come from the same sources as the step-down regulator. The currents through the inductor and switch are higher than the output current; therefore, the losses are higher. The efficiency of the boost regulators are slightly less than the buck, but at least 80 percent efficiencies should be obtained.

These step-up and step-down switching or chopping circuits are DC-DC converters. These come prepackaged with a wide range of powers and voltages. If the requirements cannot be met with a prepackaged DC-DC converter, these units can be simply designed and made.

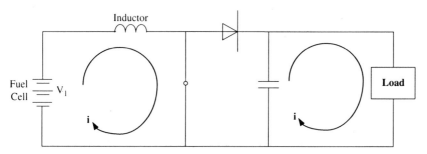

Figure 14-22 Current path when the switch is on.

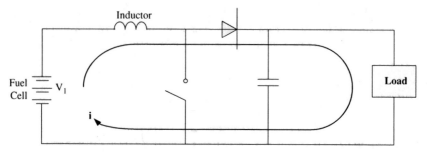

Figure 14-23 Current path when the switch is off.

DC-to-DC converters. A DC-to-DC converter is used to regulate the voltage because the fuel cell varies with the load current. Many fuel cell systems are designed for a lower voltage; therefore, a DC-DC boost converter is often used to increase the voltage to higher levels. Several types of DC-DC converters are presented in the next few paragraphs.

Single-switch boost converter. The boost converter has two modes of operation: the continuous conduction mode (CCM) and the discontinuous-conduction mode (DCM). In CCM, the inductor current is greater than zero after each switching cycle, and in DCM the inductor current is zero after each switching cycle [22]. The boost converter consists of a switch, diode, inductor, capacitor and a control block, as illustrated in Figure 14-24.

During operation, the switch goes "on" and "off" to boost the input voltage to the desired voltage level. In the "on" state, the energy is stored in the inductor, and the load current is supplied by the capacitor. In the "off" state, the energy in the inductor is transferred to the load.

Push-pull DC-DC converter. The push-pull DC-DC converter uses a transformer to increase the output voltage. The power switches do not conduct simultaneously, but alternate back and forth [22]. When this

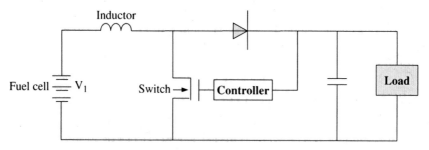

Figure 14-24 A single switch boost converter.

converter is properly designed, it is capable of generating many hundreds of watts of power output.

Full bridge DC-to-DC converters. The full bridge DC-DC converter is similar to the push-pull converter. There are several deficiencies associated with this converter type, such as transient overvoltages due to transformer leakage and output rectifier recovery [22]. The four switches are turned on and off in order to apply a positive and negative voltage at the transformer.

14.2.4 Inverters

Fuel cells generate their electricity as direct current (DC), which is an advantage for many systems. They can be used in both homes and businesses as the main power source. However, these fuel cell systems will have to connect to the AC grid. Therefore, the fuel cell output will need to be converted from DC to AC for grid-connected systems.

The main components of a single-phase converter are shown in Figure 14-25. It has four switches labeled A through D, and across each switch is a diode. The resistor and the inductor represent the load that the AC will be driven. When operating the inverter, the switches A and D are turned on, and the current flows to the load. These switches are then turned off. The current will continue to flow in the same direction, through the diodes and across the switches B and C, and then back to the supply. Switches B and C are then turned on, and the current flows to the left.

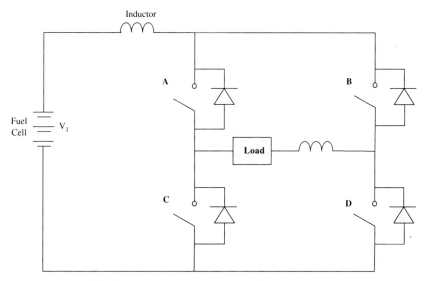

Figure 14-25 An H-bridge inverter circuit for producing single-phase alternating current.

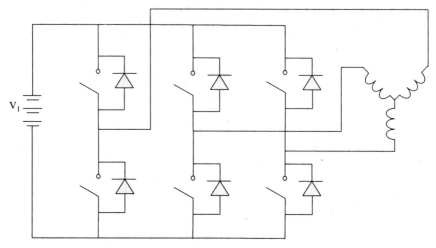

Figure 14-26 A three-phase inverter circuit.

Three-phase. Electricity is generated using three parallel circuits, and most industrial establishments use all three phases, although most homes are supplied with a single phase. The basic circuit is shown in Figure 14-26. Six switches with diodes are connected to the three-phase transformer on the right.

The signals used to turn the switches on and off are obtained from the microprocessor. Signals are taken from the three phases to turn the switches on or off. Digital signals from other sensors can be used, and instructions and information can be sent to, and received from, other parts of the system. Switches, like most electronics, are hardware that can be programmed for each application, and can be used for several systems including fuel cells, solar panels, wind-driven generators, and many other applications.

14.2.5 Supercapacitors

Supercapacitors have recently been used as a viable energy storage device. Some of the features of current commercial supercapacitors include:

- Energy densities as high as 100 times that of conventional capacitors
- Fast discharge times
- A capacitance range of 0.43 to 2700 F and up, with a nominal voltage between 2.3 and 100 V, and current ranging from 3 to 600 A
- Maintenance-free operation

Supercapacitors can be used to meet high power spikes created by the load. The stored energy in a supercapacitor is given by:

$$W_j = \frac{1}{2*CV^2} \qquad (14\text{-}34)$$

This equation explains that the energy stored in a supercapacitor is directly proportional to the square of its voltage. This means that if there is a significant voltage drop, there will also be a release of a large amount of the stored energy. Also, several losses throughout the system need to be accounted for—for example, the amount of power required for other electronics components such as the DC-DC boost converter, and the inter losses due to electrostatic resistance. If these two losses are considered into the equation, it becomes:

$$\frac{1}{2}\left[CV_{sup}^2 - C(0.7*V_{sup})^2\right]*\kappa = P_{in-rush}*t \qquad (14\text{-}35)$$

where C is the capacitance of the supercapacitor, κ is the efficiency, $P_{in-rush}$ is the amount of inrush power (watts), and the "t" is the specified duration for the inrush current.

14.2.6 Power electronics for cellular phones

The load provided by a cell-phone can range from 3 W (during analog operation of a dual-mode phone) to a few megawatts (while the phone is idle) [1]. Power converter efficiency at heavy loads, and standby current at light loads, are both necessary to achieve a usage time that exceeds Li-ion batteries.

A typical micro-fuel cell generates approximately 0.2 V to 0.4 V, depending on the fuel cell type, design, and load [2]. A single-cell fuel cell is ideal from a fuel cell design standpoint, because it eliminates many issues associated with the stacking of cells, water management, fuel distribution, and heat management. From a power conversion standpoint, such a system is troublesome.

Several parameters describe the power converter. Some power is required for the logic and gate drive, P_{ov}. P_{ov} is generally small, but it's a significant factor at light load. The MOSFET switching loss factor is [1]:

$$K = \frac{fV_{out}t_{switch}}{2} \qquad (14\text{-}36)$$

The factor, K, has units of volts and is multiplied by the input current to get switching loss, so this loss term increases linearly with load. Similarly, losses in the equivalent series resistance of the output capacitor, r_c, increase linearly with load [1]. Most notably, total resistance, R, causes losses to increase quadratically with input current. R includes fuel cell parasitic resistance, inductor resistance, MOSFET on-state resistance, and PCB trace resistance.

Input current can be determined by solving the following equation [1]:

$$RI_m^2 + \left(K + r_c\frac{P_{out}}{V_{out}} - V_{in}\right)I_{in} + P_{ov} + P_{out} - r_c\left(\frac{P_{out}}{V_{out}}\right)^2 = 0$$

Knowing input current, the efficiency is [1]:

$$\eta = \frac{P_{out}}{V_{in}I_{in}}$$

If these equations are put together, it becomes clear that efficiency drops dramatically with input voltage [1], and Resistance, R, dominates useful converter designs [1].

For an input of 1 V and a load of 1 W, efficiency is greater than 85 percent [1]. For an input of 0.4 V at the same load, efficiency is only 67 percent. The fuel cell system is affected by reduced efficiency in the power converter in a variety of ways. In order for the fuel cell to generate enough power to carry the load, it needs to be larger, and be able to reach a point of voltage collapse, just over 1.5 W at 0.4-V input in the sample converter [3].

Voltage collapse is an occurrence seen in many power applications. In the high-voltage power transmission grid, when the losses exceed 50 percent at any point, localized blackouts occur. In a boost converter, controllers assume that increasing duty cycle increases the boost ratio. As efficiency drops, a point is reached where increasing duty cycle instead decreases the boost ratio by increasing losses in the inductor and low-side switch. The controller quickly reaches the maximum duty cycle, dropping the load and possibly damaging the fuel cell or the converter [1].

To avoid voltage collapse, the fuel cell stack needs to be designed so that the boost ratio (the ratio between the desired output voltage and the actual input voltage) is as close to 1 as possible. Typical fuel cell characteristics will mandate that the boost ratio varies significantly with load. Thus, the fuel cell and converter designer must consider all of these operating points carefully [1].

14.2.7 DC-DC converters for automotive applications

The automotive fuel cell system has two voltage buses. The fuel cell's high voltage bus (~ 2000–4000 V) supplies the power for large loads, such as the air compressor drive, and the air conditioner compressor drive.

The low voltage bus is a conventional 12 V sulfuric acid battery, which is used for starting the fuel cell system, and powering small loads such as the lights and radio. The DC-DC converter manages the energy flow between the two buses in the fuel cell vehicle.

The fuel cell/battery/DC-DC converter system has two modes of operation: start-up and normal mode. The hydrogen fuel is stored in a high-pressure cylinder, and the oxygen is supplied to the fuel cell by an air compressor. During start-up, the energy is transferred from the 12 V battery to the high-voltage bus to turn on the air compressor to begin pumping oxygen into the fuel cell. The hydrogen and oxygen is supplied to the fuel cell to produce the DC electricity for the high-voltage bus.

When the fuel cell establishes a stable DC-voltage output, the vehicle will enter normal drive mode. During this mode, the fuel cell supplies all of the vehicle's energy. The DC-DC converter now works as a buck converter, and transfers energy from the high voltage bus to the low-voltage bus.

Designing a 3 kW bidirectional DC-DC converter is challenging because of the high battery current. If the converter is in boost mode, and the battery voltage is 10 V, in order to obtain the 3 kW output, the stack input current has to be 350 A if a 85% efficiency is assumed. The high current makes designing the power stage for the low-voltage side of the converter a nontrivial task. The main source of power loss is the switching device, and conduction loss is dominant when using MOSFETs for this application.

14.2.8 Multilevel converters for larger applications

Multilevel converters are of interest in the distributed energy resources area because several batteries, fuel cells, solar cells, wind turbines, and so on can be connected through a multilevel converter to feed a load or grid without voltage-balancing problems [9]. The general function of the multilevel inverter is to synthesize a desired AC voltage from several levels of DC voltages. For this reason, multilevel inverters are ideal for connecting either in series or in parallel an AC grid with renewable energy sources such as photovoltaics or fuel cells, or with energy storage devices such as capacitors or batteries. Multilevel converters also have lower switching frequencies than traditional converters, which means reduced switching losses and increased efficiency.

Advances in fuel cell technology require similar advances in power converter technology. By considering power conversion design parameters early in the system design, a small, inexpensive converter can be built to accompany a reasonably sized fuel cell for high system power and energy density.

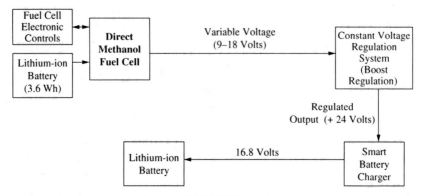

Figure 14-27 A diagram of the overall DMFC/Li-ion charger system.

14.3 Fuel Cell Hybrid Power Systems

Fuel cell hybrid power systems can be a long-term portable or stationary power solution for many types of systems. When a fuel cell is used with batteries, capacitors, solar panels, and other types of power equipment, the fuel cell can act as the primary, secondary, or backup power source. If total system power requirements are low, the energy can be taken from the battery or capacitor, or vice versa. This is especially useful if the system power requirements are variable, such as with data-logging equipment and systems that provide power to multiple devices. If the power required by the system is constant, a hybrid power system may not offer any advantages.

An example hybrid power system is shown in Figure 14-27. The direct methanol fuel cell/lithium-ion battery charger system will include the following major components: the direct methanol fuel cell, the lithium-ion battery, constant voltage regulation system, and a smart battery charger.

A 6.1-watt-hour rechargeable lithium-ion battery can be located inside the fuel cell unit to maintain the microcontroller in a low-power standby or programmed-timer sleep state for several days. The battery will also allow the system immediate startup and save the system during shutdown. The battery will be automatically charged whenever the fuel cell is running. The internal battery charging circuit will stop charging the Li-ion battery once it has reached 4.2 volts or has been charging for two hours.

14.4 System Efficiency

As described in Chapters 6-9, the actual fuel cell efficiency has to take into account many types of losses such as heat, electrode kinetics, electric

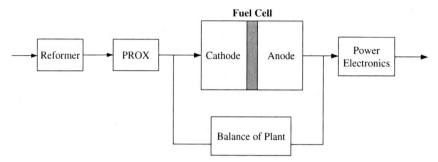

Figure 14-28 A block diagram of generic fuel cell system.

and ionic resistance, and mass transport. Additional system losses are generated by the fuel processor, compressors, pumps, blowers, fans, and the electrical subsystem. Figure 14-28 shows a block diagram of a generic fuel cell system with the flows of energy used to define the efficiencies of the individual steps.

The system efficiency is the ratio of the output electrical energy, E_{net}, and the energy in the fuel fed to the system, F_{in}:

$$\eta_{sys} = \frac{E_{net}}{F_{in}} \qquad (14\text{-}37)$$

The efficiency is actually the product of several system efficiencies:

$$\eta_{sys} = \frac{H}{F_{in}} \frac{H_{cons}}{H} \frac{E_{FC}}{H_{cons}} \frac{E_{net}}{E_{FC}} \qquad (14\text{-}38)$$

where H is the energy of hydrogen produced by the fuel processor, H_{cons} is the energy of hydrogen consumed in the fuel cell electrochemical reaction, and E_{FC} is the power produced by the fuel cell (W).

Fuel cell efficiency has two parts: fuel efficiency and voltage efficiency. A common method of defining the fuel cell efficiency is with the stoiometric ratio. Hydrogen is usually supplied to the fuel cell stack in a stoichiometric amount with ratios ranging from 1.1 to 1.2. The fuel efficiency is the ratio between the hydrogen consumed and supplied to the fuel cell:

$$\eta_{fuel} = \frac{H_{cons}}{H} = \frac{H_{cons}}{S_{H_2} H_{cons}} = \frac{1}{S_{H_2}} \qquad (14\text{-}39)$$

Any hydrogen that is unused is usually used somewhere else in the stack to create a greater overall fuel cell system efficiency. It is sometimes used to generate power in the turbine/expander or in the fuel processing process. The fuel consumption is

$$F_{in} = \frac{S_{H_2} H_{cons}}{\eta_{ref}} - (S_{H_2} - 1) H_{cons} \qquad (14\text{-}40)$$

The voltage efficiency of the fuel cell is

$$\eta_{FC} = \frac{E_{FC}}{H_{cons}} = \frac{V_{cell}}{1.482} \qquad (14\text{-}41)$$

The power produced by the fuel cell is brought to a constant voltage and current type by the power electronics subsystem. Some of the power is used to run the system's plant components. The resulting power conversion and parasitic loss efficiency is:

$$\eta_{FC} = \frac{E_{net}}{E_{FC}} = (\eta_{DC} - \xi) \qquad (14\text{-}42)$$

where $\xi = \frac{E_{aux}}{E_{FC}}$, which is the ratio between the power needed to run the parasitic load and the gross power. This equation is valid when the parasitic load runs on the same voltage and current as the main load. The overall system efficiency is then:

$$\eta_{sys} = \eta_{sref} \eta_{PROX} \eta_{fuel} \eta_{FC} \eta_{PC} \qquad (14\text{-}43)$$

If the system does not include the reforming step, $\eta_{sref}\eta_{PROX}$ can be excluded.

Chapter Summary

A fuel cell system can be very efficient with a simple plant and electrical subsystem – or a very complex one. Typically, the larger the fuel cell stack(s), the more complex the fuel cell plant subsystem will be. The number of ways to design and optimize the fuel cell plant and electrical subsystems are endless. The plant components reviewed in this chapter include humidifiers, fans, blowers, compressors, turbines, and pumps. A well-designed electrical subsystem is required for fuel cell systems because the power output of a fuel cell is unstable, and often the voltage output is very low. The power and control electronics can boost and regulate the power to produce a more desirable and efficient power system. The power electronics covered in this chapter include diodes, SCRs, MOSFETS, IGBTs, IGCTs, converters, and inverters.

Other useful topics covered in this chapter are supercapacitors, hybrid power systems, and the calculation of system efficencies.

Problems

1. A 50 kW fuel cell operates at 0.7 V per cell at 65°C at ambient pressure with an oxygen stoichiometry of 2.0. Liquid water is separated from the cathode exhaust. Calculate the amount of water that needs to be stored for three days of operation.

2. Calculate the temperature in a compressor that provides air for a 50 kW fuel cell system. Assume that air is dry at 20°C and 1 atm, and the delivery pressure is 2 atm. The fuel cell generates 0.75 volts per cell and operates with an oxygen stoiometric ratio of 2. The compressor efficiency is 0.65.

3. For the fuel cell described in Problem 2, calculate the amount of power that can be recovered if the exhaust air is run through the turbine? The exhaust gas has 100 percent humidity with a pressure drop of 0.3 atm. The turbine efficiency is 65 percent.

4. For the fuel cell system described in Problem 2, calculate the amount of water (g/s) needed to fully saturate the air at the fuel cell entrance at 75°C.

5. For the fuel cell system described in Problem 2, calculate the amount of heat (W) needed for the air humidification process.

Bibliography

[1] Kimball, Jonathan W. "Power Converters for Micro Fuel Cells." *Power Electronics Technology, October*, 2004.
[2] Lin, Bruce. "Conceptual Design and Modeling of a Fuel Cell Scooter for Urban Asia." Princeton University, Masters Thesis, 1999.
[3] Mench, M.M., Z.H. Wang, K. Bhatia, and C.Y. Wang. "Design of a Micro-direct Methanol Fuel Cell," Electrochemical Engine Center, Department of Mechanical and Nuclear Engineering, Pennsylvania State University, 2001.
[4] Mikkola, Mikko. "Experimental Studies on Polymer Electrolyte Membrane Fuel Cell Stacks." Helsinki University of Technology, Department of Engineering Physics and Mathematics, Masters Thesis, 2001.
[5] Brunstein, Vladimir. "DC-DC Converters Condition Fuel Cell Outputs." Advanced Power Associates Corp., New Milford, NJ. November 01, 2002.
[6] Makuch, Gregg. "Micro Fuel Cells Strive for Commercialization." Neah Power Systems, Bothel, Washington. February 01, 2004.
[7] Dinsdale, Julian. "New Energy Conservation Technologies." Ceramic Fuel Cells Limited ESAA Residential School Queensland University of Technology and University of Queensland, January 2004.
[8] Li, Xianguo. *Principles of Fuel Cells*. 2006. New York: Taylor & Francis Group.
[9] Ozpineci, Burak, Leon M. Tolbert, and Zhong Du. *Multiple Input Converters for Fuel Cells*. Industry Applications Conference, 2004. 39th IAS Annual Meeting. Conference Record of the 2004 IEEE. Volume 2, Issue , 3–7 Oct. 2004 p. 791–797.
[10] "Fuel Cell Origins." Smithsonian Institution, 2001. http://fuelcells.si.edu.
[11] Tse, Lawrence and Duane Bong. "A Brief History of Fuel Cells." http:// www.visionengineer.com.

[12] "Power Electronics for Fuel Cells Workshop." National Fuel Cell Research Center. August 8–9, 2003.
[13] Barbir, Frano. *PEM Fuel Cells: Theory and Practice*. 2005. Burlington, MA: Elsevier Academic Press.
[14] Izenson, Michael G. and Roger W. Hill. "Water and Thermal Balance in PEM Fuel Cells." *Journal of Fuel Cell Science and Technology*. Vol. 1, 2004, pp. 10–17.
[15] Chen, F.C., Z. Gao, R.O. Loutfy, and M. Hecht. "Analysis of Optimal Heat Transfer in a PEM Fuel Cell Cooling Plate." *Fuel Cells*. Vol. 3, No. 4, 2003.
[16] Zhang, Yangjun, Minggao Ouyang, Qingchun Lu, Jianxi Luo, and Xihao Li. "A Model Predicting Performance of Proton Exchange Membrane Fuel Cell Stack Thermal System." *Applied Thermal Engineering*. Vol. 24, 2004, pp. 501–513.
[17] Incropera, Frank P. and David P. Dewitt. *Introduction to Heat Transfer*, 4th ed. 2002. John Wiley & Sons, Inc.
[18] Larminie, James and Andrew Dicks. *Fuel Cell Systems Explained*. 2000. John Wiley & Sons, Ltd, Chichester, England.
[19] Rayment, Chris and Scott Sherwin. "Introduction to Fuel Cell Technology." Department of Aerospace and Mechanical Engineering, University of Notre Dame. May 2, 2003.
[20] Li, Xianguo. *Principles of Fuel Cells*. 2006. New York: Taylor & Francis Group.
[21] Perry, Robert H. and Don W. Green. *Perry's Chemical Engineers' Handbook*, 7th ed. 1997. McGraw-Hill: New York.
[22] *Fuel Cell Handbook*, 5th ed. EG&G Services. Parsons Incorporated. Science Applications International Corporation. U.S. Department of Energy, October 2000.
[23] Mohan, Ned, T. Undeland, and William Robbins. *Power Electronics: Converters, Applications and Design*, 2nd ed. 1995. New York: John Wiley & Sons, Inc.
[24] Robinson, Randall N. Chemical Engineering Reference Manual for the PE Exam. 5th ed. 1996. Belmont, CA: Professional Publications, Inc.
[25] GeoCool Lab. Department of Mechanical Engineering, University of Alabama. http://bama.ua.edu/~geocool/PsychWB.doc Last Accessed March 24, 2007.

Chapter

15

Fuel Types, Delivery, and Processing

Hydrogen is the most commonly used fuel for fuel cells, but several other types of fuels can also be used. The cleanest fuel is pure hydrogen because it generates zero emissions. Many fuel cell designs utilize other fuels because of the availability or perceived safety of those fuels. Fuels for fuel cells range from hydrogen to fossil fuels to biofuels. Several types of reformers and chemical conversion methods can convert all of these fuel types to hydrogen. When reformers are used, another consideration is the heat generated and the effect on the fuel cell stack. Other methods of producing hydrogen include the electrolysis of water and producing hydrogen through the use of biological methods. The use of fuels other than hydrogen may help to commercialize fuel cells in the near term, but the goal of fuel cell technology is to use pure hydrogen from renewable sources of energy other than fossil fuels. A brief comparison of five hydrogen technologies is shown in Table 15-1.

When selecting a fuel, the most important consideration is to have the stored energy in an efficient, safe and cost effective system. The method of hydrogen generation and storage is application-dependent, so several factors must be considered. In addition to the parameters listed in Table 15-1, other important considerations when selecting a fuel type is the weight, volume, dimensions and the total complexity of the system.

Some fuel types and methods of storage may be better suited for certain applications than others. For example, metal hydrides present a good choice for portable applications due to their inherent safety, good hydrogen gravimetric density, and excellent volumetric density. Compressed gas cylinders are a good option for portable, automotive and stationary applications because they can carry large amounts of

TABLE 15-1 Hydrogen Storage Comparison

	Compressed hydrogen (5000 psi – 350 bar)	Compressed hydrogen (10,000 – 700 bar)	Liquid hydrogen	Chemical hydrides	Metal hydrides
Current Volumetric Storage Density (g H_2/L)	~12.5 g H_2/L = 1.5 MJ/L	~24.2 g H_2/L = 2.9 MJ/L	~37.0 g H_2/L = 4.4 MJ/L	>22 g H_2/L >2.5 MJ/L	67 g H_2/L
Current Gravimetric Storage Density (wt%)	~2.7	~3.3	~5.0	>4.0	>1.0
Energy Density (System Volumetric Capacity) (kWh/L)	0.6 – 0.9	1.0 – 1.3	1.2 – 1.6	0.5 – 1.4	0 – 1.8
Specific Energy (Gravimetric Capacity) (kWh/kg)	2.0 – 2.5	1.5 – 2.0	2.0 – 4.5	1.2 – 1.6	0.1 – 0.5
Temperature (°C)	25	25	−253	25	25
Pressure (bar)	350	700	4	—	50
System Cost per kWh ($/kWh)	$12	$16	$6	$8	$16

hydrogen. The main disadvantage of compressed gas cylinders is that they have a poor perception of safety by the public. Refueling can easily be done from hydrogen tanks with much higher pressures. Liquid hydrogen is sometimes used for automotive and stationary applications, but the main disadvantage of this fuel storage type is the equipment required to keep the cylinder cool.

There are many other fuels and fuel storage types, such as chemical hydrides, but these are mostly still in the research and development stage. The uses for chemical hydrides and other novel hydrogen technologies will become more apparent as the research progresses on these fuel types.

The other commonly used option for supplying fuel to fuel cells is reforming various fuels to produce hydrogen before it enters the fuel cell. A few commonly used reformer technologies are steam, internal, methanol and partial oxidation reforming. Many of these reforming technologies are well established, but the final cleanup step to reduce carbon monoxide to a few ppm required for fuel cell intake is not sufficiently developed in many of these systems. Reformer technology is complex in terms of integrating the various heat and chemical flows, and the equipment required can sometimes be bulky and costly.

The topics on fuel cell fuel types, options, storage, delivery, and processing that will be covered in this chapter include:

- Hydrogen Storage Methods
- Other Common Fuel Types

- Fuel Processing
- Other Methods of Producing Hydrogen

These topics are covered in detail in the remainder of this chapter.

15.1 Hydrogen

The problem of hydrogen storage and delivery is a roadblock to mass commercialization of fuel cells. Some possibilities for hydrogen storage include compressed hydrogen, liquefied hydrogen, storage in a metal hydride, and storage in a hydrocarbon until release through a reformation process.

Compressed hydrogen is considered the best future hydrogen storage technology for portable and transportation applications. In the past, 24.8 MPa (3600 psig) metal tanks have been used, although higher pressure 34 MPa (5000 psig) lightweight polycarbonate tanks are now also in use. The drawback of compressed hydrogen storage is the low energy storage density of 31 g/L at 34 MPa [2,17].

Another potential storage medium is metal hydrides. Hydrogen can bond with certain metal compounds as a metal hydride, releasing hydrogen when heated. Conventional hydride technology allows 1 to 2 weight percent (wt.%) of H_2 stored in the metal. If higher storage percentages can be reached in operating systems, hydride technology could become the dominant method of hydrogen storage for portable.

15.1.1 Gas

Storing hydrogen as a compressed gas is the most widely used method of distributing hydrogen, and offers zero-emissions for fuel cells. Hydrogen has many unusual characteristics compared to other elements. It is the lightest and most abundant element, and can burn with oxygen to release large amounts of energy. It also has high energy content by weight. Hydrogen is highly of flammable, and burns when it makes up 4 to 75 percent of air by volume [4]. It has a low energy density by volume at standard temperature and atmospheric pressure. The volumetric density can drastically be lowered by storing compressed hydrogen under pressure, or converting it to liquid hydrogen. Table 15-2 compares the relevant properties of hydrogen, methanol, ethanol, propane, and gasoline—which all can be used as fuel for fuel cells. Details on these other fuel types can be found in Section 15.2. Hydrogen does not exist in its natural form on earth; therefore, it must be manufactured through one of various ways: steam reforming of natural gas, gasification of coal, electrolysis, or the reforming/oxidation of other hydrocarbons or biomass.

For many reasons, hydrogen is a good choice for a future energy source. Some of these reasons are [3]:

TABLE 15-2 Hydrogen Compared with Other Fuels (Compiled from [14] and [15])

Property	Hydrogen	Methanol	Ethanol	Propane	Gasoline
Hydrogen Weight Fraction	1	0.12	0.13	0.18	0.16
Molecular Weight (g/mol)	2.016	32.04	46.0634	44.10	~107.0
Density (kg/m^3) 20°C and 1 atm	0.08375	791	789	1.865	751
Normal Boiling Point (°C)	−252.8	64.5	78.5	−42.1	27 to 225
Flash Point (°C)	<−253	11	13	−104	−43
Flammability Limits in Air (Volume %)	4.0 to 75.0	6.7 to 36.0	3.3 to 19	2.1 to 10.1	1.0 to 7.6
CO_2 Production per Energy Unit	0	1.50			1.80
Autoignition Temperature in Air (°C)	585	385	423	490	230 to 480
Higher Heating Value (MJ/kg)	142.0	22.9	29.8	50.2	47.3
Lower Heating Value (MJ/kg)	120.0	20.1	27.0	46.3	44.0
Volumetric Energy Density (Wh/L)	405	4600	6100	6600	9700
Gravimetric Energy Density (Wh/kg)	39,000	6400	7850	13,900	12,200

- Hydrogen can be made from various sources. It is completely renewable. The most abundant and cleanest precursor for hydrogen is water.
- Hydrogen can be stored in many forms, from gaseous to liquid to solid. It can also be stored in various chemicals and substances such as methanol, ethanol, metal hydrides, chemical hydrides and carbon nanotubes.
- It can be produced from, and converted to, electricity with high efficiencies.
- It can be transported and stored as safely as any other fuel.

Hydrogen and other fuels for fuel cells can be made cheaply and easily by processing hydrocarbons. Hydrogen has the potential to provide energy to all parts of the economy: industry, residences, transportation, and mobile applications. It can eventually release oil-based fuels used for automobiles, and may provide an attractive solution for remote communities that cannot be supplied electricity through the grid. The fundamental attraction for hydrogen is its environmental advantage over fossil fuels—but the hydrogen is only as clean as the technologies used to produce it. The production of hydrogen can be pollutant free if it is produced by one of three methods:

- Through electrolysis using electricity derived solely from nuclear power, solar panels or renewable energy sources.
- Through steam reforming of fossil fuels combined with new carbon capture and storage technologies.
- Through thermochemical or biological techniques based on renewable biomass [4].

A major disadvantage of processing hydrocarbons is the pollution and carbon dioxide—which eliminates one of the main reasons for using hydrogen in the first place. The best low-pollution alternative for creating hydrogen is a process involving electrolysis of water by electricity.

Safety. Hydrogen is thought of as a dangerous fuel. This is partly true because, unlike gasoline and most hydrocarbons that ignite over a narrow range of fuel-to-air ratios (1.3 to 7.1 percent for gasoline), hydrogen can ignite over a wide range of concentrations (4 to 75 percent in air). Hydrogen also has relatively low ignition energy; a low-energy spark can begin an almost invisible flame (0.2 mJ at stoichiometric conditions in air, which is less than 10 percent of the ignition energy of methane, propane, or isooctane) [4,5,17].

On the other hand, hydrogen is a very light atom, and leaks tend to disperse quickly. Being lighter than air, hydrogen also tends to diffuse upwards rather than accumulate near the ground. The lower limit of flammability is higher for hydrogen than it is for gasoline, so greater concentrations of hydrogen have to be built up before ignition is reached. Slow leaks in enclosed areas were defined as the greatest risk by a thorough hydrogen safety study done by DTI for Ford and the Department of Energy.

The greatest concern about using pure hydrogen as a fuel source is that it will explode from a collision. Safety devices can be designed to shut down power to the batteries and cut off hydrogen flow in the case of a collision. In the automobile, the risk of fire or explosion of hydrogen is the same as gasoline.

Natural gas cylinders are typically designed with pressure release devices (PRDs) to discharge the contents of the cylinder in case of fire. Because failure occurs by material degradation rather than pressure increase, most eutectic switches are designed to release when a certain temperature is reached. The concept of rapid discharge of hydrogen (under five minutes) is not a feasible option considering that these devices are designed to operate when engulfed in a flame. A controlled release of hydrogen is a better solution than an abrupt failure and explosion.

Compressed gas storage. The most practical method of storing hydrogen is to compress it as a gas. The major issues with compressed gas storage are the large volume required to store the gas even when compressed, and the ability of the container to resist impact. Storage conditions are set at 3600 psi (standard for gas cylinders) at ambient temperature. The Redlich-Kwong equation of state (equation 15.1) estimates a molar volume of 0.117 L/mol under these conditions (16 percent worse than the 0.101 L/mol calculated by the ideal gas law) [17]:

$$\left(P + \frac{a}{V_m(V_m + b)T^{1/2}}\right)(V_m - b) = RT \tag{15-1}$$

where V_m is the volume per mole, R is the ideal gas constant, P is the pressure and a and b are empirical constants that can be estimated by the formulas

$$a = 0.42748 \, R^2 * T_c^{2.5}/P_c$$

$$b = 0.08664 \, R \, T_c/P_c$$

where T_c is the critical temperature, P_c is the critical pressure, and R is the ideal gas constant.

The amount of work required to compress the hydrogen gas into the cylinder equates to an energy penalty of approximately 5 to 10 percent [17]. The temperature increases when the hydrogen cylinder is filled with compressed hydrogen, and the pressure is higher than the nominal operating pressure until the cylinder has a chance to cool. The decrease in temperature during usage due to expansion of hydrogen is not as great a concern because the release rate is much slower [17].

Current hydrogen gas storage cylinders are made from steel alloys that are resistant to hydrogen embrittlement. More advanced containers are made from aluminum and wrapped with carbon fiber laminate for stiffness and lightness [17]. Fully composite cylinders made from carbon fiber impregnated with resin or some other binder are also currently being researched. Figure 15-1 shows the cross-section of a compressed hydrogen cylinder.

Several companies are developing carbon fiber reinforced 5000-psi and 10,000-psi hydrogen gas tanks. The inner liner of the tank is made of a high molecular weight polymer that prevents hydrogen gas from permeating. The middle shell is constructed of a carbon fiber epoxy

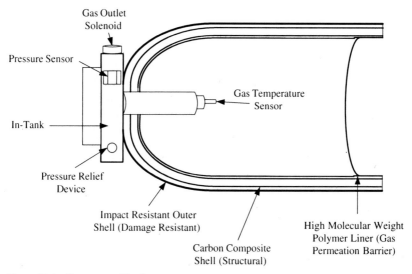

Figure 15-1 Compressed hydrogen storage.

resin composite shell. The outer shell is placed on the tank for impact and damage resistance. A temperature sensor monitors the temperature inside the tank [2]. These tanks can have hydrogen gravimetric densities as high as 9.5 percent due to their light weight [17]. Some examples of commercial hydrogen compressed gas cylinders are shown in Table 15-3.

Current issues with compressed hydrogen gas tanks are the cost, weight, and volume required. Current research involves finding lower-cost carbon fibers that are capable of meeting tank thickness without compromising weight and volume. Two approaches are currently being

TABLE 15-3 Examples of Hydrogen Compressed Gas Cylinders

Property	Compressed hydrogen 1	Compressed hydrogen 2	Compressed hydrogen 3	Compressed hydrogen 4 [5, 7]	Compressed hydrogen 4 [6, 7]
Cylinder Height	31 in (79 cm)	51 in (130 cm)	21 in (53 cm)	—	—
Cylinder Diameter	7 in (18 cm)	9 in (23 cm)	4 in (10 cm)	—	—
Cylinder Weight	65 lb (30 kg)	133 lb (60 kg)	12 lbs (5 kg)	81 lbs (37 kg)	68 lbs (31 kg)
Cylinder Volume	65 ft^3 (1840 L)	196 ft^3 (5550 L)	11 ft^3 (330 L)	8 ft^3 (237 L)	7 ft^3 (200 L)
Cylinder Pressure at Room Temperature	2000 psi (137 bar)	2000 psi (137 bar)	1800 psi (124 bar)	5000 psi (344 bar)	4351 psi (300 bar)
Total H_2 Mass	—	—	—	11 lb (5 kg)	11 lb (5 kg)

used to increase the gravimetric and volumetric storage capacities of the tanks. The first approach uses cryo-compressed tanks, and works as the gas tank temperature decreases, its volumetric capacity increases by a factor of four. The total system capacity does not decrease as much because there is an increased volume needed for the cooling system.

The second approach involves the development of conformable tanks. Gasoline tanks are conformable to take advantage of the available vehicle space, and are based upon the location of structural supporting walls [2].

15.1.2 Liquid

Liquid hydrogen is sometimes used with automotive and stationary fuel cell systems because the energy density of hydrogen is improved by storing hydrogen in the liquid form over the compressed gas form. Liquid hydrogen tanks can store more hydrogen for a particular volume than compressed gas tanks. The volumetric capacity of liquid hydrogen is 0.070 kg/L, compared to 0.030 kg/L for 10,000-psi gas tanks [2]. Current issues with this type of tank are hydrogen boil-off, the energy required for hydrogen liquefaction, volume, weight, and tank cost. The energy requirement for hydrogen liquefaction is high, and new approaches are needed to lower these energy requirements, and, therefore, the costs. Hydrogen boil-off must also be eliminated to reduce cost, and for safety considerations.

If liquid hydrogen is used, cryogenic technology and expensive well-insulated cylinders are required. At 20 K and 0.1 MPa vapor pressure, 5.3 wt% H_2 is achievable [1,2]. At this temperature, liquid hydrogen density is 70 g/L, so carrying 240 grams of hydrogen at such low temperatures is extremely difficult with very good insulation, and the liquid-nitrogen-cooled heat equipment that is typically required [2]. The energy of reducing hydrogen to 20 K and then liquefying it is an important factor when considering liquid hydrogen storage. This energy can amount to an extra 33 to 40 percent of the total energy content of the hydrogen [1,2].

Another problem is that as heat leaks slowly, the liquid hydrogen is warmed up, and over time, it is converted into gas over the liquid. Unless this gas is allowed to escape, hydrogen buildup would eventually create leaks in the tank. Therefore, a minimum boil-off rate is required.

New research is being conducted on a hybrid tank concept which combines both high-pressure gaseous and cryogenic storage. These hybrid tanks are lighter than hydrides and more compact than ambient temperature vessels. Since the temperatures are not as low as for liquid hydrogen, there is less of an energy penalty for liquefaction and less evaporative losses than regular liquid hydrogen tanks [2].

Figure 15-2 shows a state-of-the-art liquid hydrogen storage system consisting of double-wall cylindrical tanks that hold about 10 kg of hydrogen. The shell materials of this tank type are typically stainless steel or aluminum alloy. These materials are commonly used due to negligible

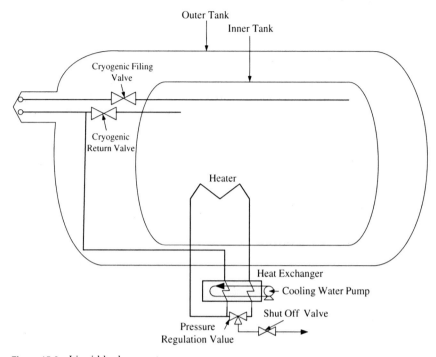

Figure 15-2 Liquid hydrogen storage.

hydrogen permeation, high modulus and strength, resistance to hydrogen brittleness, and low specific weight. Aluminum is sometimes used instead of stainless steel because of the low specific weight with the high coefficient of thermal expansion. The wall thickness for stainless steel is typically 2 to 4 mm, and the typical weight of the entire system is about 150 kg [2,6]. Space between the inner and outer vessel is used for insulation. The support structures are made of glass or carbon fiber reinforced plastics. The typical boil-off rate after the system has not been used for more than three days is 1 to 3 percent per day [2,7]. A comparison of two liquid H_2 tanks designed for an automobile are shown in Table 15-4.

TABLE 15-4 Liquid Hydrogen Tanks Designed for 640 km range in a 34 km/L Gasoline Equivalent Car

Vessel type	Liquid hydrogen 1[5, 7]	Liquid hydrogen 2[6, 7]
Volume (L)	135	121
Weight (kg)	31.3	28.0
H_2 (kg)	5	5
E/V (MJ/L)	4.4	5.6
E/m (MJ/kg)	19	24

15.1.3 Carbon nanofibers

Carbon nanotubes can store small amounts of hydrogen at ambient temperatures and pressures. Several groups are currently researching this technology in an attempt to store enough hydrogen to reach the U.S. Department of Energy benchmarks of 62 kg H_2/m^3 and 6.5 percent weight of stored hydrogen to total system weight for hydrogen fuel cell vehicles [6,7].

Carbon nanotubes have dimensions in nanometers, and can be made from several processes. The structure becomes apparent when examined under an electron microscope. Currently, three types of carbon nanotubes are being investigated for hydrogen storage: graphitic nanofibers, single-walled carbon nanotubes (SWNTs), and multiwalled carbon nanotubes (MWNTs). Three forms of graphic nanotubes are shown in Figure 15-3.

Graphitic nanofibers are made by breaking hydrocarbons or carbon monoxide over suitable catalysts. Many types of structures have been observed, such as platelets, ribbons, and herringbone arrangements. Graphitic nanofibers range from 50 nm to 100 nm in length, and 5 to 100 nm in diameter [8]. Graphitic nanofibers have a wide range of hydrogen absorption values.

Single-walled or multiwalled carbon nanotubes have highly ordered, smooth graphitic surfaces as shown in Figure 15-4. These are usually

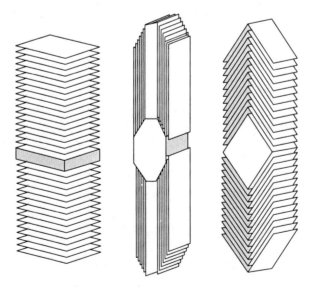

Figure 15-3 Schematic representation of the three forms of graphitic nanofibers: (a) platelet, (b) ribbon, (c) herringbone structure (Bessel, C.A. et al. J. Phys. Chem B. 2001, 105 (6), 1115 [8]).

Fuel Types, Delivery and Processing 323

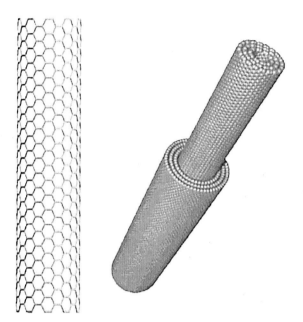

Figure 15-4 Schematic representation of a single-walled nanotube (SWNT) and a multiwalled nanotube (MWNT) [37].

prepared using an electric arc drawn between two carbon electrodes, lasar ablation, or chemical vapor deposition. No consensus has yet been made between scientists on how much hydrogen graphitic nanofibers, SWNTs and MWNTs can hold. Experimental results are often difficult to interpret due to material purity and minute differences in experimental technique. More recent studies on carbon nanotubes are beginning to show some consistency and reproducibility in experiments [6,8,37]. This topic is somewhat controversial, and the practical feasibility of hydrogen storage with carbon nanotubes still needs to be proven. Table 15-5 shows the estimated storage of carbon-based absorbents for an automobile.

TABLE 15-5 Estimated Storage of Carbon-Based Absorbants Designed for a 640-km Range in a 34 km/L Gasoline Equivalent Car

Vessel type	Li-doped carbon nanofibers [7, 9]	K-doped carbon nanofibers [7, 9]	30 wt% carbon nanofibers [6, 7]
Volume (L)	33	51	24
Weight (kg)	43	66	35
H_2 (kg)	5	5	5
E/V (MJ/L)	15	10	28
E/m (MJ/kg)	11	7.8	19

15.2 Other Common Fuel Types

The preferred fuel for all fuel cell types is hydrogen, but there are also many other types of fuels that can be used to power fuel cells. Other fuel types are typically not as efficient as pure hydrogen because the hydrogen is "stripped" from the fuel molecule, and then used in the fuel cell. The hydrogen can be "stripped" from the fuel molecule outside or inside the fuel cell. Some of the fuels that can be processed inside the fuel cell include methanol, ethanol, and ammonia. These are fed directly into the fuel cell, and the hydrogen is stripped from the fuel molecule by the platinum catalyst layer. These fuels are perceived to be safer by the public than hydrogen, and easy to transport and replenish when they are empty. One of the major disadvantages of these fuels is that they slowly poison the catalyst layer over time. Fuels that are processed outside of the fuel cell (by a fuel processor) include natural gas, propane, gasoline and other petroleum-based fuels. Methanol, ethanol, and ammonia can also be processed through a reformer outside of the fuel cell. Fuel processing by a reformer helps to avoid poisoning the fuel cell catalyst layer. It is easier in some fuel cell systems to replace the reformer if necessary instead of the catalyst layers inside the fuel cell. There are also many other methods of generating hydrogen, such as chemical and metal hydrides. All of these fuel types are described in more detail in Sections 15.2.1–15.7.1.

15.2.1 Methanol

Methanol (CH_3OH) is an alcohol fuel and one of the most important chemicals today. Methanol is an attractive fuel for fuel cells because the current supply chain can easily provide methanol, and the energy density is orders of magnitude higher than compressed hydrogen. About 90 percent of the methanol made in the world is based on synthesis gas from natural gas. Methanol can also be produced from nonpetroleum feedstocks such as coal and biomass. Approximately 75 percent of methanol is used for chemicals. The remaining 25 percent is used for fuels [12].

Methanol has been a popular choice as an alternative fuel, for portable applications and fuel cell vehicles for a number of reasons:

- Methanol has a higher energy density than hydrogen.
- It is perceived to be a safer fuel than hydrogen.
- It may lower hydrogen infrastructure costs (see Chapter 2).
- It has high octane and performance characteristics. The higher octane provides an increase in engine horsepower of about 7 to 10 percent, or more, depending on the vehicle.

- Only minor modifications are needed to allow gasoline engines to use methanol.
- There is a significant reduction of reactive emissions when using methanol.

Some studies indicate that methanol powered fuel cells (DMFCs) are the best option for rapidly commercializing fuel cells because the current infrastructure does not have to change as much as it would for hydrogen. There is also no adverse perception of methanol, and a large amount is manufactured already. Other studies indicate that there will need to be as much spent for properly setting up a methanol infrastructure as the hydrogen infrastructure [36,38].

15.2.2 Ethanol

Ethanol is an attractive fuel for fuel cells because the supply chain is already in place, and consumers may be more comfortable using this fuel than hydrogen. Ethanol is a hydrogen-rich liquid that has a high energy density compared to methanol (8.0 kWh/kg verses 6.1 kWh/kg). Ethanol can be produced from a variety of feedstocks, such as corn, sugarcane and other types of plants. In the United States, it is produced mainly from corn by a cooking, fermentation, and distillation process. Brazil uses sugarcane as the primary feedstock. Plants that have higher energy yields, such as switchgrass and sugarcane are more effective in producing ethanol than corn. Future candidates for ethanol production include cellulose feedstocks, such as wood and agricultural wastes [45].

Like methanol, ethanol should be considered as a fuel for fuel cells because:

- Ethanol has a high energy density compared to hydrogen.
- It is perceived to be a safer fuel than hydrogen.
- It may lower hydrogen infrastructure costs (see Chapter 2).
- Only minor modifications are needed to allow gasoline engines to use ethanol.

Since ethanol can be derived from a variety of biological sources (as well as through fossil fuels), this fuel is a good option for overcoming the storage and infrastructure challenges for fuel cells.

15.2.3 Metal hydrides

Due to their inherent safety, and good hydrogen gravimetric and volumetric density, metal hydrides present a good choice for portable military applications, and reversible aboard hydrogen storage [17]. Metal hydrides are a good candidate for reversible onboard hydrogen storage because they have a lower weight and less volume than compressed and liquid hydrogen options, and can release hydrogen at relatively low temperatures and pressures. Most fuel cell system designs use the waste heat generated by the fuel cell to "release" the hydrogen from the metal hydride. New packs can replace old ones, or old ones can be refueled. Refueling a small metal hydride container only takes about 5 to 15 minutes [17]. Figure 15-5 shows an illustration of metal hydride tanks, which typically look like smaller versions of compressed gas cylinders.

Metal hydrides are formed when metal atoms bond with hydrogen to form stable compounds. A large amount of hydrogen per unit volume can be extracted, so the storage density is good despite the fact they are rather heavy. They are typically used as powders in order to maximize the surface area–to-mass ratio. Issues with metal hydrides include high alloy cost, sensitivity to gaseous impurities, and low gravimetric hydrogen density [17].

Metal hydrides are safer than other alternatives because they are endothermic when releasing hydrogen, and are kept under a relatively low pressure of 1 to 10 atm within the metal hydride container [17]. The metal hydride adsorption reaction is

$$M + (x/2)\ H_2 \rightarrow MH_x\ (\text{exothermic},\ (\Delta H < 0)$$

where the number of hydrogen atoms, x, per metal atom, M, is a function of the chemistry of the metal [1,17]. Heat causes the equilibrium to shift towards free hydrogen gas, and the higher partial pressure of H_2 causes a shift towards adsorption and metal hydride formation [17]. The law of mass action shows that the equilibrium constant is $K_{eq} = [H_2]^{(-x/2)}$, and substituting this into the free energy equation:

$$\Delta G = -RT \ln K_{eq} = (x/2)\ RT \ln (P_{H_2}) \qquad (15\text{-}2)$$

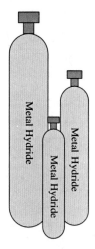

Figure 15-5 Metal hydride tanks (typical portable sizes: 20–700 L)

A hydride with high Gibbs free energy has a lower equilibrium pressure of hydrogen in the metal hydride system at a given temperature, and a stronger metal–hydrogen bond. A metal is a useful storage medium for hydrogen if it strongly bonds to hydrogen so it can be charged up. However, if the bond is too strong, the metal will not give up its hydrogen under heating or depressurization.

Hydrides are also sensitive to contaminants (some are poisoned by oxygen or water vapor) so care must be taken to only introduce pure hydrogen to the hydride.

The use of metal hydrides as a storage medium for hydrogen is dependent on thermodynamics and kinetics. The intrinsic kinetics of hydrogen dissociation are fast. The rate-determining step is heat transport into the powder. Powders generally do not conduct heat well with a thermal conductivity in the range of 1–2 W/mK [17]. For comparison, copper (one of the best thermal conductors) has a conductivity of 401 E/mK, window glass is at 1.0 W/mK, and fiberglass (a thermal insulator) has a thermal conductivity of 0.05 W/mK [16,17].

Heat typically needs to be transferred between the walls of the pressure vessel and the powder. While high surface area means fast hydrogen adsorption and desorption, it also means smaller powder particles. Conduction can be improved by embedding the metal hydride in aluminum foam, or running channels holding hot liquid through the powder. These processes can increase net thermal conductivity to as much as 7 to 9 W/mK [17].

Hydrogen tends to make the particles brittle and cause them to crack into smaller pieces. This increases the total surface area of the powder, increasing the hydrogen desorption/adsorption rate, but the smaller hydride particles can be entrained in the gas flow, requiring filtering to keep the particles out of hydrogen output. This leads to concerns about metal hydrides' long term usage [17].

Having a hydrogen storage container between the metal hydride and the fuel cell would be useful in order to stabilize the hydrogen consumption rate which is unstable due to the heat transfer rate into the metal hydride.

Complex metal hydrides have the potential for higher gravimetric hydrogen capacities than simple metal hydrides. One example is alanate (AlH4), which can store and release hydrogen reversibly when catalyzed with titanium dopants according to the following reactions:

$$NaAlH_4 = 1/3 Na_3AlH_6 + 2/3 Al + H_2$$
$$Na_3AlH_6 = 3\,NaH + Al + 3/2 H_2$$

The first reaction can release 3.7 wt.% hydrogen at 1 atm pressure, and temperatures above 33°C, and the second reaction can release 1.8 wt.% hydrogen above 110°C [1,16,17]. The amount of hydrogen that can be released is more important than the amount that the material can hold and is the key parameter used to determine system (net) gravimetric and volumetric capacities [2].

Some current issues with metal hydrides include low hydrogen capacity, slow uptake, release kinetics, and cost. The metal hydrides' gravimetric capacity 2010 DOE system target of 6 wt.% has not been reached yet, and the hydrogen release kinetics are too slow for vehicular applications. The packing density of these powders is also low (about 50 percent). Continuous research of metal hydrides will hopefully result in the design and development of complex metal hydrides for use with fuel cells in the future [18].

A new complex hydride system based upon lithium amide has been developed, and the reaction occurs at 285°C and 1 atm:

$$Li_2NH + H_2 = LiNH_2 + LiH$$

This reaction allows 6.5 wt.% hydrogen to be reversibly stored, with the potential for 10 wt.% [1,2]. One issue with using this system is that if it is utilized for portable applications with a PEMFC, the reaction temperature is quite high and needs to be lowered [18]. Further R&D on this system may lead to additional improvements in capacity and operating conditions. Table 15-6 summarizes the storage system characteristics of some types of metal hydrides.

15.2.4 Chemical hydrides

Chemical hydride storage refers to technology where hydrogen is generated through a chemical reaction. These reactions are typically chemical hydrides, water or alcohols. These reactions are not reversible, and the byproducts must be discarded. Hydrogen fuel can also be produced by chemical reaction with solid "chemical hydrides." This technology lies somewhere between metal hydrides and reforming. Chemicals such as lithium hydride, lithium aluminum hydride, and sodium borohydride can be combined with water to create hydrogen gas exothermically [32]:

$$LiH + H_2O \rightarrow LiOH + H_2 \; (\Delta H = -312 \text{ kJ/mol LiH})$$
$$LiAlH_4 + 4\,H_2O \rightarrow LiOH + Al(OH)_3 + 4\,H_2 \; (\Delta H = -727 \text{ kJ/mol LiH})$$
$$NaBH_4 + H_2O \rightarrow NaOH + H_3BO_3 + 4H_2$$

TABLE 15-6 Metal Hydride Storage System Characteristics [7, 18]

Alloy material	TiVMnTiF	TiCrMnM	LaNi5	FeMnTi	FeTi
System Volume (L)	568	365	450	433	563
System Weight (kg)	6.0	5.0	3.4	5.0	5.44
Capacity (kg H_2)	848	705.7	478.9	705.7	766.8
Capacity (MJ H_2)	1.492	1.933	1.064	1.629	1.362
Energy Density (MJ/kg)	4.988	5.227	3.683	3.713	5.321

These compounds are lighter than reversible metal hydrides and release more hydrogen because hydrogen is liberated from the water reactant. Three example chemical hydrides and their hydrogen storage capabilities are shown in Table 15-7.

A rule of thumb in the literature is adding an extra 20 percent "for the weight of the hydrogen and water cylinders, mixing device, and control valves." For example, for the LiH to produce 250 grams of hydrogen, 0.98 kg of powered hydride and a 2.2 kg tank of water would be needed. An additional 200 g of control and mixing systems would be needed. Chemical hydrides can be safely contained in small subunits. The hydrides have to be stored under nitrogen or argon, and protected from water [32].

Problems with the chemical hydrides include heat generation and cost. The less exothermic of the two reactions ($LiAlH_4$) produces 182 kJ of heat per mole of H_2 released. At maximum fuel cell power, several kilowatts of heat would be generated. This amount exceeds the heat produced by the fuel cell itself!

The costs of metal hydrides are currently high because they are not made in large quantities, and have to be greater than 95 percent purity for experimental purposes. The waste solution left after the reaction would need to be reprocessed and the cost would be expected to be lower than that for manufacturing a fresh batch of chemical hydride powder.

Hydrolysis reactions. Hydrolysis reactions are chemical hydride reactions with water to produce hydrogen. One of the most commonly studied reactions is sodium borohydride and water, as follows:

$$NaBH_4 + 2H_2O = NaBO_2 + 4H_2$$

A common method for activating the reaction is having an inert slurry that protects the hydride from contact with moisture and make the hydride pumpable. When fuel is needed, the slurry is mixed with water, and the resulting reaction produces high purity hydrogen. These reactions can be carefully controlled in an aqueous medium via pH and by

TABLE 15-7 Chemical Hydride Comparison [1,17,32]

	LiH	LiAlH	NaBH
Hydrogen-to-hydride ratio (wt%)	25.2%	21.1%	21.3%
Hydrogen-to-hydride plus stoichiometric water ratio (wt%)	7.7%	7.3%	7.3%
Hydrogen-to-system ratio (wt%), assuming 20% additional weight	6.4%	6.1%	6.1%
Mass of hydride powder needed for 250 g hydrogen output	980 g	1177 g	1173 g
Cost for mass of hydride given above	$268	$503	$178

using a catalyst. Although the material hydrogen capacity is high, and the release kinetics are fast, this reaction requires water and fuel to be carried onboard separately, and the sodium borohydride to be regenerated off-board. Many aspects of this hydrogen generation method are being investigated, such as cost, life-cycle impact, and regeneration energy requirements [32].

Another hydrolysis reaction that is being researched is the reaction of MgH_2 with water to form $Mg(OH)_2$ and H_2. Particles of MgH_2 are contained in a non-aqueous slurry to inhibit reaction when hydrogen is not required. Material capacities for $Mg(OH)_2$ can be as high as 11 wt.%, but this chemical hydride system also faces similar challenges to the sodium borohydride system.

Hydrogenation/dehydrogenation reactions. Hydrogenation and dehydrogenation reactions have been studied for many years, and there are several systems that have excellent potential for hydrogen storage. One of these reactions is the decalin-to-naphthalene reaction, which can release 7.3 wt.% hydrogen at 210°C via the reaction:

$$C_{10}H_{18} = C_{10}H_8 + 5H_2$$

The kinetics of hydrogen evolution is enhanced by a platinum-based or noble metal supported catalyst. Research is currently directed on lowering dehydrogenation temperatures. An advantage of the dehydrogenation systems are that they do not require water. The reactions are endothermic, which means there is no waste heat to be removed while producing hydrogen onboard [1,32].

New chemical approaches. New approaches being researched include reacting lightweight metal hydrides such as LiH, NaH, and MgH_2 with methanol and ethanol (alcoholysis). The hydrogen production from these reactions is controlled, and can be produced at room temperature. Like hydrolysis reactions, alcoholysis reaction products must be recycled off-board the vehicle. Alcohol must be carried onboard the vehicle, which impacts system-level weight, volume, and complexity. Figure 15-6 shows an example of a chemical hydride system.

Another new chemical approach may be hydrogen generation from ammonia-borane materials by the following reaction:

$$NH_3BH_3 = NH_2BH_2 + H_2 = NHBH + H_2$$

This reaction occurs at less than 120°C and releases 6.1 wt.% hydrogen. Hydrogen release kinetics and selectivity are improved by incorporating ammonia-borane nanosized particles in a mesoporous scaffold.

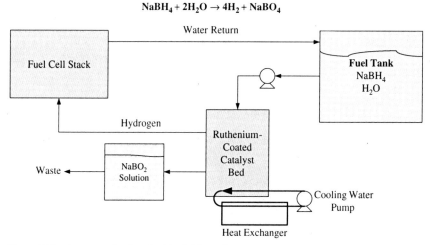

Figure 15-6 An example chemical hydride system.

Table 15-8 shows a comparison of a few types of chemical hydride storage materials.

15.2.5 Ammonia

Ammonia reforming is different than the standard fossil-fuel reforming because it offers clean combustion from a chemical feedstock that is commercially available as a fertilizer. Ammonia is manufactured from natural gas and nitrogen through steam reformation. The reaction takes place at high temperatures and has to be compressed to a high pressure (about 100 bar) to react with nitrogen in the Haber process. The process efficiencies and costs are very similar to producing methanol.

Ammonia contains 17.6 percent hydrogen atoms by weight. This is as good as the partial oxidation reformation of methanol, with the advantage that the only products of ammonia reforming are hydrogen and

TABLE 15-8 Chemical Hydride Storage Materials (Adapted from [1])

Name	Formula	Percent hydrogen	Density (kg/L)	Volume (L) to store 1 kg H_2
Lithium hydride	LiH	12.68	0.82	6.5
Beryllium hydride	BeH_2	18.28	0.67	8.2
Sodium hydride	NaH	4.3	0.92	25.9
Aluminum hydride	AlH_3	10.8	1.3	7.1
Potassium hydride	KH	2.51	1.47	27.1
Calcium hydride	CaH_2	5.0	1.9	1.1
Lithium borohydride	$LiBH_4$	18.51	0.67	8.0
Sodium borohydride	$NaBH_4$	10.58	1.0	9.5
Aluminum borohydride	$Al(BH_4)_3$	16.91	0.545	11
Palladium hydride	Pd_2H	0.471	10.78	20

nitrogen gas. Ammonia is easily liquefied under pressure with a liquid density of 601 g/L at 300 K (equivalent to an H_2 volumetric density of 55 g/L). The liquefaction requires a pressure of only 10 bar at 300 K [1,17].

The recovery of hydrogen from ammonia involves a simple dissociation reaction:

$$NH_3 \rightarrow 1/2\ N_2 + 3/2\ H_2\ (\Delta H = +\ 46.4\ kJ/mol\ NH_3)$$

The reaction takes place at over 600°C, therefore, an external heat source may be required depending upon the fuel cell type used. For example, the exhaust from a traditional PEM exits at only 80°C [1,17,19]. There are several ways to obtain the required heat for the reformer and the reaction. One method is burning the hydrogen in the reformer's output stream to provide the necessary temperature for the reformer and to provide the heat needed for the reaction. Another method is designing the anode of the fuel cell so that the exhaust stream has enough heating value from the unconsumed hydrogen to supply the required heat if burned. Temperatures of 900°C are needed if the output of the converter needs to have ppm levels of ammonia in the output hydrogen.

One of the major problems with undissociated ammonia is with the product gas. Although the concentration is less than 50 ppm, this is still enough to damage fuel cells with acid and polymer electrolytes, so an acid scrubber is needed to remove the final traces of ammonia gas from the converter. Another issue is that ammonia is toxic; spills and evaporative emissions could damage the lungs. An example ammonia/fuel cell system is shown in Figure 15-7.

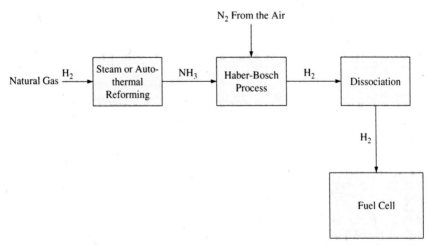

Figure 15-7 Hydrogen supply through ammonia to a fuel cell.

15.2.6 Natural gas

Natural gas is a mixture of hydrocarbons in which methane is the component of greatest concentration, with small amounts of ethane, propane, and other hydrocarbon-based fuels. Natural gas can be stored in a compressed gaseous state (CNG) or in a liquefied state (LNG). In the liquefied state, it is super-cooled to a temperature of minus 260°F and stored in insulated, pressurized tanks [1,46]. There are also trace elements of nitrogen, carbon dioxide, helium, and sulfur. Natural gas is obtained from extracted wells, or with crude oil. It can also be produced as a product of landfill operations. Natural gas has low CO emissions compared with gasoline because it contains less carbon. The type of processing that the natural gas is put through depends upon the geographic region. In certain parts of the world, the natural gas is blended with ethane, propane, and butane to meet the needs of the location. Since natural gas is odorless, natural gas companies also add odor to the gas for safety reasons. Common mixtures are different proportions of thiophenes and mercaptans.

15.2.7 Propane

Propane is a relatively simple molecule (C_3H_8) compared to gasoline, and it undergoes more complete combustion when used as a fuel. It is a byproduct of natural gas processing and petroleum refining. Propane is a "cleaner" fuel than gasoline, and has about one-third less reactive hydrocarbons, 20 percent less NOx, and 60 percent less CO. Propane is used in many end-use sectors as a heating fuel, engine fuel, cooking fuel, and chemical feedstock. Under standard conditions, propane is a gas, but with moderate pressure (100 to 300 psi), it is easy to store and transport. Propane alone has been used as a road fuel for decades, and was widespread in fleet automotive use prior to the enactment of the Energy Policy Act of 1992 (EPACT). Propane has about 73 percent of the energy density of gasoline, by volume. It requires a slightly larger tank, or more frequent fill-ups in order for an automobile to obtain the same mileage [1,46].

15.2.8 Gasoline and other petroleum-based fuels

Petroleum is a mixture of hydrocarbon-based chemical compounds that provide high-value liquid feeds, solvents, and lubricants. Most of the hydrocarbons contain 4 to 12 carbon atoms and have a boiling point between 80 and 437°F [1]. Various types of fossil fuels can be derived from petroleum, tar sands, oil shales, gas hydrates, coal and natural gas. Gasoline is produced from the refining of crude petroleum. Petroleum-based fuels account for at least 50 percent of the world's total energy supply, and include paraffins, nathas, gasoline, diesel fuel, and light and heavy oils. Gasoline is doped with chemical compounds to improve engine lubrication, reduce corrosion, and improve combustion characteristics. For fuel cell systems, different types of reformers are used depending upon the fuel type selected. The trace compounds found in the fuel are a very important consideration when contemplating the type of fuel to select. The common trace compounds found in fossil fuels are sulfur, nitrogen, or oxygen, CO, CO_2, NOx, SOx, VOCs, OH^-, and organic–metallic compounds [1,46]. Some of these emissions are known as probable human carcinogens, including benzene, formaldehyde, acetaldehyde, and 1,3-butadiene. Gasoline can also impact the environment if spilled since it spreads on water surfaces and quickly penetrates porous soils and groundwater. Gasoline is extremely flammable, is highly volatile, and has a corresponding risk of fire and explosion. It also contains a relatively high proportion of hydrocarbons, which can affect the human nervous system. Gasoline evaporates quickly and leaves little residue when spilled. The complex chemical nature of gasoline requires a complex production process, but also allows for significant flexibility in adjusting fuel specifications to meet different performance and emission standards. About 95 percent of the gasoline consumed in the U.S. is used for fuel for automobiles and motorcycles. Currently arguments abound regarding the use of petroleum-based fuels for fuel cells due to the existing infrastructure that already exists. Many studies also contradict this argument, and condone using gasoline in fuel cell vehicles [1,46].

15.2.9 Bio-fuels

Biomass describes the natural organic material associated with living organisms from algae to trees, and animal manure to municipal waste. This is an important source of renewable fuel and can be obtained by combustion or conversion to biogas, ethanol, syngas, or liquid hydrocarbons. The biogas composition varies widely depending upon the source of the biofuel, and how it is produced. Biogases are typically combinations of methane, carbon dioxide, and nitrogen combined with many other organic materials. Biogases are especially attractive for use in MCFCs, SOFCs, and PAFCs because of the low heating values. Methanol and ethanol can also be produced from biofuels and are attractive for some fuel cell systems [1].

15.3 Fuel Processing

Each type of fuel cell stack has different tolerances towards the types of fuels that can be utilized. The lower the operating temperature of the stack, the stricter the requirements are on a purified fuel. The complexity of the fuel processing system is dependent on the type of fuel cell and fuel used. The fuel processing system usually consists of a series of catalytic chemical reactors that convert the fuel into hydrogen. Fuel fed to a PAFC should be hydrogen-rich and contain less than 0.5 percent carbon monoxide. Fuels used for a PEMFC should be essentially carbon monoxide free. Other fuel cell types such as MCFCs and SOFCs operate at temperatures high enough to allow internal reforming. Since PEMFC and PAFCs have stricter fuel processing requirements, these systems will be looked at in more detail. Most of these subsystems consist of three processes: fuel reforming, water shift reaction, and carbon monoxide cleanup [1,17].

Liquid hydrocarbon fuels contain more energy per unit volume than hydrogen and are cheaply available. Methanol produces less carbon dioxide than gasoline, and reduces long-term oil dependence. A comparison of gravimetric and volumetric energy densities of hydrocarbon-based fuels are shown in Table 15-9.

Another important issue is that the fuel cell must be optimized to accept an anode fuel stream that may have as little as 40 percent hydrogen for partial oxidation, or 75 percent hydrogen for steam reforming. The lower amount of hydrogen means that the fuel cell efficiency is lowered. Also, water must be either carried to supply the water-gas shift reaction (and any steam reformer), or recirculated from the exhaust. The total water needed is on the order of 3 grams per gram of H_2 for partial oxidation, and 4.5 grams per gram of H_2 for steam reforming [1,46].

CO poisoning is an important issue for polymer electrolyte membrane fuel cells. CO poisons the platinum on the electrode, reducing voltage at a given current density. For the same power output, a fuel cell running off reformed hydrogen must be sized larger. It is difficult to completely eliminate CO from the reformer exhaust, and fuel cells can only tolerate, at most, 50 ppm CO before efficiency drops dramatically, so a final clean-up step is required even after the water-gas shift. A preferential oxidizer (PROX) is needed to perform CO removal [1,17].

The amount of hydrogen that can be produced for reforming various hydrocarbons is listed in Table 15-10 for both steam reforming and partial oxidation. Both processes include a water gas shift reaction to create more hydrogen from shifting carbon monoxide.

Excellent overall weight fractions and hydrogen densities can be achieved in the fuel, but this does not include the additional weight of reformer equipment required, or the extra water needed.

TABLE 15-9 Fuel Gravimetric and Volumetric Energy Densities, Lower Heating Value Basis [17]

Fuel	By mass	By volume
Hydrogen Gas, atmospheric pressure	120 MJ/kg	11 kJ/L @STP
Compressed Hydrogen Gas, 3600 psi	120 MJ/kg	2700 kJ/L @3600 psi
Gasoline	44 MJ/kg	31800 kJ/L in liquid
Methanol	20 MJ/kg	15900 kJ/L liquid
Natural Gas (pure methane)	50 MJ/kg	36 kJ/L @STP
Compressed Methane, 3600 psi	50 MJ/kg	8700 kJ/L @ 3600 psi

TABLE 15-10 Hydrogen Produced from Hydrocarbon Reformation [17]

Fuel	Formula	Steam reforming			Partial oxidation		
		Wt% H_2	g H_2 per L_{fuel}	Moles H_2O per mol fuel	Wt% H_2	g H_2 per L_{fuel}	Moles H_2O per mol fuel
Methanol	CH_3OH	19%	150	1.0	13%	100	0.0
Ethanol	C_2H_5OH	26%	209	3.0	22%	168	2.0
Methane (LNG)	CH_4	50%	205	2.0	38%	151	1.0
Gasoline	$C_8H_{15.4}$	43%	301	16.3	28%	200	8.1
Diesel Fuel	$C_{14}H_{25.5}$	42%	357	28.3	28%	231	14.2

To simply compare the effectiveness of reforming processes, the hydrogen percent and the steam-to-carbon ratio (S/C) can be used. The H_2 percent is the molar percentage of H_2 in the reformate stream at the outlet of the fuel reformer:

$$y_{H_2} = \frac{n_{H_2}}{n_{tot}} \quad (15\text{-}3)$$

where n_{H_2} is the number of moles produced by the fuel reformer, and n_{tot} is the total number of moles of all gases at the outlet. The steam-to-carbon ratio is the number of moles of molecular water (n_{H2O}) to the moles of atomic carbon (n_c) in a fuel:

$$\frac{S}{C} = \frac{n_{H_2O}}{n_c} \quad (15\text{-}4)$$

The efficiency of the fuel reformer can be calculated in terms of the higher heating value (HHV):

$$\eta_{FR} = \frac{\Delta H_{HHV,H_2}}{\Delta H_{HHV,Fuel}} \quad (15\text{-}5)$$

Each reforming process produces varying amounts of hydrogen, requires different steam-to-carbon ratios, and has different advantages and disadvantages.

15.3.1 Desulfurization

Most fossil fuel–based liquids have sulfur compounds that need to be removed before any further fuel processing can be performed. Levels as low as 1 ppb are enough to permanently poison a PEM fuel cell anode catalyst. Gasoline typically has about 300 to 400 ppm of sulfur compounds. To ensure adequate lifetimes of fuel processors, the desulphurization step is very important. Over the past few years, limits have been put on the amount of sulfur to help reduce emissions from vehicles. The design of any kind of desulphurization system for a fuel cell stack must be carefully considered. It is common practice to use a desulphurization (HDS) reactor where organic sulfur compounds are converted into hydrosulfide through nickel-molybdenum oxide or cobalt-molybdenum oxide catalysts through a reaction of the type [1,17,45]:

$$(C_2H_5)_2S + 2H_2 \rightarrow 2C_2H_6 + H_2S$$

The rate of hydrogenolysis increases with increasing temperatures between 300 to 400°C [1]. Sulfur compounds such as thiophene (C_4H_4S) and tetrahydrothiophene (THT) ($C_4H_8O_2S$) have a slower reaction rate and are absorbed onto a bed of zinc oxide forming zinc sulfide:

$$H_2S + ZnO \rightarrow ZnS + H_2O$$

HDS as a means of removing sulfur is suited to PEM or PAFC systems. This system cannot be applied easily to SOFC or MCFC systems because there is not enough excess hydrogen for HDS.

15.3.2 Steam reforming

Steam reforming is an endothermic process that combines the fuel with steam to produce products. The steam-reforming equivalent of the reaction for methane and a generic hydrocarbon are the following equations:

$$CH_4 + H_2O \rightarrow CO + 3H_2 \; (\Delta H = +206.2 \text{ kJ/mol})$$
$$C_nH_m + nH_2O \rightarrow nCO + (m/2 + n)\,H_2$$

The next step is the water-gas shift reaction. Most of the remaining carbon monoxide reacts with water to produce additional hydrogen. A typical conversion is from 7.1 percent CO in a steam reformer's output (or 46.1 percent from a partial oxidation reactor), to 0.5 percent coming out of the water-gas shift reactor [17]:

$$CO + H_2O \rightarrow CO_2 + H_2 \; (\Delta H = -41.2 \text{ kJ/mol})$$

These reactions are usually carried out using a nickel support catalyst at temperatures over 500°C [1,17]. The product gas is a mixture of carbon monoxide, carbon dioxide, hydrogen, methane, and steam. The product composition is determined by the temperature of the reactor, the operating pressure, the composition of the feed gas, and the amount of steam fed into the reactor. Steam reforming systems are efficient because waste heat from the later processes can be recycled as input into the endothermic steam reforming process. Steam reformers also produce large amounts of hydrogen because some comes from the water. In fuel cell systems that require a low amount of CO, further processing of the fuel will be required. When natural gas is reformed, the reaction eventually becomes exothermic as the temperature is lowered. The reverse of the reaction becomes favored, and the formation of methane begins to dominate [45]. An issue with steam reforming is that the fuel cell has to be optimized to accept an anode fuel stream that may have as little as 75 percent hydrogen. The lower amount of hydrogen means that the fuel cell efficiency is lowered. Also, water must be either carried to supply the water-gas shift reaction and steam reformer, or recirculated from the exhaust. The total water needed is 4.5 grams per gram of H_2 for steam reforming. Figure 15-8 shows an example of a methanol steam reforming/PEMFC system.

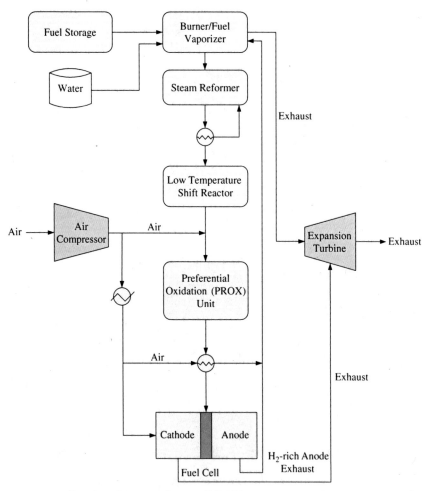

Figure 15-8 A methanol steam reforming/PEMFC system.

Example 15-1 In a steam reforming process, the reformer is consuming methane (CH_4) which will also undergo a water-gas-shift reaction. What is the hydrogen yield and steam-to-carbon ratio for the combined reactions?

The combined reaction for the steam-reforming reaction of CH_4 and the water gas shift reaction is:

$$CH_4 + 2H_2O \text{ (g)} + CO_2 + 4H_2$$

This reaction has a hydrogen percentage of:

$$y_{H_2} = \frac{4 mol H_2}{4 mol H_2 + 1 mol CO_2} = 0.80$$

The steam-to-carbon ratio is:

$$\frac{S}{C} = \frac{n_{H_2O}}{n_c} = 2$$

15.3.3 Carbon formation

Carbon formation is a potential risk in fuel gas from reforming systems. In the absence of air or steam, natural gas will decompose when heated above 650°C, yielding reactions of the type [1,17]:

$$CH_4 \rightarrow C + 2H_2 \;(\Delta H = +75.0 \text{ kJ/mol})$$

Higher molecular weight hydrocarbons tend to decompose more rapidly than methane; therefore, the risk of carbon formation is greater. The Boudouard reaction from a disproportionate amount of carbon monoxide is:

$$2CO \rightarrow C + CO_2$$

The risk of extra carbon from the last two reactions can be reduced by adding steam to the fuel stream. Steam can also lead to the following reaction:

$$C + H_2O \rightarrow CO + H_2$$

The minimum amount of steam that needs to be added to the stream can be calculated. Carbon formation on steam reforming catalysts has been the subject of tremendous study. The carbon that poisons the catalyst attaches to the nickel crystallites in the catalyst. It can take only seconds to poison the catalyst and cause the reactor to be plugged.

15.3.4 Internal reforming

Many internal reforming concepts have been created and applied for use with molten carbonate or solid oxide fuel cell stacks. The heat required to reform low molecular weight hydrocarbons can be provided via the heat generated by the stack. There are two main approaches to internal reforming in fuel cells: direct and indirect reforming. The advantages of internal reforming compared with external reforming is [1]:

- The system cost is reduced because an external reformer is not needed.
- Less steam is required.
- Hydrogen is more evenly distributed in the cell.
- The conversion of methane is high.
- System efficiency is higher.

Indirect internal reforming. Indirect internal reforming involves the conversion of methane by reformers in close contact with the stack. For example, in certain designs, there are plate reformers next to each cell. A variation of this type of reforming is placing the reforming catalyst in the gas distribution path of each cell. An illustration of indirect internal reforming is shown in Figure 15-9.

Direct internal reforming. In direct internal reforming, the fuel reforming takes place in the anode compartment of the fuel cell. This can be accomplished by placing the catalyst directly in the fuel channels, or having the reaction take place directly in the anode. This method utilizes the heat from the anode and the steam from the electrochemical reaction in the MCFC or the SOFC. Gases that have been used in direct reforming stacks include natural gas, naptha, kerosenes, and coal gases. Figure 15-10 shows a diagram of direct internal reforming.

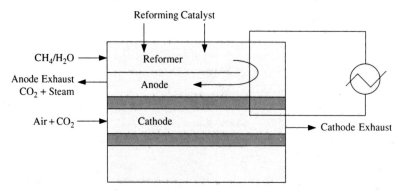

Figure 15-9 Indirect internal reforming.

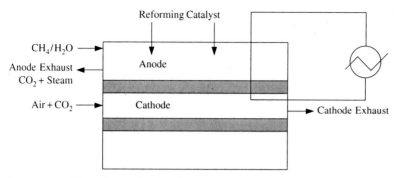

Figure 15-10 Direct internal reforming.

15.3.5 Direct hydrocarbon oxidation

Direct hydrocarbon oxidation converts methane to carbon dioxide and water with high efficiency. The major issue with direct hydrocarbon oxidation is carbon formation. This method has been investigated on novel ceramic anodes for certain types of SOFCs. Development of this process is still in the research and development stages.

15.3.6 Partial oxidation

Partial oxidation is a simpler system than steam reforming because of fewer heat integration and water management issues. An example of this type of system is shown in Figure 15-11. It has lower capital costs for this reason. Partial oxidation reformers also have superior startup times, fuel flexibility, and may have faster response times.

Methane and other hydrocarbons may be converted to hydrogen for fuel cells through the following reaction:

$$CH_4 + 1/2 O_2 \rightarrow CO + 2H_2 \; (\Delta H = -247 \text{ kJ/mol})$$

Partial oxidation is carried out at 1200–1500°C without a catalyst. An advantage over other catalytic processes is that the impurities such as sulphur compounds do not need to be removed. This process can also handle much heavier hydrocarbons that other catalytic processes, therefore, it is ideal for processing diesels, and other heavy petroleum-based fuels. This reaction is successful on a large scale, but is problematic if scaled down. Unlike steam reforming, heat from the fuel cell cannot be utilized in this reaction which decreases the system efficiency. Another important issue is that the fuel cell must be optimized to accept an anode fuel stream that may have as little as 40 percent hydrogen [1,45]. The lower amount of hydrogen means that the fuel cell efficiency is lowered. Also, water must be either carried for the reaction or recirculated from the fuel cell exhaust.

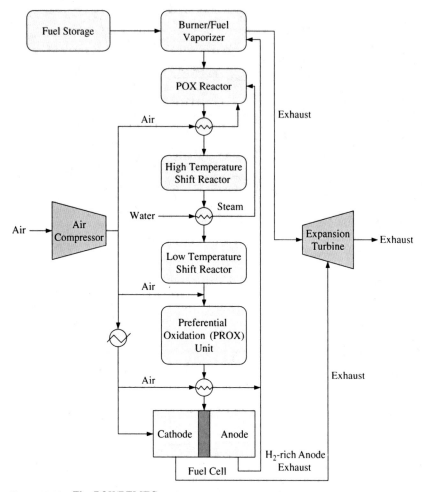

Figure 15-11 The POX/PEMFC system.

The total water needed is on the order of 3 grams per gram of H_2 for the partial oxidation process.

Example 15-2 A partial oxidation reforming reactor consumes methane and air. What is the maximum hydrogen yield, and the reformer efficiency? The HHV of methane is 55.5 MJ/kg (880 MJ/mol) and the HHV of H_2 is 142 MJ/kg (284 MJ/kmol) at standard pressure and temperature.

For air, the total reaction is

$$C_xH_y + 1/2x(O_2 + 3.76\ N_2) \leftrightarrow xCO + 1/2yH_2 + 1.88xN_2$$

For methane, the total reaction is

$$C_xH_y + 1/2x(O_2 + 3.76\ N_2) \leftrightarrow xCO + 1/2yH_2 + 1.88xN_2$$

The hydrogen yield is

$$y_{H_2} = \frac{2\ molH_2}{2\ molH_2 + 1\ molCO + 1.88\ molN_2} = 0.41$$

The fuel reformer efficiency is in terms of HHV:

$$\eta_{FR} = \frac{\Delta H_{HHV,H_2}}{\Delta H_{HHV,Fuel}} = \frac{2\ kmolH_2(284\ MJ/kmolH_2)}{1\ kmolC_xY_y(880\ MJ/kmolC_xY_y)} \approx 65\%$$

15.3.7 Pyrolysis

Pyrolysis is the process of heating hydrocarbons in the absence of air. The hydrocarbon is cracked into hydrogen and solid carbon. The process usually uses light hydrocarbons, and the hydrogen produced is usually very pure. One of the challenges in pyrolysis is the removal of carbon from the reactor. A commonly used method is turning off the reactor, and allowing air into it to form carbon dioxide. Precise control of this process is critical to producing large amounts of hydrogen, and for preventing too much carbon buildup—which can poison the catalyst. This can plug the reactor, which means that no amount of air flow can remove the carbon from the system [1].

15.3.8 Methanol reforming

Methanol reforming is the leading reforming technology candidate for PEM fuel cells, because as a liquid it has a high efficiency and energy density, and because it is easier to reform than gasoline. An example reformer performance is shown in Table 15-11. It contains 12.5 percent hydrogen by weight. Johnson Matthey predicts an additional 20-percent volume for a PROX on top of that required for the partial oxidizer system, and 95-percent efficiency for this second stage, with overall efficiencies of 89 percent [1]. Fuel cells running on reformate cannot be dead-ended,

TABLE 15-11 Example Reformer Performance [1]

Reformer volume	7.2 L
Reformer weight	9.5 kg + cleanup unit weight
Hydrogen output after cleanup	1.36 L/s
Hydrogen output	0.061 mol/s
Fraction of hydrogen in output	43%
Fraction of CO in output	8 +/− 5 ppm
Overall efficiency	76%
Output per mole of methanol	2.3 mol H_2

so hydrogen utilization at the anode decreases to about 85 percent, for a total efficiency of 76 percent. The reduced hydrogen content in the reformate output (when compared to pure hydrogen) reduces the voltage of the fuel cell by approximately 0.128 volts per A/cm^2, or roughly 20 percent at maximum power output [1,17].

At this production rate, 55 moles of methanol would be needed for the required 250 grams of hydrogen gas output. This equals 1.8 kg of methanol, and a volume of 2.2 L—much less than a compressed gas cylinder or metal hydride hydrogen absorption device. The total volume is only 9.4 L, and the total weight of the system is 11.3 kg plus the weight of the PROX [1,17,23].

15.4 Bioproduction of Hydrogen

Several bio-fuel processing systems are currently in the research stage. There are currently three main categories of biological processes that create hydrogen:

- Photosynthesis using unicellular organisms that use hydrogenase or nitrogenase reactions.
- Bacteria that produce hydrogen anaerobically.
- Processes that use a combination of bacteria to break down complex organic molecules into a compound that can be transformed to hydrogen using hydrogen-producing organisms.

Sections 15.41–15.42 describe the most popular photosynthesis and digestion processes that are currently being researched.

15.4.1 Photosynthesis

Photosynthesis consists of the conversion of light energy to biochemical energy by a photochemical reaction and the reduction of atmospheric carbon dioxide to organic compounds such as sugars. Certain plants and algae absorb solar energy, and several types of algae and cyanobacteria produce hydrogen instead of sugars during photosynthesis. These organisms have hydrogenase or nitrogenase enzymes to catalyze hydrogen formation [1].

Several types of green algae produce hydrogen. Some common ones in the literature are Scenedesmus, Chlamydomonas, C. reinhardtii, Anabaena Cylindrica, Synechococcus, Rhodobacter sp, Rubrivivax gelatinosus, Rhodovulum sulfidophilum, R. sphaeroides as well as many others. Scenedesmus produces hydrogen on light exposure after being kept in the dark under anaerobic conditions. This green alga is a "water splitting" organism, and performs biophotolysis using hydrogenase to reduce water to hydrogen. Much research work has been conducted with a mutant of Chlamydomonas to produce a continuous flow reactor using this type of algae. Although high efficiencies of light to hydrogen have been obtained, the hydrogenase reaction is very sensitive to oxygen, which has prohibited large advances in development with this type of bacteria. New methods have recently been developed by the National Renewable Energy Laboratory (NREL) to improve hydrogen by controlling oxygen and sulfur in the environment of C. reinhardtii. The algae culture is allowed to grow under normal conditions, and then is deprived of both oxygen and sulfur, which makes it switch to an alternate metabolism to produce hydrogen. This cycle can be repeated several times to generate hydrogen [1].

A nitrogen-fixing cyanobacterium Anabaena Cylindrica produces hydrogen and oxygen simultaneously in an argon atmosphere for several hours. Hydrogen production occurs under normal atmospheric conditions, but at a much slower rate than nitrogen fixation. The hydrogen production can be increased by starving the algae of nitrogen.

Most of the photosynthetic systems being studied have low efficiencies. However, recent advances in genetic engineering may help create more efficient systems, such as R. Sphaeroides, which has a photoenergy conversion of 7 percent, thus approaching the efficiency of photovoltaic systems at 10 percent [1].

15.4.2 Digestion processes

Hydrogen can also be produced by the microbial digestion of organic matter in the absence of light. Many of these bacteria types produce hydrogen with acid and many other low molecular weight acids. Unfortunately, the reaction rates are usually low, and large amounts of hydrogen are not produced due to the inhibition of microbial hydrogenases, along with the reaction of hydrogen with other species. The hydrogen-generating enzymes hinder hydrogen produced in the build-up of hydrogen. Hydrogen also typically reacts with organic species or carbon dioxide, which produces methane. One of the main challenges in digestion processes is inhibiting methane production to favor hydrogen production. Increasing the temperature in these types of systems is not an effective method of increasing the process efficiency because this usually makes the organic enzymes and bacteria ineffective. Some of the most promising bacteria being researched include ecscherichica coli, aerobacter aerogenes, aerobactor cloacae, pseudomonas sp, and Clostridium butyricum. One of the best types is the Clostridium butyricum, which has yielded 35 mmol (784 mls) of hydrogen per hour from 1 g of microorganism at 37°C [1]. The use of biological processes for hydrogen production is at the point of technical system development, and the photosynthesis-algae-bacteria systems currently seem to be the most promising [1].

15.5 Electrolyzers

Electrolyzers use electricity to break water into hydrogen and oxygen. The reactions are very similar to fuel cells, except the reactions go the opposite way. There are numerous ways to build and configure an electrolyzer, and different electrolytes can be used just as in fuel cells. However, it is not practical to use high-temperature systems since the water would have to be supplied as steam. Commonly used electrolytes are proton exchange membranes and alkaline liquids. Potassium hydroxide was a commonly used electrolyte in the past, but recently PEM membranes are more typical. The basic structure of a PEM-based electrolyzer is shown in Figure 15-12.

At the negative electrode, the protons are removed from the electrolyte, and electrons are provided by the external electrical supply:

$$4H^+ + 4e^- \rightarrow 2H_2$$

At the positive electrode, the water is oxidized, and the oxygen is made via:

$$2H_2O \rightarrow O_2 + 4H^+ + 4e^-$$

PEM electrolyzers are popular because many of the typical problems of PEM fuel cells are not applicable. Water supplied to the cathode can also be easily used to cool the cell, and water management is much simpler since the positive electrode must be flooded with water. Hydrogen produced by this type of electrolyzer is of high purity. The only issue is the presence of water vapor in the system. Water diffuses through the electrolyte as in fuel cells; therefore, electrolyte designers use various

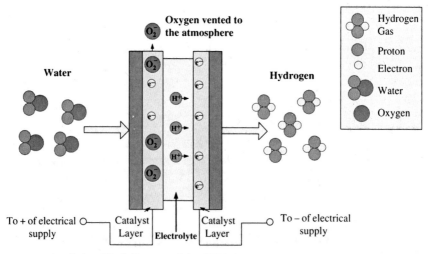

Figure 15-12 A simplified diagram of the PEM electrolyzer.

techniques to avoid this occurrence. A common method is to use thicker electrolytes than those used in fuel cells [1].

Electrolyzers at first may not seem like the ideal method of generating hydrogen since electricity is needed to obtain the hydrogen, but this type of system may be useful, and ideal, in certain configurations. The near-term solution for electrolyzers would be in portable systems. For boats, soldiers, and fuel cell–powered cars, it could be advantageous to plug-in an electrolyzer to generate a sufficient amount of hydrogen is to be used on a trip. Certain studies prove that this method is more effective than gas reformer/fuel cell systems. There are also systems that do not simply use electricity from the grid, but from solar and wind power. These systems can be charged at night and used in the day for various applications. There are many commercial and prototype systems that generate electricity from the sunlight, for generating hydrogen by electrolysis. There has been much success with this type of system, and work needs to be continued to increase the efficiency and the cost of these systems for integration into portable and stationary devices. Currently, the majority of electrolyzers are used to make hydrogen for applications that are not used with fuel cells. Hydrogen has many other uses, and some of the advantages of using electrolyzers are (1) the hydrogen produced is very pure, (2) it is produced in-situ as needed, and does not have to be stored, and (3) it is a much cheaper method than gas supplied in high-pressure cylinders [1].

15.5.1 Electrolyzer efficiency

The electrolyzer efficiency is calculated in the same way as a fuel cell. As stated previously, the fuel cell efficiency is given by the formula [1]:

$$\eta_{fuel\_cell} = \frac{V_{cell}}{1.48} \tag{15-6}$$

And the inverse of this formula is the efficiency of the electrolyzer:

$$\eta_{EL} = \frac{1.48}{V_{el\_cell}} \tag{15-7}$$

The losses in electrolyzers are the same as fuel cells, and typical values for V_{cell} and V_{el_cell} are 1.6–2.0 V depending upon the current density. Typical operating efficiencies of commercial electrolyzer units are about 60 to 70 percent [1].

15.5.2 High pressure in electrolyzers

When designing an electrolyzer, one of the questions is whether to run the electrolyzer at high pressure. The gas in the electrolyzer is released when pressure is built up inside the electrolyzer. The hydrogen can be directly released into the high-pressure cylinder. Another approach is to operate at air pressure, and then the hydrogen gas is pumped to a higher pressure and stored in a cylinder. The energy cost for the pumping option is the work done to compress the gas. The thermodynamic formula for isothermally compressing a gas from P_1 to P_2 is [1]:

$$\Delta W = P_1 V_1 \ln\left(\frac{P_2}{P_1}\right) = nRT \ln\left(\frac{P_2}{P_1}\right) \qquad (15\text{-}8)$$

where n is the number of moles and R is the universal gas constant. For the other approach mentioned, the voltage rise from an increase in hydrogen pressure is [1]:

$$\Delta V = \frac{RT}{2F} \ln\left(\frac{P_2}{P_1}\right) \qquad (15\text{-}9)$$

The increase in voltage can be converted into work as follows:

$$\Delta W = \Delta V \times Q \qquad (15\text{-}10)$$

The charge produced by one mole of hydrogen is 2F. The work done in producing n moles of hydrogen at high pressure is [1]:

$$\Delta W = \Delta V \times 2F \times n \qquad (15\text{-}11)$$

Substituting this equation into the previous equations yields [1]:

$$\Delta W = \frac{RT}{2F} \ln\left(\frac{P_2}{P_1}\right) \times 2F \times n = nRT \ln\left(\frac{P_2}{P_1}\right) \qquad (15\text{-}12)$$

Neither design is better than the other, and the specific design used must be determined by the designer, the system costs, and the materials available.

Chapter Summary

Many types of fuels can be used for different fuel cell types. In addition, several types of reformers and chemical conversion methods can convert all of the fuels mentioned in this chapter to hydrogen. The cleanest fuel type is hydrogen, but there are also many fuel cell designs that utilize other fuels because of the availability or perceived safety of other fuels. Common fuels include hydrogen, methanol, ethanol, ammonia, natural gas, and gasoline. Many forms of hydrogen can also be used, including compressed gas, liquid, metal hydrides, and carbon nanotubes. Hydrogen can also be made through various chemical reactions (chemical hydrides).

Several types of fuel processing systems can also be used to obtain hydrogen. Some common methods of processing hydrogen include steam reforming, internal reforming, partial oxidation, and methanol reforming, as well as many other types. When reformers are used, another consideration is the heat generated, and the effect on the fuel cell stack. Other methods of producing hydrogen include using electricity to electrolyze water, and producing hydrogen through the use of biological methods. The use of fuels other than hydrogen may be beneficial for the commercialization of hydrogen in the near term, but the goal of fuel cell technology is to use pure hydrogen from renewable sources of energy other than fossil fuels.

Problems

1. A partial oxidation reformer obtains its heat from the combustion of methane. Compare the ratio of hydrogen produced per mole of methane consumed for a steam reformer, and a partial oxidation reformer. Assume that the reactants enter and leave the reactor at 950 K.

2. A partial oxidation reformer consumes gasoline and air. What is the maximum H_2 yield, and the steam-to-carbon ratio?

3. A steam reformer uses propane and air. Calculate the steam-to-carbon ratio and the amount of hydrogen produced per mole of propane consumed if the reactants and products enter and leave the reactor at 1000 K.

4. Autothermal reforming combines the steam reforming, partial oxidation, and the water-gas-shift reaction into a single process. The total reaction is

$$C_xH_y + zH_2O(l) + \left(x - \frac{1}{2}z\right)O_2 \leftrightarrow xCO_2 + \left(z + \frac{1}{2}y\right)H_2$$

Using methane and liquid water, estimate the hydrogen yield, reformer efficiency, and the steam-to-carbon ratio.

Bibliography

[1] Larminie, James and Andrew Dicks. *Fuel Cell Systems Explained*, 2nd ed. West Sussex, England: John Wiley & Sons, 2003.
[2] "Gaseous and Liquid Hydrogen Storage." Hydrogen Fuel Cells, Infrastructure Technologies Program. U.S. Department of Energy. http://www.eere.energy.gov/hydrogenandfuelcells/storage/hydrogen_storage.html. Last updated 8/11/06.
[3] Barbir, Frano. *PEM Fuel Cells: Theory and Practice*. 2005. Burlington, MA: Elsevier Academic Press.
[4] "The Hydrogen Economy: A Non-Technical Review." United Nations Environment Program E. 2006.
[5] Berry, G.D. and S.M. Aceves. "Onboard Storage Alternatives for Hydrogen Vehicles." *Energy & Fuels*. Vol. 12, No. 1, 1998, pp. 49–55.
[6] Klos, H. "Technical and Economic Practicability of Carbon Nanostructures Hydrogen Storage Systems." 12th World Hydrogen Energy Conference. Vol. 2, 1998, pp. 893–898.

[7] Pettersson, Joakim and Ove Hjortsberg. "Hydrogen Storage Alternatives–A Technological and Economic Assessment." December 1999. Volvo Teknisk Utveckling AB.
[8] C.A. Bessel, K. Laubernds, N.M. Rodriguez, and R.T.K. Baker. "Graphite Nanofibers as an Electrode for Fuel Cell Applications." *J. Phys. Chem. B.* Vol. 105, No. 6, 2001, pp. 1115–1118.
[9] Chen, P, X. Wu, J. Lin, and K.L. Tan. "High H_2 Uptake by Alkali-Doped Carbon Nanotubes under Ambient Pressure and Moderate Temperatures." *Science.* Vol. 285, No. 5424, 1999, pp. 91–93.
[10] "Alternative Fuels for Fleet Vehicles." Pacific Northwest Pollution Prevention Resource Center, 2004. www.pprc.org. Last accessed November 20, 2006.
[11] Wittcoff, Harold A., Bryan G. Reuben, and Jeffrey S. Plotkin. *Industrial Organic Chemicals*, 2nd ed., New York: John Wiley & Sons, Inc, 2004.
[12] "Petroleum and Coal. Organic Chemistry: Structure and Nomenclature of Hydrocarbons." http://chemed.chem.purdue.edu/genchem/topicreview/bp/1organic/coal.html. Last accessed September 2, 2006.
[13] "Ethanol Fuel." http://en.wikipedia.org/wiki/Ethanol_fuel Last Updated September 2, 2006. Last accessed October 2, 2006.
[14] U.S. Department of Energy Hydrogen Analysis Resource Center. "Comparative Properties of Hydrogen and Fuels." http://hydrogen.pnl.gov.
[15] *Perry's Chemical Engineers' Handbook*, 7th ed., 1997, McGraw-Hill.
[16] Halliday, David, Robert Resnik, and Jearl Walker. *Fundamentals of Physics*, 5th ed. 1997. John Wiley & Sons, Inc, New York. pp. 470.
[17] Lin, Bruce. "Conceptual Design and Modeling of a Fuel Cell Scooter for Urban Asia." Princeton University, Masters Thesis. 2000. http://www.spineglass.net/scooters/.
[18] T-Raissi, A., A. Banerjee, and K. Sheinkopf. "Metal hydride Storage Requirements for Transportation Applications." 31st Intersociety Energy Conversion Engineering Conference. Vol. 4, 1996, pp. 2280–2285.
[19] Saika, T., Mitsuhiro Nakamura, Tetsuo Nohara, and Shinji Ishimatsu. "Study of Hydrogen Supply System with Ammonia Fuel." *JSME International Journal Series B*, Vol. 49, No. 1, 2006, pp.78–83 .
[20] Sangtongkitcharoen, W., S. Assabumrungrat, V. Pavarajarn, N. Laosiripojana, and P. Praserthdam. "Comparison of Carbon Formation Boundary in Different Modes of Solid Oxide Fuel Cells Fueled by Methane." *Journal of Power Sources.* Vol. 142, 2005, pp. 75–80.
[21] Brown, Lee. "A Comparative Study of Fuels for Onboard Hydrogen Production for Fuel Cell Powered Automobiles." *International Journal of Hydrogen Energy.* Vol. 26, 2001, pp. 381–397.
[22] Sarker, Arindam and Rangan Banerjee. "Net Energy Analysis of Hydrogen Storage Options." *International Journal of Hydrogen Energy.* Vol. 30, 2005, pp. 867–877.
[23] Choi, Yongtaek and Harvey G. Stenger. "Kinetics, Simulation, and Optimization of Methanol Steam Reformers for Fuel Cell Applications." *Journal of Power Sources.* Vol. 142, 2005, pp. 81–91.
[24] Raissi, Ali-T and David Block. "Hydrogen: Automotive Fuel of the Future." *IEEE Power and Energy Magazine.* Nov/Dec, 2004.
[25] Ghosh, P.C., B. Emonts, H. Janben, J. Mergel, and D. Stolten. "Ten years of Operational Experience with a Hydrogen-Based Renewable Energy Supply System." *Solar Energy.* Vol. 75, 2003, pp. 469–478.
[26] Kamarudin, S.K., W.R.W. Daud, A. Md. Som, A.W. Mohammad, S. Takriff, and M.S. Masdar. "The conceptual design of a PEMFC system via simulation." *Chemical Engineering Journal.* Vol. 103, 2004, pp. 99–113.
[27] Ogden, Joan M., Margaret M. Steinbugler, and Thomas G. Kreutz. "A Comparison of Hydrogen, Methanol, and Gasoline as Fuels for Fuel Cell Vehicles: Implications and Infrastructure Development." *Journal of Power Sources.* Vol. 79, 1999, pp. 143–168.
[28] Wu, S. H., D.B. Kotak, and M.S. Fleetwood. "An Integrated System Framework for Fuel Cell–Based Distributed Green Energy Applications." *Renewable Energy.* Vol. 30, 2005, pp. 1525–1540.
[29] Agbossou, Kodjo, Kolhe Mohanlal, Jean Hamelin, and Tapan K. Bose. "Performance of a Stand-Alone Renewable Energy System Based Upon Energy Storage as Hydrogen." *IEEE Transactions on Energy Conversion.* Vol. 19, No. 3, September 2004.

[30] Barbir, Frano. "PEM Electrolysis for Production of Hydrogen from Renewable Energy Sources." *Solar Energy.* Vol. 78, 2005, pp. 661–669.
[31] Balat, Mustafa and Nuray Ozdemir. "New and Renewable Hydrogen Production Processes." *Energy Sources,* Vol. 27, 2005, 1285–1298.
[32] Aiello, R., M.A. Matthews, D. L. Reger, and J.E. Collins. "Production of Hydrogen Gas from Novel Chemical Hydrides." *International Journal of Hydrogen Energy.* Vol. 23, No. 12, 1998, pp. 1103–1108.
[33] Burke, Andrew and Monterey Gardiner. "Hydrogen Storage Options: Technologies and Comparisons for Light Duty Vehicle Applications." Institute of Transportation Studies, University of California, Davis. UCD-ITS-RR-05-01. January 2005.
[34] Eberle, U., G. Arnold, and R. von Helmolt. "Hydrogen Storage in Metal-Hydride Systems and Their Derivatives." *Journal of Power Sources.* Vol. 154, 2006, pp. 456–460.
[35] Zuttel, Andreas. "Materials for Hydrogen Storage." *Materials Today.* September 2003.
[36] Ananthachar, Vinay, and John J. Duffy. "Efficiencies of Hydrogen Storage Systems Onboard Fuel Cell Vehicles." *Solar Energy.* Vol. 78, 2005, pp. 687–694.
[37] Atkinson, Kaylene, Siegar Roth, Michael Hirscher, and Werner Grunwald. "Carbon Nanostructures: An Efficient Hydrogen Storage Medium for Fuel Cells?" *Fuel Cells Bulletin.* Vol. 4, Issue 38, November 2001, pp. 9–12.
[38] Thomas, C.E., Brian D. James, Frank D. Lomax, Jr., and Ira F. Kuhn. "Fuel Options for the Fuel Cell Vehicle: Hydrogen, Methanol, or Gasoline?" *International Journal of Hydrogen Energy.* Vol. 25, 2000, pp. 551–567.
[39] Maeland, Arnulf J. "Approaches to Increasing Gravimetric Hydrogen Storage Capacities of Solid Hydrogen Storage Materials." *International Journal of Hydrogen Energy.* Vol. 28, 2003, pp. 821–824
[40] Zhou, Li, Yaping Zhou, and Yan Sun. "Enhanced Storage of Hydrogen at the Temperature of Liquid Nitorgen." *International Journal of Hydrogen Energy.* Vol. 29, 2004, pp. 319–322.
[41] Lee, Jung Woo, Hyun Seok Kim, Jai Young Lee, and Jeung Ku Kang. "Hydrogen Storage and Desorption Properties of Ni-Dispersed Carbon Nanotubes." *Applied Physics Letters.* Vol. 88, No. 143126, April, 2006.
[42] Wolf, Joachim. "Liquid Hydrogen Technology for Vehicles." MRS Bulletin. September 2002.
[43] Wee, Jung-Ho. "Performance of a Unit Cell Equipped with a Modified Catalytic Reformer in Direct Internal Reforming Molten Carbonate Fuel Cell." *Journal of Power Sources.* Vol. 156, 2006, pp. 288–293.
[44] Bove, Roberto and Piero Lumghi. "Experimental Comparison of MCFC Performance Using Three Different Biogas Types and Methane." *Journal of Power Sources.* Vol. 145, 2005, pp. 588–593.
[45] Ethanol. Wikipedia. http://en.wikipedia.org/wiki/Ethanol Last Updated November 26, 2006. Last Accessed December 10, 2006.
[46] Fuel Cell Handbook, 5th ed. EG&G Services. Parsons Incorporated. Science Applications International Corporation. U.S. Department of Energy, October 2000.
[47] Browning, D., P. Jones, and K. Packer. "An investigation of hydrogen storage methods for fuel cell operation with man-portable equipment" J. Power Sources 65 (1–2) March–April 1997, pp. 187–195.

Chapter

16

Fuel Cell Operating Conditions

Fuel cell operating conditions depend upon the fuel cell and stack design. The optimal conditions will vary from cell to cell (unless identical fuel cells are mass produced); therefore, it is advantageous to calculate the optimal parameters required to operate the cell. Table 16-1 shows typical operating conditions for fuel cells. The operating conditions that will be discussed in this chapter are:

- Operating Pressure
- Operating Temperature
- Flow Rates of Reactants
- Humidity of Reactants
- Fuel Cell Mass Balance

Using the correct operating conditions for each of these parameters is critical in obtaining good performance from a fuel cell.

16.1 Operating Pressure

A fuel cell can be operated at ambient pressure or in a pressurized state. Fuel cell performance will improve with increased pressures, but the need for gas compression and storage may not make the fuel cell system more efficient. Pressurization of the fuels also changes the water management in the cell; therefore, the fuel cell operating conditions must be analyzed from a system perspective. Fuel cell parameters such as inlet fuel compositions, the diffusivity of the porous electrode or gas diffusion layer (GDL), and the flow-field plate designs may change with reactant

TABLE 16-1 Typical Operating Conditions for Fuel Cells

Fuel cell system	Proton exchange membrane fuel cell (PEMFC)	Direct methanol fuel cell (DMFC)	Solid oxide fuel cell (SOFC)	Alkaline fuel cell (AFC)	Phosphoric acid fuel cell (PAFC)	Liquid molten carbonate fuel cell (MCFC)
Fuel	H_2	$CH_3OH + H_2O$	CO, H_2	H_2	H_2	H_2/CO/Reformate
Oxidizer	O_2, air	O_2, air	O_2, air	O_2, air	O_2, air	CO_2, O_2, air
Operating temperature	Room temperature to 100°C	Room temperature to 100°C	600 to 1000°C	Room temperature to 250°C	150 to 220°C	620 to 660°C
Operating pressure (atm)	1 to 3	1 to 3	1 atm	1 to 4	3 to 10	1 to 10
Major contaminants	CO <100 ppm, sulfur, dust	CO <100 sulfur, dust	<100 ppm sulfur	CO_2	CO <100 ppm, sulfur, dust, NH_3	H_2S, HCl, As, H_2Se, NH_3, AsH_3, dust

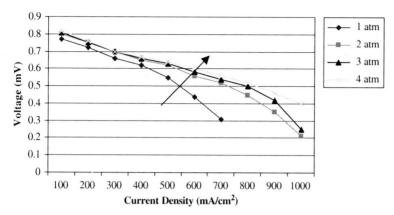

Figure 16-1 Example of a PEMFC at several pressures with anode: Pt/C 1 mg/cm^2, cathode: Pt/C 1 mg/cm^2, electrolyte: Nafion 117, temp: 80°C, fuel: H$_2$/Air 2/2, humidity: 100 percent [25, 35].

gas pressure. Since the saturation pressure does not vary for constant operating temperature, the mole fraction of an oxidant such as oxygen may increase with increasing operating pressure. As shown in Figures 16-1 and 16-2, as the operating pressure was increased, the fuel cell performance improved.

Many fuel cell types are fed reactant gases from the pressurized tank, and controlled by pressure and back-pressure regulators. The inlet pressure is always higher than the outlet pressure due to the pressure drop in the channels. The fan, blower, or compressor must be capable of delivering the required flow rate at the desired pressure.

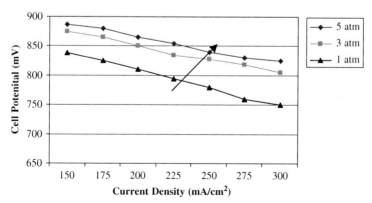

Figure 16-2 Example of an MCFC at several pressures with anode: Ni-4 percent Al, Cr, cathode: NiO (Ni-3% MgCO$_3$), electrolyte: LiCO$_3$/NO$_2$CO$_3$ (60/40), temp: 650°C, fuel: H$_2$/CO$_2$:80/20 percent, oxidant: air/CO$_2$ = 70/30 percent [19].

16.2 Operating Temperature

As in the case of the pressure, a higher operating temperature also creates higher fuel cell efficiency. For each fuel cell design, there is an optimal temperature. The operating temperature must be chosen based upon the fuel cell system. A fuel cell generates heat as a by-product of the electrochemical reaction, and this heat must often be removed from the cell to maintain the desired temperature. The designed operating temperature of the fuel cell affects many factors. In the case of PEMFCs and DMFCs, higher operating temperature means that more of the product water is vaporized so more waste heat goes into the latent heat of vaporization, and less liquid water is left to be pushed out of the fuel cell. In high temperature fuel cells, it means greater difficulty in finding lowcost, highly manufacturable materials for fuel cell stacks. Higher temperatures also mean faster kinetics and a voltage gain that exceeds the voltage loss from the negative thermodynamic relationship between the open-circuit voltage and temperature. Lower temperatures mean shorter warm-up times for the system, and lower thermomechanical stresses. Corrosion and other time- and temperature-dependent processes are lessened, and much less water is required for the saturation of input gases. The upper limit of operation for PEMFCs is about 90°C because water evaporates from the membrane and performance drops quickly.

Figure 16-3 illustrates the polarization curves of a PEMFC at several operating temperatures between 65 and 85°C. The increase in temperature is due to the increase in gas diffusivity and membrane conductivity at the higher temperatures. At lower temperatures, water is in its liquid state, and may flood the catalyst and gas diffusion layers, which may decrease gas diffusivity. The gas diffusivity of the fuel cell improves at

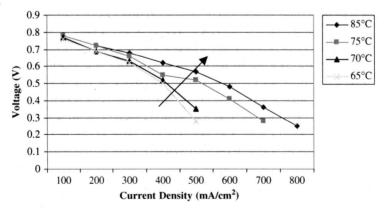

Figure 16-3 Example of a PEMFC at several temperatures with anode: Pt/C 1 mg/cm^2, cathode: Pt/C 1 mg/cm^2, electrolyte: Nafion 117, fuel: H$_2$/Air 2/2, humidity: 100 percent [25, 35].

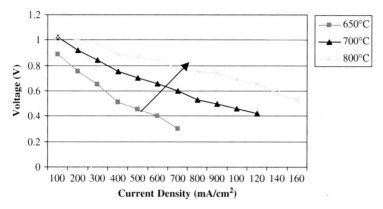

Figure 16-4 Example of an SOFC at several temperatures with anode: YSZ + NiO (0.5 mm thick), cathode: (La0.85Sr0.14)0.98MnO$_3$, electrolyte: YSZ (10 microns), fuel: H$_2$/Air 2/2 [10, 11].

higher fuel cell temperatures, while membrane conductivity may decrease at high temperatures due to the reduction of water content in the membrane. Figure 16-4 illustrates the affect of temperature on an SOFC. Increasing the temperature increases the reaction kinetics in the SOFC, as in all fuel cell types.

For smaller fuel cells, sometimes heaters are required to warm the fuel cell to the required operating temperature as explained in Chapter 9. The cell operating temperature is often not uniform; therefore, it is necessary to monitor several places in the fuel cell with thermocouples to obtain an idea of the actual fuel cell temperature.

16.3 Flow Rates of Reactants

The flow rate of the reactants must be equal to, or greater than, the rate at which those reactants are consumed within the cell. Oxygen and hydrogen are introduced into the fuel cell system at the appropriate flow rate necessary for the current at any given moment. This requires a variable–flow system if the stoichiometry is to be kept constant. Even in an "atmospheric pressure" system, some pressure over atmospheric is needed in order to push the gases through the flowfields in the flow plates, and to force the liquid water out of the same flow fields. The additional pressure required is 0.1 to 2.0 psi (0.7 to 13.8 kPa) above atmospheric.

According to the literature, a minimum flow speed of 0.35 m/s is needed to eliminate product water. Due to the fact that the cathode reaction is much slower than the anode reaction, oxygen is often supplied at a higher stoichiometric flow rate. The ratio of the air flow rate to the minimum flow rate required for stoichiometric oxygen–hydrogen reaction is often 2.0 or higher to maintain a high concentration of oxygen at the cathode, and to push product water generated at the cathode out of the fuel cell. A standard rule-of-thumb used for fuel cells is to have a flow rate that is 2.5 times the stoichiometric flow rate, where the hydrogen consumption is 100 percent [4].

The flow rates of the reactants are also dependent upon the stack size. Low efficiency implies there has to be a greater reactant flow into the fuel cell. The sizes and costs of the reactant management equipment are dependent upon the flow rates [4].

Figure 16-5 shows the polarization curves for a PCFC fuel cell with various oxygen flow rates and a constant hydrogen flow rate of 50 cc/min. As the oxygen flow rate decreased, the cell performance decreased. The polarization curve in the low current density region did not change with the decreasing oxygen flow rate because the O_2 flow rate did not have a large effect on the activation process of the cell in the measured flow rate range.

Figure 16-6 shows the results of a PEMFC at three normal stoichiometry sets: (1) a standard stoichiometry of 1.2/2.0, (2) an anode rich stoichiometry of 2.4/2.0, and (3) a cathode rich stoichiometry of 1.2/3.0. The stoichiometry corresponds to a flow rate that was 1.2 times greater for the hydrogen and 2.0 times greater than required on the cathode for air. The graph indicates that the cathode rich stoichiometry performed the best. If pure oxygen was used, the stoichiometry of hydrogen to oxygen would be closer to equal for this fuel cell.

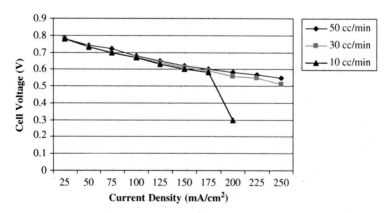

Figure 16-5 Example of a PCFC at several O_2 flow rates with anode: Pt/C + carbon paper (0.2 mm thick), cathode: Pt/C + carbon paper (0.2 mm thick), electrolyte: electrolyte matrix: SiC/PTFE, electrolyte: 105 wt% H_3PO_4, fuel: H_2/O_2, temperature: 190°C [36].

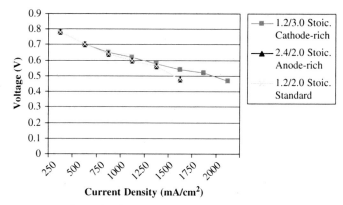

Figure 16-6 Example of a PEMFC as a function of stoichiometry, with anode: Pt/C + carbon paper (0.4 mg/cm²), cathode: Pt/C + carbon paper (0.4 mg/cm²), electrolyte: 25 um thickness, Fuel: H₂/air, temperature: 70°C [37].

The rate at which hydrogen and oxygen are consumed and water is generated is determined by Faraday's law:

$$N_{H_2} = \frac{I}{2F} \quad (16\text{-}1)$$

$$N_{O_2} = \frac{I}{4F} \quad (16\text{-}2)$$

$$N_{H_2O} = \frac{I}{2F} \quad (16\text{-}3)$$

where N is the consumption rate (mol/s), I is the current (A), and F is Faraday's constant (C/mol).

The mass flow rates of the reactants in consumption (g/s) are:

$$m_{H_2} = \frac{I}{2F} M_{H_2} \quad (16\text{-}4)$$

$$m_{O_2} = \frac{I}{4F} M_{O_2} \quad (16\text{-}5)$$

$$m_{H_2O} = \frac{I}{2F} M_{H_2O} \quad (16\text{-}6)$$

where M_{H_2}, M_{O_2}, and M_{H_2O} are the molecular weights of the hydrogen, oxygen, and water, respectively. The reactants may be supplied in excess of consumption. The ratio of the actual flow rate of a reactant at the cell inlet and the consumption rate of the reactant is the stoichiometric ratio(S) [3]:

$$S = \frac{N_{act}}{N_{cons}} = \frac{m_{act}}{m_{cons}} \quad (16\text{-}7)$$

The fuel utilization is the inverse of the stoichiometric ratio:

$$\eta_{fuel\_util} = \frac{1}{S} \quad (16\text{-}8)$$

In fuel cells that use pure hydrogen, the hydrogen must be purged because of the accumulation of fuels or water. The frequency and duration of purges depends upon the purity of hydrogen, the rate of nitrogen permeation, and the water net transport through the membrane. The loss of hydrogen due to purging must be taken into account through fuel utilization [3]:

$$\eta_{fuel\_util} = \frac{N_{cons}}{N_{cons} + N_{loss} + N_{prg}\tau_{prg}f_{prg}} \quad (16\text{-}9)$$

where N_{loss} is the rate of hydrogen loss (mol/s), N_{prg} is the rate of hydrogen purge (mol/s), T_{prg} is the duration of hydrogen purge (s), and F_{prg} is the frequency of purges (1/s).

Hydrogen may be supplied in excess (S > 1) in flow-through mode. As mentioned previously, air is usually supplied in flow-through mode with a stoichiometry of S = 2 or higher. Unused gas can be recirculated by a pump or a compressor. Fuel utilization in various modes of operation are summarized next [3]:

$$\eta_{fuel\_util} = \frac{N_{cons}}{N_{cons} + N_{loss}} \quad (16\text{-}10)$$

Dead-end mode with purging [3]:

$$\eta_{fuel\_util} = \frac{N_{cons}}{N_{cons} + N_{loss} + N_{prg}\tau_{prg}f_{prg}} \quad (16\text{-}11)$$

Flow-through mode [3]:

$$\eta_{fuel\_util} = \frac{N_{cons}}{N_{act}} \quad (16\text{-}12)$$

When designing a fuel cell, the flow rates must be calculated. Pure hydrogen can be supplied stoichiometrically, or with a slightly higher stoichiometry of 1.1 to 1.2. If pure hydrogen is not used, higher stoichiometries are required and need to be calculated. For hydrocarbon-based hydrogen gaseous streams, the stoichiometric ratio can be 1.5 and greater.

For a pure oxygen flow rate, the stoichiometry is between 1.2 and 1.5, but if air is used, the stoichiometry is 2 or higher. Higher flow rates result in better fuel cell performance, but air is supplied by a blower or compressor whose power consumption was proportional to the flow rate. Fuel cell efficiency improves with flow rate because the higher flow helps remove product water from the fuel cell, and high flow rates keep the oxygen concentration high.

16.4 Humidity of Reactants

In PEM fuel cell stacks, the reactant gases need to be humidified before entering the cell for high ionic conductivity of the membrane as mentioned in Chapter 8. The polarization curves for different operating levels of air and hydrogen relative humidity are shown in Figure 16-7. The best performance was obtained when hydrogen was at 100-percent humidity. At high current densities, the transport from the anode by electro-osmotic drag exceeds transport to the anode by back diffusion from the cathode, which leads to membrane dehydration and performance degradation [25, 35]. Air that has a low humidity can exacerbate this effect by reducing the rate of back diffusion from the cathode. Humidifying the anode gases leads to higher levels of fuel cell performance. When air has high humidity levels, the performance only improves slightly. This could be due to the back diffusion of the water at the cathode.

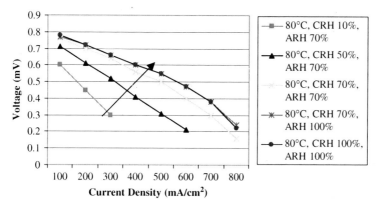

Figure 16-7 Example of a PEMFC as a function of humidity, anode: Pt/C 1 mg/cm^2, cathode: Pt/C 1 mg/cm^2, electrolyte: Nafion 117, fuel: H_2/Air 2/2, humidity: 100 percent [25, 35].

The humidity mass ratio, the ratio between the amount of water vapor present in the gas stream and the amount of dry gas is:

$$x = \frac{G_v}{G_a} \qquad (16\text{-}13)$$

The humidity molar ratio is:

$$X = \frac{N_v}{N_a} \qquad (16\text{-}14)$$

The molar ratio of gases is the same as the ratio of partial pressures:

$$X = \frac{p_v}{p_a} = \frac{p_v}{P - p_v} \qquad (16\text{-}15)$$

where P is the total pressure and p_v and p_a are the partial pressures of vapor and gas, respectively. Relative humidity is a ratio between the water vapor partial pressure; p_v and p_a are the partial pressures of the vapor and gas, respectively.

Relative humidity is the ratio between the water vapor partial pressure, p_v, and the saturation pressure, p_{vs}, which is the maximum amount of water vapor that can be present in gas for given conditions [3,39]:

$$\varphi = \frac{p_v}{p_{vs}} \qquad (16\text{-}16)$$

The enthalpy of dry gas is [39]:

$$h_g = c_{pg} t \qquad (16\text{-}17)$$

where h_g is the enthalpy of dry gas, J/g, C_{pg} is the specific heat of gas, J/gK, and T is the temperature in °C.

Enthalpy of the water vapor is [39]:

$$h_v = c_{pv} t + h_{fg} \qquad (16\text{-}18)$$

where h_{fg} is the heat of evaporation, 2500 J/g at 0°C. Enthalpy of the moist gas is then [3,39]:

$$h_{vg} = c_{pg} t + x(c_{pv} t + h_{fg}) \qquad (16\text{-}19)$$

And the unit is J/g dry gas.

The enthalpy of liquid water is:

$$h_w = c_{pw} t \qquad (16\text{-}20)$$

If the gas contains both water and vapor, the enthalpy can be found by [3,39]:

$$h_{vg} = c_{pg}t + x_v(c_{pv}t + h_{fg}) + x_w c_{pw}t \qquad (16\text{-}21)$$

where x_v is the water vapor content and x_w is the liquid water content. The total water content thus is

$$x = x_v + x_w \qquad (16\text{-}22)$$

Air going into the cell is typically humidified to prevent drying of the membrane near the cell inlet. Air enters the cell relatively dry. At lower temperatures, smaller amounts of water are required to humidify the cell than at higher temperatures. As the cell's air is heated up and the pressure increases, it needs more and more water.

16.5 Fuel Cell Mass Balance

In order to properly determine the fuel cell mass flow rates, the mass that flows into and out of each process unit in the fuel cell subsystems, and in the fuel cell itself needs to be accounted for in order to determine the overall fuel, oxidant, and (if applicable) water requirement(s) for the process. The procedure for formulating a mass balance is as follows:

1. A flowchart must be drawn and labeled. Enough information should be included on the flowchart to have a summary of each stream in the process. This includes known temperatures, pressures, mole fractions, mass flow rates, and phases.
2. Write the appropriate mass balance equation in order to determine the flow rates of all stream components, and solve for any desired quantities.

An example flowchart is shown in Figure 16-8. Hydrogen enters the cell at temperature, T, and pressure, P with the mass flow rate, m_{H_2}. Oxygen enters the fuel cell from the environment at a certain T, P, and

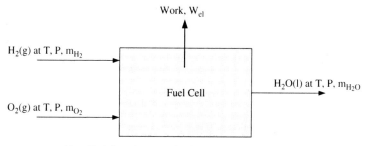

Figure 16-8 Detailed flowchart to obtain mass balance equation.

m_{O_2} (mass flow rate). The hydrogen and oxygen react completely in the cell to produce water, which exits at a certain T, P, and m_{H_2O} (mass flow rate). This reaction can be described by:

$$2H_2(g) + O_2(g) \rightarrow 2H_2O(l)$$

W_{el} in Figure 16-8 is the work available due to chemical availability. The generic mass balance for the fuel cell in this example is:

$$m_{H_2} + m_{O_2} = m_{H_2O} + W_{el} \qquad (16\text{-}23)$$

The formal definition for material balances in a system can be written as:

Input (enters through system boundaries)	+	Generation (produced within the system)	−	Output (leaves through system boundaries)	−	Consumption (consumed within the system)	=	Accumulation (buildup within the system)

Generally, the fuel cell mass balance requires that the sum of all of the mass inputs is equal to the mass outputs, which can be expressed as:

$$\sum (m_i)_{in} = \sum (m_i)_{out} \qquad (16\text{-}24)$$

Where i is the material going into and out of the cell (examples include hydrogen, oxygen, carbon dioxide, propane etc). The flow rates at the inlet are proportional to the current and number of cells. The cell power output is:

$$W_{el} = n_{cell} V_{cell} I \qquad (16\text{-}25)$$

All of the flows are proportional to the power output and inversely proportional to the cell voltage:

$$I \cdot n_{cell} = \frac{W_{el}}{V_{cell}} \qquad (16\text{-}26)$$

The inlet flow rates for a fuel cell that uses hydrogen and oxygen are shown in equations 16-27 to 16-30.
The hydrogen mass flow rate is:

$$m_{H_2,in} = S_{H_2} \frac{M_{H_2}}{2F} I \cdot n_{cell} \qquad (16\text{-}27)$$

The oxygen mass flow rate (g/s) is:

$$m_{O_2,in} = S_{O_2} \frac{M_{O_2}}{4F} I \cdot n_{cell} \quad (16\text{-}28)$$

The air mass flow rate (g/s) is:

$$m_{air,in} = \frac{S_{O_2}}{x_{O_2}} \frac{M_{air}}{4F} I \cdot n_{cell} \quad (16\text{-}29)$$

The nitrogen mass flow rate (g/s) is:

$$m_{N_2,in} = S_{O_2} \frac{M_{N_2}}{4F} \frac{1 - x_{O_2,in}}{x_{O_2,in}} I \cdot n_{cell} \quad (16\text{-}30)$$

The additional inlet flow rates for a PEMFC or a DMFC are shown in equations 16-31 to 16-34.

Water vapor in the hydrogen inlet is [3]:

$$m_{H_2O\,in\,H_2,in} = S_{H_2} \frac{M_{H_2O}}{2F} \frac{\varphi_{an} P_{vs(T_{an,in})}}{P_{an} - \varphi_{an} P_{vs(T_{an,in})}} I \cdot n_{cell} \quad (16\text{-}31)$$

Water vapor in the fuel inlet (g/s) is [3]:

$$m_{H_2O,fuel,in} = \frac{S_{H_2}}{x_{H_2}} \frac{M_{H_2O}}{2F} \frac{\varphi_{an} P_{vs(T_{an,in})}}{P_{an} - \varphi_{an} P_{vs(T_{an,in})}} I \cdot n_{cell} \quad (16\text{-}32)$$

Water vapor in the oxygen inlet is [3]:

$$m_{H_2O\,in\,O_2,in} = S_{O_2} \frac{M_{H_2O}}{4F} \frac{\varphi_{ca} P_{vs(T_{an,in})}}{P_{ca} - \varphi_{ca} P_{vs(T_{an,in})}} I \cdot n_{cell} \quad (16\text{-}33)$$

Water vapor in the air inlet (g/s) is [3]:

$$m_{H_2O\,in\,air\,in} = \frac{S_{O_2}}{x_{O_2}} \frac{M_{H_2O}}{4F} \frac{\varphi_{ca} P_{vs(T_{an,in})}}{P_{ca} - \varphi_{ca} P_{vs(T_{an,in})}} I \cdot n_{cell} \quad (16\text{-}34)$$

The outlet flow rates for a fuel cell that uses hydrogen and oxygen are shown in equations 16-35 to 16-39.

The unused hydrogen flow rate is:

$$m_{H_2,out} = (S_{H_2} - 1) \frac{M_{H_2}}{2F} I \cdot n_{cell} \quad (16\text{-}35)$$

The oxygen flow rate at the outlet is equal to the oxygen supplied at the inlet minus the oxygen consumed in the fuel electrochemical reaction:

$$m_{O_2,out} = (S_{O_2} - 1)\frac{M_{O_2}}{4F} I \cdot n_{cell} \qquad (16\text{-}36)$$

The nitrogen flow rate at the exit is the same as the inlet because nitrogen does not participate in the fuel cell reaction [3]:

$$m_{N_2,out} = m_{N_2,in} = S_{O_2}\frac{M_{N_2}}{4F}\frac{1 - x_{O_2,in}}{x_{O_2,in}} I \cdot n_{cell} \qquad (16\text{-}37)$$

The depleted air flow rate is then simply a sum of the oxygen and nitrogen flow rates [3]:

$$m_{air,out} = \left[(S_{O_2} - 1)M_{O_2} + S_{O_2}\frac{1 - x_{O_2,in}}{x_{O_2,in}} M_{N_2}\right]\frac{I \cdot n_{cell}}{4F} \qquad (16\text{-}38)$$

The oxygen volume fraction at the outlet is much lower than the inlet volume fraction [3]:

$$x_{O_2,out} = \frac{S_{O_2} - 1}{\frac{S_{O_2}}{x_{O_2,in}} - 1} \qquad (16\text{-}39)$$

The additional outlet flow rates and balances for a PEMFC or DMFC are shown in equations 16-40 to 16-41.

The water in the hydrogen exhaust can be in vapor or liquid form. The water vapor content at the anode outlet is the smaller of the total water flux [3]:

$$m_{H_2Oin,H_2Oout,V} = \min\left[(S_{H_2} - 1)\frac{M_{H_2O}}{2F}\frac{P_{vs(T_{out,an})}}{P_{an} - \Delta P_{an} - P_{vs(T_{out,an})}}\right.$$

$$\left. I \cdot n_{cell}, m_{H_2Oin,H_2Oout}\right] \qquad (16\text{-}40)$$

where ΔP_{an} is the pressure drop on the anode side. The amount of liquid water is the difference between the total water present and the water vapor:

$$M_{H_2Oin,H_2out,L} = m_{H_2Oin,H_2out} - m_{H_2Oin,H_2out,V}$$

Water content in the cathode exhaust is equal to the amount of water brought into the cell, generated in the cell, and transported across the membrane:

$$m_{H_2Oin,Airout} = m_{H_2Oin,Airin} + m_{H_2Ogen} + m_{H_2OED} - m_{H_2OBD}$$

The water in the hydrogen exhaust can be in vapor or liquid form. The water vapor content at the cathode outlet is the smaller of the total water flux [3]:

$$m_{H_2Oin,Airout,V} = \min\left[\left(\frac{S_{O_2} - x_{O_2,in}}{x_{O_2,in}}\right)\frac{M_{H_2O}}{4F}\frac{P_{vs(T_{out,an})}}{P_{ca} - \Delta P_{ca} - P_{vs(T_{out,an})}}\right.$$
$$\left. I \cdot n_{cell}, m_{H_2Oin,Airout}\right] \quad (16\text{-}41)$$

Example 16-1 A hydrogen–air PEM fuel cell generates 500 watts at 0.7 V. Dry hydrogen is supplied in a dead-end mode at 20°C. The relative humidity of the air at the fuel cell inlet is 50% at a pressure of 120 kPa. Liquid water is injected at the air inlet to help cool the fuel cell. The oxygen stoichiometric ratio is 2, and the outlet air is 100% saturated at 80°C and atmospheric pressure. What is the water injected flow rate (g/s)?

The water mass balance is

$$m_{H_2O\_air\_in} + m_{H_2O\_Inject} + m_{H_2O\_gen} = m_{H_2O\_in\_air\_out}$$

In order to calculate the amount of water in air, the saturation pressure needs to be calculated. To calculate the saturation pressure (in Pa) for any temperature between 0°C and 100°C:

$$p_{vs} = e^{aT^{-1}+b+cT+dT^2+eT^6+f\ln(T)}$$

where a, b, c, d, e, and f are the coefficients.

a = −5800.2206, b = 1.3914993, c = −0.048640239,
d = 0.41764768 × 10^{-4}, e = −0.14452093 × 10^{-7}, and f = 6.5459673.

With T = 293.15, and p_{vs} = 2.339 kPa, the amount of water in air can be calculated:

$$m_{H_2O,airin} = \frac{S_{O_2}}{x_{O_2}} \frac{M_{H_2O}}{nF} \frac{\phi_{ca} P_{vs(T_{ca,in})}}{P_{ca} - \phi_{ca} P_{vs(T_{ca,in})}} I \cdot n_{cell}$$

$$m_{H_2O,airin} = \frac{2}{0.2095} \frac{18.015}{(4*96{,}485 \text{ As/mol})} \frac{0.50*2.339}{120 \text{ kPa} - 0.50*2.339} \frac{500 \text{ W}}{0.7 \text{ V}} *1$$

$$m_{H_2O,airin} = 0.00313 \text{ g/s}$$

Water generated is

$$m_{H_2O,gen} = \frac{I}{nF} M_{H_2O} = \frac{714.29}{2 \times 96485} 18.015 = 0.0667 \text{ g/s}$$

Water vapor in air out is

$$m_{H_2Oin,H_2out,V} = \left[\left(\frac{S_{O_2} - x_{O_2,in}}{x_{O_2,in}}\right) \frac{M_{H_2O}}{4F} \frac{P_{vs(T_{out,ca})}}{P_{ca} - \Delta P_{ca} - P_{vs(T_{out,ca})}} I \cdot n_{cell}\right]$$

First calculate the saturation pressure:

$$T = 80°C = 353.15 \text{ K}, P_{vs} = 47.67 \text{ kPa}, P_{ca} - \Delta P_{ca} = 101.325 \text{ kPa}$$

$$m_{H_2Oin,H_2out,V} = \left[\left(\frac{2 - 0.2095}{0.2095}\right) \frac{18.015}{(4*96,485)} \frac{47.67}{(101.325 - 47.67)} \frac{500}{0.7} * 1\right]$$

$$m_{H_2Oin,airout} = 0.253 \text{ g/s}$$

The water balance is therefore:

$$m_{H_2O\_air\_in} + m_{H_2O\_Inject} + m_{H_2O\_gen} = m_{H_2O\_in\_air\_out}$$

$$0.00313 + m_{H_2Oinject} + 0.0667 = 0.253$$

$$m_{H_2Oinject} = 0.22317 \text{g/s}$$

Example 16-2 A PEM fuel cell with 100 cm² of active area is operating at 0.50 A/cm² at a voltage of 0.70. The operating temperature is 75°C and 1 atm with air supplied at a stoichiometric ratio of 2.5. The air is humidified by injecting hot water (75°C) before the stack inlet. The ambient air conditions are 1 atm, 22°C and 70-percent relative humidity. Calculate the air flow rate, the amount of water required for 100-percent humidification of air at the inlet, and the heat required for humidification.

The oxygen consumption is

$$N_{O_2} = \frac{I}{4F} = \frac{0.50 \text{ A/cm}^2 \times 100 \text{ cm}^2}{4 \times 96,485} = 0.129 \times 10^{-3} \text{ mol/s}$$

The oxygen flow rate at the cell inlet is

$$N_{O_2} = SN_{O_2,cons} = 2.5 \times 0.129 \times 10^{-3} = 0.324 \times 10^{-3} \text{ mol/s}$$

$$N_{air} = N_{O_2,act} \frac{1}{x_{O_2}} = \frac{0.324 \times 10^{-3}}{0.21} = 1.54 \times 10^{-3} \text{ mol/s}$$

$$m_{air} = N_{air} m_{air} = 1.54 \times 10^{-3} \text{ mol/s} \times 28.85 \text{ g/mol} = 0.0445 \text{ mol/s}$$

Fuel Cell Operating Conditions

The amount of water in air at the cell inlet where $\varphi = 1$ is

$$m_{H_2O} = x_s m_{air} \quad \text{and} \quad x_s = \frac{m_{H_2O}}{m_{air}} \frac{p_{vs}}{P - p_{vs}}$$

where p_{vs} is the saturation pressure at 348.15 K, and P is the total pressure (101.325 kPa).

$$p_{vs} = e^{aT^{-1}+b+cT+dT^2+eT^3+f\ln(T)} = 38.6 \text{ kPa}$$

$$x_s = \frac{m_{H_2O}}{m_{air}} \frac{p_{vs}}{P - p_{vs}} = \frac{18}{28.85} \frac{38.6}{(101.325 - 38.6)} = 0.384 \, g_{H_2O}/g_{air}$$

$$m_{H_2O} = x_s m_{air} = 0.384 \, g_{H_2O}/g_{air} \times 0.128 \, g_{air}/s = 0.0491 \, g_{H_2O}/s$$

The amount of water in ambient air at 70% RH and 295.15 K is

$$x_s = \frac{m_{H_2O}}{m_{air}} \frac{\varphi p_{vs}}{P - \varphi p_{vs}} = \frac{18}{28.85} \frac{0.70 \times 2.645}{101.325 - 0.7 \times 2.645} = 0.0116 \, g_{H_2O}/g_{air}$$

$$m_{H_2O} = x_s m_{air} = 0.0116 \, g_{H_2O}/g_{air} \times 0.128 \, g_{air}/s = 0.001486 \, g_{H_2O}/s$$

The amount of water needed for humidification of air is

$$m_{H_2O} = 0.0491 - 0.001486 = 0.047614 \, g_{H_2O}/s$$

The heat required for humidification can be calculated from the heat balance.

$$H_{air,in} + H_{H_2O,in} + Q = H_{air,out}$$

The enthalpy of wet/moist air is

$$h_{vair} = c_{p,air}t + x(c_{p,v}t + h_{fg})$$

Humidified air:

$$h_{vair} = 1.01 \times 75 + 0.384 \times (1.87 \times 75 + 2500) = 1089.61 \text{ J/g}$$

Ambient air:

$$h_{vair} = 1.01 \times 22 + 0.0116 \times (1.87 \times 22 + 2500) = 51.70 \text{ J/g}$$

Water: $h_{H_2O} = 4.18 \times 75 = 313.5 \text{ J/g}$

$$Q = 1089.61 \text{ J/g} \times 0.128 \text{ g/s} - 51.70 \text{ J/g} \times 0.128 \text{ g/s}$$
$$-313.5 \text{ J/g} \times 0.0157 \text{ g/s} = 127.93 \text{ W}$$

Chapter Summary

Fuel cell operating conditions depend upon the fuel cell and stack design. The optimal conditions will vary from cell to cell (unless the fuel cells are mass produced); therefore, it is necessary to calculate the required parameters for efficient fuel cell operation. Increasing the pressure of the reactants will improve fuel cell performance, but may lower the efficiency of the fuel cell system as a whole due to the equipment required for the pressure increase. Higher fuel cell operating temperatures mean faster reaction kinetics and a voltage gain, however, there is a optimal temperature for each fuel cell type. For example, PEM fuel cells are limited to operating from 20°C to 100°C due to the evaporation of water from the membrane. The flow rate of the reactants must minimally match the rate of reactant consumption in the fuel cell. If air is used as the oxidant, a stoichiometry of 2 or higher is needed to maintain a high concentration of hydrogen at the cathode. For PEMFCs, the level of humidity is also an important operating parameter. Often, humidifying one or both of the reactant streams will result in better fuel cell performance. Like fuel cell efficiency increases due to temperature or pressure, the balance-of-plant must be considered. When calculating the operating conditions, performing mass balances on the reactants, products and electricity generated will help the designer to estimate the optimal pressures, temperatures, flow rates, and humidity for a particular fuel cell system.

Problems

1. A PEM fuel cell with 50 cm^2 of active area is operating at 0.60 A/cm^2 at a voltage of 0.70. The operating temperature is 60°C at 3 atm with air supplied at a stoichiometric ratio of 2. The air is humidified by injecting hot water (60°C) before the stack inlet. The ambient air conditions are 1 atm, 22°C, and 60-percent RH. Calculate the air flow rate, the amount of water required for 100-percent humidification of air at the inlet, and the heat required for humidification.

2. A fuel cell has a current density of 0.7 A/cm^2 with a voltage of 0.65, and an active area of 250 cm^2. The hydrogen flow rate is 2 LPM and the air flow rate is 6 LPM, with 80-percent saturation at 70°C and 2 atm. Calculate the hydrogen and oxygen stoichiometric ratios and the amount of water vapor present in the hydrogen and air streams.

3. For the fuel cell described in Problem 2, calculate the amount of liquid water in the hydrogen and air outlets.

4. For the fuel cell described in Problem 2, calculate the fuel cell efficiency and the rate of heat generated (W).

5. A hydrogen–air fuel cell generates 50 watts at 0.65 V. Dry hydrogen is supplied in a dead-end mode at 90°C. Liquid water is injected at the air inlet to help cool the fuel cell. What is the water-injected flow rate (g/s)?

Bibliography

[1] Sousa, Ruy, Jr. and Ernesto Gonzalez. "Mathematical Modeling of Polymer Electrolyte Fuel Cells." *Journal of Power Sources.* Vol. 147, 2005, pp. 32–45.
[2] O'Hayre, Ryan, Suk-Won Cha, Whitney Colella, and Fritz B. Prinz. *Fuel Cell Fundamentals.* 2006. New York: John Wiley & Sons.
[3] Barbir, Frano. *PEM Fuel Cells: Theory and Practice.* 2005. Burlington, MA: Elsevier Academic Press.
[4] Lin, Bruce. 2000. "Conceptual Design and Modeling of a Fuel Cell Scooter for Urban Asia." Princeton University, Masters Thesis. 1999.
[5] Mench, M.M., Z.H. Wang, K. Bhatia, and C.Y. Wang. "Design of a Micro-direct Methanol Fuel Cell." Electrochemical Engine Center, Department of Mechanical and Nuclear Engineering, Pennsylvania State University. 2001.
[6] Mench, Matthew M., Cao-Yang Wang, and Stephan T. Tynell. "An Introduction to Fuel Cells and Related Transport Phenomena." Department of Mechanical and Nuclear Engineering, The Pennsylvania State University. Draft. http://mtrl1.mne.psu.edu/Document/jtpoverview.pdf Last Accessed March 4, 2007.
[7] You, Lixin, and Hongtan Liu. "A Two-Phase Flow and Transport Model for PEM Fuel Cells." *Journal of Power Sources.* Vol. 155, 2006, pp. 219–230.
[8] Springer et al. "Polymer Electrolyte Fuel Cell Model." *J. Electrochem. Soc.* Vol. 138, No. 8, 1991, pp. 2334–2342.
[9] Li, Xianguo. *Principles of Fuel Cells.* 2006. New York: Taylor & Francis Group.
[10] Subhash, Singhall C. "Solid Oxide Fuel Cells: Fundamentals and Applications." Pan American Advanced Studies Institute. Rio de Janeiro, Brazil. Pacific Northwest National Laboratory, U.S. Department of Energy.
[11] Chung, Brandon W., Christopher N. Chervin, Jeffrey J. Haslam, Ai-Quoc Pham, and Robert S. Glass. "Development and Characterization of a High Performance Thin-Film Planar SOFC Stack." *Journal of the Electrochemical Society.* Vol. 152, No. 2, 2005, pp. A265–A269.
[12] Zhu, Bin. "Proton and Oxygen Ion-Mixed Conducting Ceramic Composites and Fuel Cells." *Solid State Ionics.* Vol. 145, 2001, pp. 371–380.
[13] Yoon, K.H., J.Y. Choi, J.H. Jang, Y.S. Cho, and K.H. Jo. "Electrode/Matrix Interfacial Characteristics in a Phosphoric Acid Fuel Cell." *Journal of Applied Electrochemistry.* Vol. 30, 2000, pp. 121–124.
[14] Song, Rak-Hyun and Dong Ryul Shin. "Influence of CO Concentration and Reactant Gas Pressure on Cell Performance in PAFC." *International Journal of Hydrogen Energy.* 26, 2001, pp. 1259–1262.
[15] Soler, J., T. Gonzalez, M.J. Escudero, T. Rodrigo, and L. Daza. Endurance test on a single cell of a novel cathode material for MCFC. Journal of Power Sources. Vol. 106, 2002, pp. 189–195.
[16] Silva, V.S., J. Schirmer, R. Reissner, B. Ruffmann, H. Silva, A. Mendes, L.M. Madeira, and S.P. Nunes. "Proton Electrolyte Membrane Properties and Direct Methanol Fuel Cell Performance." *Journal of Power Sources.* Vol. 140, 2005, pp. 41–49.
[17] Rowshanzamir, S. and M. Kazemeini. "A New Immobilized-Alkali H_2/O_2 Fuel Cell." *Journal of Power Sources.* Vol. 88, 2000, pp. 262–268.
[18] Verma, A. and S. Basu. "Direct Use of Alcohols and Sodium Borohydride as Fuel in an Alkaline Fuel Cell." *Journal of Power Sources.* Vol. 145, 2005, pp. 282–285.

[19] Morita, H., M. Komoda, Y. Mugikura, Y. Izaki, T. Watanabe, Y. Masuda, and T. Matsuyama. "Performance Analysis of Molten Carbonate Fuel Cell Using a Li/Na Electrolyte." *Journal of Power Sources.* Vol. 112, 2002, pp. 509–518.
[20] Bove, Roberto and Piero Lumghi. "Experimental Comparison of MCFC Performance Using Three Different Biogas Types and Methane." *Journal of Power Sources.* Vol. 145, 2005, pp. 588–593.
[21] Nakagawa, Nobuyoshi and Yikun Xiu. "Performance of a Direct Methanol Fuel Cell Operated." *Journal of Power Sources.* Vol. 118, 2003, pp. 248–255.
[22] Lim, C. and C. Y. Wang. "Development of High Power Electrodes for a Liquid Feed Direct Methanol Fuel Cell." *Journal of Power Sources.* Vol. 113, 2003, pp. 145–150.
[23] Wong, C.W., T.S. Zhao, Q. Ye, and J.G. Liu. "Experimental Investigations of the Anode Flow Field of a Micro Direct Methanol Fuel Cell." *Journal of Power Sources.* Vol. 155, 2006, pp. 291–296.
[24] Liu, J.G., T.S. Zhao, Z.X. Liang, and R. Chen. "Effect of Membrane Thickness on the Performance and Efficiency of Passive Direct Methanol Fuel Cells." *Journal of Power Sources.* Vol. 153, 2006, pp. 61–67.
[25] Yan, Qiangu, Hossein Toghiani, and Junxiao Wu. "Investigation of Water Transport Through Membrane in a PEM Fuel Cell by Water Balance Experiments." *Journal of Power Sources.* Vol. 158, 2006, pp. 316–325.
[26] Fabian, Tibor, Jonathan D. Posner, Ryan O'Hayre, Suk-Won Cha, John K. Eaton, Fritz B. Prinz, and Juan G. Santiago. "The Role of Ambient Conditions on the Performance of a Planar, Air-Breathing Hydrogen PEM Fuel Cell." *Journal of Power Sources.* Vol. 161, 2006, pp. 168–182.
[27] Nguyen, Nam-Trung and Siew Hwa Chan. "Micromachined Polymer Electrolyte Membrane and Direct Methanol Fuel Cells – A Review." *J. Micromech. Microeng.* Vol. 16, 2006, pp. R1–R12.
[28] U.S. Patent No. 7,029,781 B2. Microfuel Cell Having Anodic and Cathodic Microfluidic Channels and Related Methods. Lo Priore, Stefano, Michele Palmieri, and Ubaldo Mastromatteo. STMicroelectronics, Inc., Carollton, TX. April 18, 2006.
[29] Cha, S.W., R. O'Hayre, Y. Saito, and F.B. Prinz. "The Influence of Size Scale on the Performance of Fuel Cells." *Solid State Ionics.* Vol. 175, 2004, pp. 789–795.
[30] Heinzel, A., C. Hebling, M. Muller, M. Zedda, and C. Muller. "Fuel Cells for Low Power Applications." *Journal of Power Sources.* Vol. 105, 2002, pp. 250–255.
[31] Lu, G.Q. and C.Y. Wang. "Development of Micro Direct Methanol Fuel Cells for High Power Applications." *Journal of Power Sources.* Vol. 144, 2005, pp. 141–145.
[32] Blum, A., T. Duvdevani, M. Philosoph, N. Rudoy, and E. Peled. "Water Neutral Micro Direct Methanol Fuel Cell (DMFC) for Portable Applications." *Journal of Power Sources.* Vol. 117, 2003, pp. 22–25.
[33] Mitrovski, Svetlana, M. Lindsay, C.C. Elliott, and Ralph G. Nuzzo. "Microfluidic Devices for Energy Conversion: Planar Integration and Performance of a Passive, Fully Immersed H_2-O_2 Fuel Cell." *Langmuir.* Vol. 20, 2004, pp. 6974–6976.
[34] Wozniak, Konrad, David Johansson, Martin Bring, Anke Sanz-Velasco, and Peter Enoksson. "A Micro Direct Methanol Fuel Cell Demonstrator." *J. Micromech. Microeng.* Vol. 14, 2004, pp. S59–S63.
[35] Yan, Qiangu, H. Toghianai, and Heath Causey. "Steady State and Dynamic Performance of Proton Exchange Membrane Fuel Cell (PEMFCs) Under Various Operating and Load Changes." *Journal of Power Sources.* Vol. 161, 2006, pp. 492–502.
[36] Song, Rak-Hyun, Chang-Soo Kim, and Dong Ryul Shin. "Effects of Flow Rate Starvation of Reactant Gases on the Performance of Phosphoric Acid Fuel Cells." *Journal of Power Sources.* Vol. 86, 2000, pp. 289–293.
[37] Kim, Sunhoe, S. Shimpalee, and J.W. Van Zee. "The Effect of Stoichiometry on Dynamic Behavior of a Proton Exchange Membrane Fuel Cell (PEMFC) During Load Change." *Journal of Power Sources.* Vol. 135, 2004, pp. 110–121.
[38] Felder, Richard M., and Ronald W. Rousseau. Elementary Principles of Chemical Processes. 2nd ed. Wiley Series in Chemical Engineering, 1986.
[39] Perry, Robert H. and Don W. Green. Perry's Chemical Engineer's Handbook. 7th ed. New York: McGraw-Hill, 1997.

Chapter

17

Fuel Cell Characterization

There are many characterization techniques that allow quantitative comparison of almost every property of every part of the fuel cell stack. Selecting the best characterization techniques will help the researcher to understand why the fuel cell is performing well or poorly. These techniques will help discriminate between activation, ohmic and concentration losses, fuel crossover, and defective materials, as well as many other properties. In-situ testing is very important for fuel cells because the performance of the fuel cell cannot be determined simply by characterizing its parts. In addition, not only can the actual parts cause performance losses, but the contact between the parts can also cause decreased performance in fuel cells. Table 17-1 summarizes the techniques examined in this chapter, and offers a quick summary of what each technique characterizes.

In addition to the techniques listed in Table 17-1, topics that will be covered in this chapter include:

- Fuel Cell Testing Setup
- Verification of the Testing Assembly
- Baseline Testing Conditions and Operating Parameters

Insuring that the fuel cell apparatus is properly setup with a set of baseline conditions helps to obtain useful test results. This is critical in obtaining accurate testing results and good fuel cell performance.

17.1 Fuel Cell Testing Setup

Figure 17-1 shows a very basic fuel cell test station for fuel cell measurements. For individuals or projects with limited testing equipment,

TABLE 17-1 Selection of Fuel Cell Characterization Techniques

Type of characterization	Technique
Fuel Cell Durability Testing and Examination	
Constant voltage/current/power and power cycling	Voltage-current measurements, cell impedance
Steady-state and transient testing	Polarization curve, cell impedance
Characterization of Membranes, Catalysts, Gas Diffusion Layers	
Structural analysis	SEM, TEM, EDS
Elemental analysis	XRF, ICP-MS
Particle size distribution	XRD
Electrochemical surface area	BET, cyclic voltammetry, impedance measurements
Carbon bonding interactions, polymer degradation	Neutron scattering

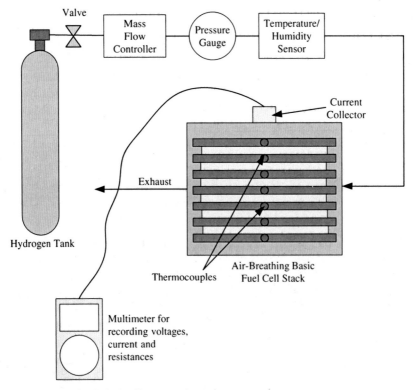

Figure 17-1 A basic fuel cell test station.

one can begin testing fuel cells with just a multimeter, a temperature/ humidity monitor for the room, and an oscilloscope. This will provide only a limited amount of information about the fuel cell system, but it will be enough to begin examining the overall system. In order to optimize the fuel cell system, a test setup which allows the monitoring of many fuel cell and fuel properties simultaneously enables the researcher to analyze how the fuel cell can be improved. Some of the parameters that can be monitored and controlled are the temperatures, pressures, flow rates, and humidity levels of the fuels at the stack entrance and exit.

Mass flow controllers, pressure gauges, and thermocouples help to monitor the fuel cell operating conditions. A potentiostat/galvanostat and impedance analyzer measurement devices can help accurately track the electrical performance of the fuel cell. Some of the experiments that can be conducted include polarization curves, current interrupt tests, and cyclic voltammetry. A test setup like the one shown in Figure 17-2 can help accurately characterize most fuel cell parameters and operating conditions.

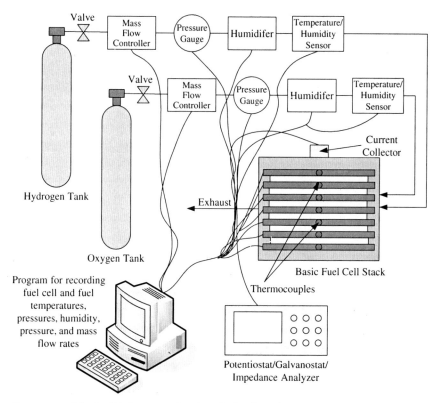

Figure 17-2 A fuel cell lab test station for a fuel cell stack.

In order to begin testing the fuel cell, the test setup needs to be verified, the cell conditioned, and baseline testing conditions and operating parameters must be established. After the testing setup has been completed, accurate tracking of the temperature, pressure, flow rates, and other operating conditions need to be monitored throughout the testing process.

17.2 Verification of the Assembly

After the fuel cell has been put into the test apparatus, the person performing the test must verify that the fuel cell is seated correctly in the test apparatus. This should be conducted before any fuel cell is tested because fuel cell testing exposes the experimenter to voltage, current, hydrogen, and oxygen and other chemical types depending upon the actual fuel cell being tested. Any hazards are usually eliminated by checking the test apparatus through visual inspection, pressure and leakage tests for the hydrogen, oxygen, coolant, and other chemicals used in the fuel cell stack. It is usually a good idea to have a hydrogen and oxygen sensor in the lab environment. The instrumentation also needs to be checked to insure it is properly installed, and correctly connected to the data acquisition system, and is generating the expected fuel cell data [7].

17.3 Fuel Cell Conditioning

Before fuel cell testing is performed, the cell must be adequately conditioned. Many fuel cells (depending upon the fuel cell type) begin providing power as indicated within seconds or minutes of supplying fuel, but in order to test accurately and consistently, a procedure should be developed and used for the startup and conditioning of the cell. There are several "standard" methods that exist for starting a fuel cell. The cell voltage, current, and other operating parameters should be recorded throughout the fuel cell testing process.

One frequently used method for fuel cell conditioning is supplying pure hydrogen to the cell or stack at open-circuit voltage (OCV) for 15 minutes, running at 600 mV/cell for 75 minutes, and then running three hold cycles at 850 mV/cell for 20 minutes followed by 600 mV/cell for 30 minutes [4]. The total conditioning time is approximately four hours.

Many techniques are discussed in the literature, and MEA manufacturers may impose additional conditions, or their own procedures for optimum performance of their products. Regardless of the technique used, detailed criteria needs to be developed to ensure that the conditioning is adequate. A commonly used rule-of-thumb used in industry is that the conditioning is complete if the quarter power and rated power voltages from the initial polarization curve are within 2 percent of the expected values [4].

17.4 Baseline Test Conditions and Operating Parameters

A baseline set of conditions should be established for each set of fuel cell tests. This baseline is a function of the hardware being tested, the test facility, and the operating conditions. A baseline set of conditions should be established for the following:

- Cell Temperature
- Fuel and Oxidant Composition
- Fuel and Oxidant Inlet Temperatures
- Fuel and Oxidant Inlet Dew Point Temperatures (relative humidity)
- Fuel and Oxidant Exit Pressures
- Fuel and Oxidant Stoichiometries (flowrates)
- Current Density (mA/cm^2) or Cell Voltage (mV/cell)
- Coolant Inlet and Exit Temperatures
- Coolant Inlet and Exit Pressures

An example of a set of baseline operating conditions is a cell that runs on hydrogen and air with an average cell temperature of 75 ±1°C. The hydrogen and air inlet temperatures should be 75 ±1°C with dew points of 73 ±1°C. The fuel and oxidant exhaust pressures should be both 150 ±2 kPa. The fuel and oxidant flow rates both correspond to a stoichiometry of 2 [7]. Usually, a current or voltage is held constant for the duration of the test. The current state of the art seems to use 200 and 1000 mA/cm^2 for the current density [7]. If a voltage needs to be selected, the commonly used values are 850 mV/cell and 700 mV/cell [7].

One common issue encountered during fuel cell testing with platinum cathodes is that a platinum hydroxide layer builds up on the cathode surface causing reversible cell voltage degradation. One method of minimizing this effect is alternating between two current densities or voltages. For example, alternating between 850 mV/cell for two hours and 700 mV/cell for ten hours minimizes the effect [7]. The next few paragraphs briefly explain the parameters that should be monitored during the startup, conditioning, and during the testing process.

17.4.1 Temperature

The fuel cell temperature must be documented and maintained throughout the fuel cell tests. At a minimum, the inlet and outlet temperatures should be monitored in addition to the fuel cell stack itself. In order to determine how the fuel cell could be improved, the temperatures across the fuel cell stack and in each cell should be monitored. Any inhomogeneities in temperature can alter the fuel cell performance.

17.4.2 Pressure

Gas pressures are typically monitored at the fuel inlets and outlets using pressure transducers to determine the amount of pressure drop that occurred within the fuel cell stack. Increasing the fuel cell pressure may improve cell performance, but the system should be studied to determine how this will affect the other fuel cell parameters, and whether the system design is optimal for increased pressures.

17.4.3 Flow rate

Flow rates can be monitored using mass flow controllers. The fuel cell performance can be studied by maintaining the same flow rate, and ramping the flow rates up or down and recording the resulting fuel cell parameters described in this section.

17.4.4 Compression force

For most bipolar fuel cell stack configurations, there is an optimal compression force that helps the fuel cell achieve the best performance. Cells with low compression forces can suffer from increased ohmic loss, while cells that have high compression forces can suffer from pressure or concentration losses [1].

17.5 Polarization Curves

A common fuel cell measurement is the fuel cell polarization curve. This curve is essential for characterizing the fuel cell since it is the most common method of testing fuel cells and can provide an easy comparison to other published data. As shown in Figure 17-3, the polarization curve shows the voltage output of the fuel cell for a given current density loading. Polarization curves are usually taken with a potentiostat/galvanostat, which draws a fixed current from the fuel cell, and measures the fuel cell output voltage. By slowly stepping the load on the potentiostat, the voltage response of the fuel cell can be determined. If a potentiostat is unavailable, a more primitive method of determining the current/voltage relationship from the fuel cell is to take several different types of small resistors (to act as a load), and measure the voltage output on a multimeter [1].

Three distinct regions of a fuel cell polarization curve are noticeable:

- At low power densities, the cell potential drops as a result of the activation polarization.
- At intermediate current densities, the cell potential drops linearly with current due to ohmic losses.
- At high current densities, the cell potential drop departs from the linear relationship with current density as a result of a more pronounced concentration polarization.

When the current (load) on a fuel cell is changed, the fuel cell heat and water balance change, and it may take several minutes to hours for the fuel cell to reach a new equilibrium point. While testing, a designated time period should be used to allow the fuel cell to reach the new equilibrium. The establishment of an equilibrium period may vary depending upon whether the fuel cell load has been increased or decreased. Relevant test data can be obtained from the fuel cell by adjusting the load in various ways: the load can be programmed to increase or decrease by a certain step-size, or the load can be randomly selected. The most typical method is to have the load increased step-to-step. The data can be taken at multiple current or voltage points. A typical method for obtaining measurements is beginning at open-circuit voltage, and collecting five to six points between 600 mV/cell and 850 mV/cell and waiting 15 minutes between each point. The data from the last five minutes should be averaged and then plotted as average current versus average voltage [7].

Reliable voltage/current curves require a stable environment, where the temperature, pressure, humidity and flow rates maintain the desired level while the test(s) are being conducted. If the conditions are fluctuating, the voltage/current characteristics may change. In addition to keeping the environment stable, the fuel cell itself may take a while to have a constant output. Depending upon the fuel cell design and its size, the fuel cell may stabilize within the first few minutes. Large fuel cell stacks (automotive and larger) may require up to 30 minutes to stabilize after a current or voltage change [1]. For large fuel cell systems, obtaining polarization curves can be a tedious process, and only a limited number of points can be taken.

Useful information can be gained from the parameters of the polarization curve if the experimental results are fitted to the reversible cell potential, exchange current density, Tafel slope, cell resistance, or limiting current. If the polarization curve is recorded while increasing and decreasing the current, it may show hysteresis. This typically

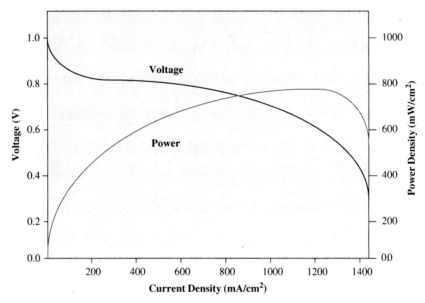

Figure 17-3 Example of a PEMFC polarization curve.

indicates a change in fuel cell conditions, such as drying or flooding of the membrane.

A difference in polarization curves between a cell operating with air and pure oxygen may be used to characterize flooding or problems in the cathode. The difference in cell voltage between pure oxygen and air results from the difference in the concentration of oxygen at the catalyst surface—this difference should be nearly constant at any current density. An increase in cell potential difference between oxygen and air at higher current densities indicates mass transport problems.

17.6 Fuel Cell Resistance

One of the most important parameters to test in a fuel cell is the cell resistance, and specifically the resistance of the electrolyte. Recall from chapter 7, that the passage of ions within a phase of finite resistance creates a voltage loss called ohmic polarization. Whether the electrolyte being characterized is the proton exchange membrane in a PEMFC, the high-temperature ionically conductive oxide in SOFC, or the acid-soaked matrix in a phosphoric acid fuel cell (PAFC), electrolyte resistance is an important metric for fuel cell performance because it quantifies internal cell losses [4,7].

Since fuel cell current densities are high compared with other electrochemical processes, small amounts of ohmic resistance (milliohms) have a significant effect on overall efficiency. The electrolyte resistance should be monitored during electrolyte development and stack manufacture because ohmic losses generate heat that must be removed from the fuel cell. Electrolyte resistivity is a function of operating conditions; therefore, it is important that the resistance be determined while the cell is operating.

The electrode / solid electrolyte interface has a large capacitance associated with it that allows an AC measurement to be used to determine an electrolyte resistance separate from the polarization resistance. The methods typically used for electrolyte resistance measurement are the current interrupt (IR) method, AC resistance, electrochemical impedance spectroscopy (ETS), and high-frequency resistance (HFR) methods. Sections 17.6.1–17.6.7 compare these methods and several others, and outlines the pros and cons of each [4].

17.6.1 Current interrupt

The polarization curve provides useful information, but additional information is needed about the fuel cell to accurately study its performance characteristics. Cell resistance provides a lot of information about a fuel cell that polarization curves do not. For example, a resistance increase may indicate the drying out of a membrane. The current interrupt method works by using a load contactor relay switch placed in series with the load bank and fuel cell. The relay is opened while the cell is operating at a certain current, and the voltage is measured. The voltage measurement over a given time period qualitatively measures the membrane impedance. This voltage response is used to calculate a slope that gives the ohmic resistance of the fuel cell components. The switching speed of the relay takes milliseconds, and an oscilloscope is required to record the waveform when the relay is open [7].

One method of measuring cell resistance in a fuel cell is to interrupt the current for a very short interval (a few milliseconds), and have the resulting voltage gain be recorded. The difference between the cell voltage before and after the current interrupt, divided by the current, is the cell resistance. The difficulty with using this simple method is determining the exact point of voltage gain. The voltage behavior may be monitored in real time using an oscilloscope, or several voltage readings may be done at different interrupt times. This method can be used with simple measurement devices and can be implemented on high-power fuel cell systems.

The test is conducted as follows:

1. The fuel cell is set at predetermined operating conditions.
2. Voltage sensing leads are attached to the positive and negative terminals of the fuel cell or fuel cell stack.

3. At a given time, the current relay is opened and the cell voltage(s) are monitored by the oscilloscope for three to ten seconds.
4. The cell voltage response is plotted versus current, and the slope of the curve estimates the total ohmic losses in the fuel cell [7].

17.6.2 The AC resistance method

The AC resistance method applies a fixed, single high-frequency sine wave (typically −1 kHz) to the fuel cell being tested to measure the total impedance magnitude of the cell and the load in parallel at that frequency. The electrolyte ohmic resistance can be calculated after correcting for the impedance of the load, which is in parallel with the fuel cell [4].

This method only provides a single data point, like the current interrupt method. An advantage is that the cell is minimally disturbed while this test is being conducted. Accurate results cannot be obtained without an exact gain-phase characterization of the impedance on the load at operating conditions. The only method of separating the actual cell impedance is to use a measured load impedance, along with the magnitude and phase data of the milliohmmeter, to calculate the impedance of the cell itself [4].

17.6.3 The high-frequency resistance (HFR) method

In the HFR method, the electrolyte resistance can be determined by applying a small AC signal to the electronic load to modulate the DC load current. The magnitude and the phase of the AC voltage and current response are measured by a frequency response analyzer. A high, single frequency is used, typically on the order of 1 kHz. The component of interest is the real (Z) component is displayed in milliohms [4].

The HFR measurement minimally disturbs the cell; therefore, it is suitable for routine testing during normal fuel cell operation. The ideal measurement frequency can vary depending upon the fuel cell type, and must be carefully selected. Selecting the proper frequency is accomplished by examining the phase difference between the applied current and measured voltage. If the difference is zero, the cell is functioning in a resistive manner. This frequency should be where the imaginary component of the impedance is zero ($Z'' = 0$). This occurs in the Nyquist plot where the impedance data crosses the real axis at high frequency (small Z'). If the plot does not cross the real axis at high frequency, the highest frequency before the plot deviates from the semicircle shape can be used. HFR frequencies range from 500 to 3 kHz. The same frequency must be used for data comparison. The method of choosing the HFR frequency also requires that the test system has Electrochemical

Impedance Spectroscopy capability. Most true frequency response analyzers can measure over a wide range of frequencies [4].

17.6.4 Electrochemical (EIS) impedance spectroscopy

Electrochemical impedance spectroscopy (EIS) is a sensitive and powerful tool that can be used in-situ to characterize the impedances of the processes that occur within the fuel cell or fuel cell stack [4,7].

Performing EIS measurements can be helpful in identifying the kinetics, ohmic resistances, the electrolytic, contact, and diffusion layer resistances, and the transport limitations in the system. This method uses a small sinusoidal perturbation potential at one or several frequencies to the fuel cell stack, and the response is an alternating current (AC) signal of the same frequency with a possible shift in phase and change in amplitude [7]. In some measurement systems, the AC current is imposed on top of the DC current generated by the fuel cell, and the voltage is recorded instead of the current. The recorded response is used to calculate the impedance using a mathematical technique. By repeating this at a number of frequencies, an Electrochemical Impedance Spectroscopy (EIS) is obtained [4,7].

The EIS measurement is obtained with an electronic load, a function generator, and a computer. The same setup can also be used for obtaining a fuel cell polarization or IV curves.

EIS is a noninvasive technique that varies the current of the fuel cell a few percent of the operating current or voltage. The magnitude of the fluctuations is dependent on the peak-peak current of the AC signal imposed and the impedance of the fuel cell at each frequency.

EIS is useful in characterizing fast and slow transport phenomena because it tests both a single and a large range of frequencies. This test helps to characterize many of the phenomena discussed in Chapters 6-8, such as mass transfer resistance, resistance to electron transfer during electrochemical reactions, and ionic resistance through the membrane [7].

The researcher then uses this data to characterize the various impedances inside the fuel cell. As a result, the interpretation of the data can be ambiguous and difficult, even for experienced users [7].

The EIS technique is an extension of the HFR technique, but differs in two ways. The HFR technique only uses a single frequency, and only examines the real component of the impedance. EIS measures a broad range of frequencies and monitors the resulting variations in magnitude and phase of the cell voltage and current with a frequency response analyzer to determine the complex impedance (Z' and Z'') of the fuel cell.

This results in a data set that can be plotted in Nyquist or Bode formats for analysis and modeling. Modeling software allows accurate analysis of the transport phenomena and kinetics of the fuel cell.

An example of the testing procedure is as follows:

1. The fuel cell is set at predetermined operating conditions.
2. Voltage sensing leads are attached to the positive and negative terminals of the fuel cell or stack. After the fuel cell has been connected to the load, set the DC operating point.
3. Set the function generator to the frequency where the impedance measurement will be made. The EIS will either apply a sinusoidal potential perturbation at each frequency and read the current response or apply a sinusoidal current perturbation and read the voltage response.
4. Look at the fuel-cell voltage and current waveforms with the load.
5. Transfer the data to the PC.
6. Perform an FFT on both the voltage and current waveform data.
7. Divide the FFT voltage data by the FFT current data to obtain the complex impedance.
8. In the complex FFT impedance data, look up the resultant measured impedance at the desired frequency.
9. Repeat the steps to obtain an EIS plot.

Performing the test is straightforward except for selecting the function generator frequency. Electrochemical impedance spectroscopy is the most widely used technique for distinguishing fuel cell losses. Impedance is a measure of a system to impede the flow of electrical current. It is given by the ratio between a time-dependent voltage and current:

$$Z = \frac{V(t)}{i(t)} \tag{17-1}$$

With known amplitude and frequency, a signal is defined as:

$$I = I_{max} \sin(\omega t) \tag{17-2}$$

where I_{max} is the signal amplitude (A), $\omega = 2\pi f$, where f is the frequency (s^{-1} or Hz), and t is time (s).

The response is then:

$$V = V_{max} \sin(\omega t - \varphi) \tag{17-3}$$

where V_{max} is the response amplitude, and φ is the phase angle.

The sinusoidal impedance response of a system is

$$Z = \frac{V_0 \cos(\omega t)}{i_0 \cos(\omega t - \phi)} = Z_0 \frac{\cos(\omega t)}{i_0 \cos(\omega t - \phi)} \qquad (17\text{-}4)$$

In the Nyquist plot, the size of the two semicircle peaks can be attributed to the magnitude of the anode and cathode activation losses. The size of the three impedances corresponds to the relative size of the η_Ω, Z_{fA}, and Z_{fC}. Figure 17-4 clearly shows that the cathode activation loss dominates the fuel cell performance, while the ohmic and the anode activation losses are small [1].

17.6.5 Stoichiometry (utilization) sweeps

A stoichiometry (utilization) sweep is a test conducted to evaluate the fuel and oxidant flow distribution through the cell. Uneven flow distribution could be a result of manifold effects, non-uniform composition of the cell, and uneven machine tolerances. This test is usually conducted for fuels other than hydrogen, and with bipolar plate designs other than serpentine [7].

A typical stoichiometry (utilization) sweep is conducted at a set current, and the stoichiometry is changed. The stack and cell performance should be continuously recorded throughout the testing process. The test is usually conducted in the following manner:

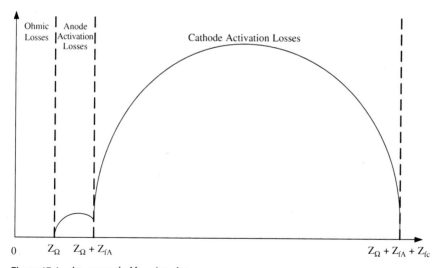

Figure 17-4 An example Nyquist plot.

1. Increase the fuel flow to the electrode not being tested to a high flow rate to insure that this does not affect the test.
2. Increase the flow rate of the electrode being tested.
3. Decrease the flow rate at the electrode being tested in small increments.
4. Return the flow of the electrode being evaluated to the initial values.
5. Return the fuel flow of the electrode not being evaluated to the initial values.
6. Compare the voltage change to the performance level that the Nernst equation predicts.

This test should be carefully conducted because it stresses the cell/stack by approaching starvation. This test is not conducted with "clean fuels" because there is no performance change as the fuel is reduced until starvation. "Dirty fuels" show a change due to the fuel partial pressure. Data should be obtained at multiple points. The "standard" test method uses about 5 to 6 points, with a 15-minute stabilization time between each point. The data from the last five minutes should be averaged and plotted as average current versus average voltage. If the cell becomes unstable during testing, the flows should be increased immediately so that the fuel cell is not damaged [7].

17.6.6 Limiting current

Limiting current measurements are conducted to investigate the mass transport limitations within the fuel cell (MEA). The two common methods for measuring limiting currents are the potentiostatic method or voltage control, and galvanostatic methods or current control. The technique selected depends upon the equipment available, but if there is an access to both methods, the potentiostatic method should be used because there is less of a risk of damaging the fuel cell (MEA).

In the potentiostatic method, the cell voltage is swept from open circuit voltage to 0.3 V, the limiting current is where the slope of the potential–current curve is infinity. In the galvanostatic method, the test must be stopped at ~0.2 V to prevent the voltage from going negative. The limiting current in this method is also when the slope of the potential–current curve is infinity. The measurements should be repeated five times. The limiting current of the anode is determined by diluting the fuel to a concentration level of 1 to 4 percent in N_2 [7]. Once the baseline limiting current level has been established, the effect of contaminants may be determined. The effect of absorbed contaminants reduces the number of available active sites, which results in a reduction of limiting current.

The gas diffusivity in the porous electrode layer or GDL and catalyst layer structure also has a large effect on the limiting current; therefore, it is necessary to use the same fuel cell layers to produce data that are comparable. This technique is a simple, noninvasive technique that is valuable in evaluating the effects of different contaminants under different operating conditions on the same fuel cell. Comparisons between different fuel cell, however, would be difficult [7].

17.6.7 Cyclic voltammetry

A cyclic voltammetry test is useful in evaluating the electrochemical area of both electrodes. The voltage response is swept back and forth between two voltage limits, while the current is measured. From this data, the catalyst surface area can be calculated. The voltage sweep is linear with time, and the typical CV waveform is provided in Figure 17-5.

Although valuable information can be obtained from sweeps of both electrodes over the full voltage range (0 to 1.2 V depending upon the fuel cell), it is recommended that the anode voltage not be raised more than 0.7 V. This is recommended to avoid corrosion of the catalyst support and dissolution of the catalyst. Removal of impurities can be complicated by these two reactions. The experiments need to be carefully thought out to avoid removing the contaminant on one electrode while cycling the opposite electrode. A typical scan rate is 50 mV/s; however, the scan rate will depend upon the electrode size and the maximum current load of the instrument being used. The electrochemical surface area can be

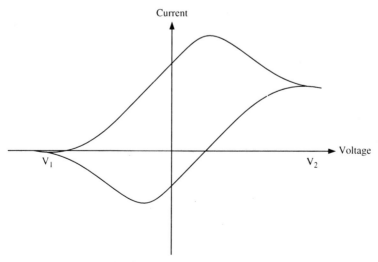

Figure 17-5 An example of a cyclic voltammagram.

determined by calculating the area under the hydrogen adsorption peaks minus the area under the baseline current [7].

17.7 Current Density Mapping

Mapping the current density distribution helps to insure that the fuel cell (catalyst layer and water and heat distribution through the membrane) is uniform. The bipolar plates in a fuel cell are good current collectors; therefore, any current that is collected quickly equalizes. In order to map the current distribution, the measurement must be taken before the current equalizes. This has been accomplished using several different methods in the past, including segmenting the cell, segmenting the current collector, using several small fuel cells or subcells, or a network of passive resistors. The different segments of the fuel cell were controlled using a multichannel potentiostat/galvano-stat [2].

17.8 Neutron Imaging

Neutron imaging provides valuable information about the water management in fuel cells. Neutron imaging is the only in-situ method for visualizing the water distribution in a "real-world" PEMFC. Unlike x-rays, whose interaction with material increases with the number density of electrons, how neutrons interact varies somewhat randomly across the periodic table. For instance, a neutron's interaction with hydrogen is approximately 100 times greater than that with aluminum, and ten times greater than that with deuterium. It is this sensitivity to hydrogen (and insensitivity to many other materials) that is exploited in neutron imaging studies of water transport in operating fuel cells. Using neutrons to characterize a fuel cell provides the necessary contrast to image the hydrogen and water in a PEMFC. Real-Time Radiography has a device that provides an immediate qualitative image of the water management system [2].

17.9 Characterization of Fuel Cell Layers

The fuel cell characterization techniques discussed thus far are used to study fuel cell electrical behavior, but there are also numerous other techniques that can help evaluate and characterize fuel cell materials and performance. These techniques typically evaluate the physical or chemical structure of the fuel cell components in order to help determine the most significant impact on fuel cell performance. Some of these methods include pore structure, catalyst surface area, electrode microstructure, and electrolyte/electrode chemistry.

17.9.1 Porosity determination

The porosity of a material is the ratio of void space to the total volume of the material. Fuel cell electrodes and catalyst layers must exhibit good porosity in order to be effective. Porosity can be determined using several methods. One popular way is to measure the mass and volume, as well as the bulk density of the material. This is an inaccurate method of determining the porosity for fuel cells because it does not take into consideration the pore space that is open to the surface, and also location of the pores in the material. Several volume infiltration techniques can help determine the effective porosity. One method is to immerse the sample in a liquid that does not enter the pores. At a low pressure, mercury will not infiltrate pore spaces because of the surface tension. The method of determining pore size is called mercury porosimetry.

The porous sample is placed in a chamber that is evacuated, mercury is injected into the porous sample. The pressure is extremely low at first, and then it is steadily increased, and the volume at each pressure is recorded. Mercury will enter the pore radius only when the pressure in the chamber is:

$$p \geq \frac{2\gamma}{r} \cos\theta \qquad (17\text{-}5)$$

where γ is the surface tension of mercury (480 dyn/cm^2), θ is the contact angle of mercury (137 degrees in the cermet), r is the intrusion radius of cylindrical pores and p is the applied pressure (typical rate is 1–1800 bar with an increase rate of 1 bar per second) [9]. Using this equation with the experimental mercury pressure data allows pore size curves to be calculated.

17.9.2 BET surface area determination

Surface area determination is an important characterization tool because the most effective catalyst layers have extremely high surface areas. The electrochemically active area can be roughly determined from cyclic voltammetry or impedance measurements. However, the most accurate technique for determining active surface area is the Brunauer-Emmett-Teller (BET) method.

The BET method absorbs an inert gas, such as nitrogen, argon, or krypton on the sample surface at low temperatures. The typical test consists of evacuating a dry sample of all gas, and cooling it to 77 K. A layer of inert gas will physically adhere to the sample surface, which lowers the pressure in the chamber. The surface area of the sample can be calculated from the measured absorption isotherm.

17.9.3 Transmission electron microscopy (TEM)

Transmission electron microscopy (TEM) is an imaging technique that focuses a beam of electrons onto the sample, which causes an enlarged version to appear in the microscope. The TEM image interacts with the electron beam based upon diffraction rather than absorption, although the beam is still affected by the volume and density of the material. The intensity of the diffraction depends upon the orientation of the planes of atoms relative to the electron beam. TEM is a useful technique for analyzing the structure of the catalyst, diffusion layers, and the electrolyte in certain fuel cell types. One of the issues with TEM is that sample preparation may take a lot of time since the sample has to be thin enough to be electron transparent. The field of view is also very small, which raises the possibility that the region analyzed may not be characteristic of the entire sample.

17.9.4 Scanning electron microscopy (SEM)

Scanning electron microscopy produces high-resolution images of a sample surface. The scanning electron microscope (SEM) creates the magnified images by using electrons instead of light waves. Samples have to be prepared carefully to withstand the vacuum inside the microscope. SEM images have a 3-D structure and are good for determining the surface structure of the sample. SEM is a useful technique for analyzing the structure of the catalyst, diffusion layers, and the electrolyte in certain fuel cell types, but like TEM, the field of view is also very small, which raises the possibility that the region analyzed may not be characteristic of the entire sample.

17.9.5 X-ray diffraction (XRD)

X-ray diffraction is a technique that looks at the pattern produced by the diffraction of x-rays through the closely spaced lattice of atoms in the crystal to reveal the nature of the lattice. This gives information about the material and molecular structure of the sample being analyzed. The spacing in the crystal lattice can be determined using Bragg's law. The electrons that surround the atoms, rather than the atomic nuclei themselves, are the entities that physically interact with the incoming x-ray photons. XRD is useful for characterizing the particle size distribution of the catalyst layer in the fuel cell. The test usually takes a lot of time, and can only be performed on a small sample set.

17.9.6 Energy dispersive spectroscopy (EDS)

When the electron beam of the SEM is scanned across a sample surface, it generates x-ray fluorescence from the atoms in its path. The energy of each x-ray photon is characteristic of the element that produced it. The energy dispersive spectroscopy (EDS) microanalysis system collects the x-rays, sorts and plots them by energy, and automatically identifies and labels the elements responsible for the peaks in this energy distribution. Data output is either this element analysis, the original spectrum showing the number of x-rays collected at each energy, or maps of distributions of elements over areas of interest. EDS can be a useful technique for analyzing the structure of the catalyst, diffusion layers, and the electrolyte for fuel cells.

17.9.7 X-ray fluorescence (XRF)

X-ray fluorescence (XRF) is used to measure the chemical analysis of materials, and is a technique that is fast and nondestructive. It is used in industry and the field for measurement and control of materials. XRF can use x-rays or other sources such as alpha particles, protons, or high-energy electron beams. The technique involves aiming an x-ray beam 2 mm in diameter at the surface of an object.

The interaction of x-rays with an object causes secondary x-rays to be generated. Each element produces x-rays of different energies, which can be detected and displayed as a spectrum of intensity against energy. This technique is accurate and fast, but it is not sensitive enough to measure low concentrations of elements (<0.1 percent). It will, however, quickly determine the alloy composition of a metal, or analyze materials such as ceramics and glass. One limitation of the technique is that only a thin layer, less than 0.1 mm, is actually analyzed. This can sometimes give misleading results on corroded or plated metals unless the surface is cleaned.

17.9.8 Inductively coupled plasma mass spectroscopy (ICP-MS)

Inductively coupled plasma mass spectroscopy (ICP-MS) is a type of mass spectroscopy that is highly sensitive and capable of analysis of many types of metals and nonmetals below one part in 10^{12}. It uses inductively coupled plasma to produce ions that can be detected and identified by comparison with an elemental standard. This technique analyzes elements with mass ranges from 7 to 250 (Li to U). An ICP-MS can detect particle levels as low as parts per trillion to 10–100 parts per million.

Chapter Summary

Fuel cell characterization techniques allow comparison of almost every property of every part of the fuel cell stack. Selecting a combination of characterization techniques will allow the researcher to determine why the fuel cell is performing well or poorly. The results of these tests will help discriminate between activation, ohmic and concentration losses, fuel crossover, defective materials, as well as many other properties. Some of the electrical measurements described in this chapter include polarization curves, the current interrupt method, the AC resistance method, the high-frequency resistance method, electrochemical impedance spectroscopy, stoichiometry sweeps, limiting current tests, cyclic voltammetry, pressure drop measurements, and current density mapping. Other types of testing for characterizing the fuel cell layers include neutron imaging, porosity determination, BET surface area determination, transmission electron microscopy, scanning electron microscopy, x-ray diffraction, energy dispersive spectroscopy, x-ray fluorescence, and inductively coupled plasma mass spectroscopy. In-situ testing is very important for fuel cells because the performance of the fuel cell cannot be determined simply by characterizing its parts. However, the actual parts and the contact between the parts can also cause decreased performance in fuel cells.

Problems

1. List several methods to check for cell resistance. Describe how to use these methods.

2. Describe how hysteresis shows up in the fuel cell polarization curve.

3. How does cell flooding show up in fuel cell testing?

4. Draw what a fuel cell IV curve looks like when there is a reduced voltage as I increases. How can the voltage be increased?

5. Draw a Nyquist diagram that has flooding in the cathode. How can this be resolved?

Bibliography

[1] O'Hayre, Ryan, Suk-Won Cha, Whitney Colella, and Fritz B. Prinz. *Fuel Cell Fundamentals*. 2006. New York: John Wiley & Sons.
[2] Barbir, Frano. *PEM Fuel Cells: Theory and Practice*. 2005. Burlington, MA: Elsevier Academic Press.
[3] Raposa, Gary. "Performing AC Impedance Spectroscopy Measurements on Fuel Cells." *Fuel Cell Magazine*. February/March 2003.
[4] Smith, Matt, Kevin Cooper, Derek Johnson, and Louie Scribner. "Comparison of Fuel Cell Electrolyte Resistance Measurement Techniques." *Fuel Cell Magazine*, April/May 2005.
[5] Pekula, N., K. Heller, P.A. Chuang, A. Turhan, M.M. Mench, J.S. Brenzier, and K. Unlu. "Study of Water Distribution and Transport in a Polymer Electrolyte Fuel Cell Using Neutron Imaging." *Nuclear Instruments and Methods in Physics Research A*. Vol. 542, 2005, pp. 134–141.

[6] Turhan, A., K. Heller, J.S. Brenizer, and M.M Mench. "Quantification of Liquid Water Accumulation and Distribution in a Polymer Electrolyte Fuel Cell Using Neutron Imaging." *Journal of Power Sources.* Vol. 160, 2006, pp. 1195–1203.

[7] "Primer on Fuel Cell Component testing: Primer for Generating Test Plans." The U.S. Fuel Cell Council's Joint Hydrogen Quality Task Force. Document No. USFCC 04-003. November 29, 2004. www.usfcc.com.

[8] Lu, G.Q.and C.Y. Wang. Development of micro direct methanol fuel cells for high power applications. *Journal of Power Sources.* Vol. 144, 2005, pp. 141–145.

[9] Lane, A. M. "Interpretation of Mercury Porosimetry Data." Ph.D Thesis. University of Massachusetts. 1984, pp. 3–7.

Appendix A

Useful Constants and Conversions

The following table lists constants and conversions that may be helpful for fuel cell design calculations. These constants are used in many equations throughout the book.

Physical Constants

Avogadro's number	N_A	6.02×10^{23} atoms/mol
Universal gas constant	R	0.08205 L atm/mol K
		8.314 J/mol K
		0.08314 bar m^3/mol K
		8.314 kPa m^3/mol K
Planck's constant	h	6.626×10^{-34} J s
		4.136×10^{-15} eV·s
Boltzmann's constant	k	1.38×10^{-23} J/K
		8.61×10^{-5} eV/K
Electron mass	m	9.11×10^{-31} kg
Electron charge	q	1.60×10^{-19} C
Faraday's constant	F	96,485.34 C/mol

Appendix B

Thermodynamic Properties of Selected Substances

The following table lists enthalpy of formation, the Gibbs function of formation, and absolute entropy for selected substances at 25°C and 1 atm.

Substance	Formula	H_f (J/mol)	G_f (J/mol)	S (J/mol∗K)
Acetylene	$C_2H_2(g)$	+226,730	+209,170	200.85
Ammonia	$NH_3(g)$	−46,190	−16,590	192.33
Benzene	$C_6H_6(g)$	+82,930	+129,660	269.20
Carbon	C(s)	0	0	5.74
Carbon Dioxide	$CO_2(g)$	−393,522	−394,360	213.80
Carbon Monoxide	CO(g)	−110,530	−137,150	197.65
Ethane	$C_2H_6(g)$	−84,680	−32,890	229.49
Ethyl Alcohol	$C_2H_5OH(g)$	−235,310	−168,570	282.59
Ethyl Alcohol	$C_2H_5OH(l)$	−277,690	−174,891	160.70
Ethylene	$C_2H_4(g)$	+52,280	+68,120	219.83
Hydrogen	$H_2(g)$	0	0	130.68
Hydrogen	H(g)	+217,999	+203,290	114.72
Hydrogen Peroxide	$H_2O_2(g)$	−136,310	−105,600	232.63
Hydroxyl	OH(g)	+38,987	+34,280	183.70
Methane	$CH_4(g)$	−74,850	−50,790	186.16
Methyl Alcohol	$CH_3OH(g)$	−200,670	−162,000	239.70
Methyl Alcohol	$CH_3OH(l)$	−238,660	−166,360	126.80
n-Butane	$C_4H_{10}(g)$	−126,150	−15,710	310.12
n-Dodecane	$C_{12}H_2(g)$	−291,010	+50,150	622.83
Nitrogen	$N_2(g)$	0	0	191.61
Nitrogen	N(g)	+472,680	+455,510	153.30
n-Octane	$C_8H_{18}(g)$	−208,450	+16,530	466.73
n-Octane	$C_8H_{18}(l)$	−249.950	+6,610	360.79
Oxygen	$O_2(g)$	0	0	205.14
Oxygen	O(g)	+249,170	+231,770	161.06
Propane	$C_3H_8(g)$	−130,850	−23,490	269.91
Propylene	$C_3H_6(g)$	+20,410	+62,720	266.94
Water	$H_2O(g)$	−241,826	228,590	188.83
Water	$H_2O(l)$	−285,826	237,180	69.92

Appendix C

Molecular Weight, Gas Constant and Specific Heat for Selected Substances

The following table lists the molecular weight, gas constants and ideal gas specific heat at constant pressure for selected substances at 300 K (Adapted from Li, Xianguo. Principles of Fuel Cells. 2006. New York: Taylor & Francis Group).

Gas	Formula	Molecular weight, W, kg/kmol	Gas constant kJ/(kg K)	Specific heat, C_p, kJ/(kg K)
Air	—	28.97	0.2870	1.005
Argon	Ar	39.948	0.2081	0.5203
Butane	C_4H_{10}	58.124	0.1433	1.7164
Carbon Dioxide	CO_2	44.01	0.1889	0.846
Carbon Monoxide	CO	28.011	0.2968	1.040
Ethane	C_2H_6	30.070	0.2765	1.7662
Ethylene	C_2H_4	28.054	0.2964	1.5482
Helium	He	4.003	2.0769	5.1926
Hydrogen	H_2	2.016	4.1240	14.307
Methane	CH_4	16.043	0.5182	2.2537
Neon	Ne	20.183	0.4119	1.0299
Nitrogen	N_2	28.013	0.2968	1.039
Octane	C_8H_{18}	114.230	0.0729	1.7113
Oxygen	O_2	31.999	0.2598	0.918
Propane	C_3H_8	44.097	0.1885	1.6794
Steam	H_2O	18.015	0.4615	1.8723

Appendix D

Gas Specific Heats at Various Temperatures

The following table lists the ideal gas specific heat at constant pressure at various temperatures (Adapted from Li, Xianguo. Principles of Fuel Cells. 2006. New York: Taylor & Francis Group).

T (K)	Air kJ/(kg K)	CO_2 kJ/(kg K)	CO kJ/(kg K)	H_2 kJ/(kg K)	$H_2O(g)$ kJ(kmol K)	N_2 kJ/(kg K)	O_2 kJ/(kg K)
250	1.003	0.791	1.039	14.051	33.324	1.039	0.913
300	1.005	0.846	1.040	14.307	33.669	1.039	0.918
350	1.008	0.895	1.043	14.427	34.051	1.041	0.928
400	1.013	0.939	1.047	14.476	34.467	1.044	0.941
450	1,020	0.978	1.054	14.501	34.914	1.049	0.956
500	1.029	1.014	1.063	14.513	35.390	1.056	0.972
550	1.040	1.046	1.075	14.530	35.891	1.065	0.988
600	1.051	1.075	1.087	14.546	36.415	1.075	1.003
650	1.063	1.102	1.100	14.571	36.960	1.086	1.017
700	1.075	1.126	1.113	14.604	37.523	1.098	1.031
750	1.087	1.148	1.126	14.645	38.100	1.110	1.043
800	1.099	1.169	1.139	14.695	38.690	1.121	1.054
900	1.121	1.204	1.163	14.822	39.895	1.145	1.074
1000	1.142	1.234	1.185	14.983	41.118	1.167	1.090

Appendix E

Specific Heat for Saturated Liquid Water at Various Temperatures

The following table lists specific heat at constant pressure for saturated liquid water H_2O (Adapted from Li, Xianguo. Principles of Fuel Cells. 2006. New York: Taylor & Francis Group).

Temperature (°C)	Specific heat, C_p kJ/(kg K)
0	4.2178
20	4.1818
40	4.1784
60	4.1843
80	4.1964
100	4.2161
120	4.250
140	4.283
160	4.342
180	4.417

Appendix F

Thermodynamic Data for Selected Fuel Cell Reactants at Various Temperatures

The following tables list thermodynamic data for hydrogen, oxygen, water, carbon monoxide, carbon dioxide, methane, and nitrogen for 200–1000 K.

Appendix F

Hydrogen Thermodynamic Data

T(K)	$\hat{g}$ (T) (kJ/mol)	$\hat{h}$ (T) (kJ/mol)	$\hat{s}$ (T) (J/mol K)	C_p(T) (J/mol K)
200	−26.66	−2.77	119.42	27.26
220	−29.07	−2.22	122.05	27.81
240	−31.54	−1.66	124.48	28.21
260	−34.05	−1.09	126.75	28.49
280	−36.61	−0.52	128.87	28.7
298.15	−38.96	0	130.68	28.84
300	−39.20	0.05	130.86	28.85
320	−41.84	0.63	132.72	28.96
340	−44.51	1.21	134.48	29.04
360	−47.22	1.79	136.14	29.1
380	−49.96	2.38	137.72	29.15
400	−52.73	2.96	139.22	29.18
420	−55.53	3.54	140.64	29.21
440	−58.35	4.13	142	29.22
460	−61.21	4.71	143.3	29.24
480	−64.08	5.3	144.54	29.25
500	−66.99	5.88	145.74	29.26
520	−69.91	6.47	146.89	29.27
540	−72.86	7.05	147.99	29.28
560	−75.83	7.64	149.06	29.3
580	−78.82	8.22	150.08	29.31
600	−81.84	8.81	151.08	29.32
620	−84.87	9.4	152.04	29.34
640	−87.92	9.98	152.97	29.36
660	−90.99	10.57	153.87	29.39
680	−94.07	11.16	154.75	29.41
700	−97.18	11.75	155.61	29.44
720	−100.30	12.34	156.44	29.47
740	−103.43	12.93	157.24	29.5
760	−106.59	13.52	158.03	29.54
780	−109.75	14.11	158.8	29.58
800	−112.94	14.7	159.55	29.62
820	−116.14	15.29	160.28	29.67
840	−119.35	15.89	161	29.72
860	−122.58	16.48	161.7	29.77
880	−125.82	17.08	162.38	29.83
900	−129.07	17.68	163.05	29.88
920	−132.34	18.27	163.71	29.94
940	−135.62	18.87	164.35	30
960	−138.91	19.47	164.99	30.07
980	−142.22	20.08	165.61	30.14
1000	−145.54	20.68	166.22	30.2

Oxygen Thermodynamic Data

T(K)	$\hat{g}(T)$ (kJ/mol)	$\hat{h}(T)$ (kJ/mol)	$\hat{s}(T)$ (J/mol K)	$C_p(T)$ (J/mol K)
200	−41.54	−2.71	194.16	25.35
220	−45.45	−2.19	196.63	26.41
240	−49.41	−1.66	198.97	27.25
260	−53.41	−1.10	201.18	27.93
280	−57.45	−0.54	203.27	28.48
298.15	−61.12	0.00	205.00	28.91
300	−61.54	0.03	205.25	28.96
320	−65.66	0.62	207.13	29.36
340	−69.82	1.21	208.92	29.71
360	−74.02	1.81	210.63	30.02
380	−78.25	2.41	212.26	30.30
400	−82.51	3.02	213.82	30.56
420	−86.80	3.63	215.32	30.79
440	−91.12	4.25	216.75	31.00
460	−95.47	4.87	218.14	31.20
480	−99.85	5.50	219.47	31.39
500	−104.25	6.13	220.75	31.56
520	−108.68	6.76	221.99	31.73
540	−113.13	7.40	223.20	31.89
560	−117.61	8.04	224.36	32.04
580	−122.10	8.68	225.48	32.19
600	−126.62	9.32	226.58	32.32
620	−131.17	9.97	227.64	32.46
640	−135.73	10.62	228.67	32.59
660	−140.31	11.27	229.68	32.72
680	−144.92	11.93	230.66	32.84
700	−149.54	12.59	231.61	32.96
720	−154.18	13.25	232.54	33.07
740	−158.84	13.91	233.45	33.19
760	−163.52	14.58	234.33	33.30
780	−168.21	15.24	235.20	33.41
800	−172.93	15.91	236.05	33.52
820	−177.66	16.58	236.88	33.62
840	−182.40	17.26	237.69	33.72
860	−187.16	17.93	238.48	33.82
880	−191.94	18.61	239.26	33.92
900	−196.73	19.29	240.02	34.02
920	−201.54	19.97	240.77	34.12
940	−206.36	20.65	241.51	34.21
960	−211.20	21.34	242.23	34.30
980	−216.05	22.03	242.94	34.40
1000	−220.92	22.71	243.63	34.49

$H_2O(l)$ Thermodynamic Data

T(K)	$\hat{g}$ (T) (kJ/mol)	$\hat{h}$ (T) (kJ/mol)	$\hat{s}$ (T) (J/mol K)	C_p(T) (J/mol K)
273	−305.01	−287.73	63.28	76.10
280	−305.46	−287.20	65.21	75.81
298.15	−306.69	−285.83	69.95	75.37
300	−306.82	−285.69	70.42	75.35
320	−308.27	−284.18	75.28	75.27
340	−309.82	−282.68	79.85	75.41
360	−311.46	−281.17	84.16	75.72
373	−312.58	−280.18	86.85	75.99

$H_2O(g)$ Thermodynamic Data

T(K)	$\hat{g}$ (T) (kJ/mol)	$\hat{h}$ (T) (kJ/mol)	$\hat{s}$ (T) (J/mol K)	C_p(T) (J/mol K)
280	−294.72	−242.44	186.73	33.53
298.15	−298.13	−241.83	188.84	33.59
300	−298.48	−241.77	189.04	33.60
320	−302.28	−241.09	191.21	33.69
340	−306.13	−240.42	193.26	33.81
360	−310.01	−239.74	195.20	33.95
380	−313.94	−239.06	197.04	34.10
400	−317.89	−238.38	198.79	34.26
420	−321.89	−237.69	200.47	34.44
440	−325.91	−237.00	202.07	34.62
460	−329.97	−236.31	203.61	34.81
480	−334.06	−235.61	205.10	35.01
500	−338.17	−234.91	206.53	35.22
520	−342.32	−234.20	207.92	35.43
540	−346.49	−233.49	209.26	35.65
560	−350.69	−232.77	210.56	35.87
580	−354.91	−232.05	211.82	36.09
600	−359.16	−231.33	213.05	36.32
620	−363.43	−230.60	214.25	36.55
640	−367.73	−229.87	215.41	36.78
660	−372.05	−229.13	216.54	37.02
680	−376.39	−228.39	217.65	37.26
700	−380.76	−227.64	218.74	37.50
720	−385.14	−226.89	219.80	37.75
740	−389.55	−226.13	220.83	37.99
760	−393.97	−225.37	221.85	38.24
780	−398.42	−224.60	222.85	38.49
800	−402.89	−223.83	223.83	38.74
820	−407.37	−223.05	224.78	38.99
840	−411.88	−222.27	225.73	39.24
860	−416.40	−221.48	226.65	39.49
880	−420.94	−220.69	227.56	39.74
900	−425.51	−219.89	228.46	40.00
920	−430.08	−219.09	229.34	40.25
940	−434.68	−218.28	230.21	40.51
960	−439.29	−217.47	231.07	40.76
980	−443.92	−216.65	231.91	41.01
1000	−448.57	−215.83	232.74	41.27

CO Thermodynamic Data

T(K)	$\hat{g}$ (T) (kJ/mol)	$\hat{h}$ (T) (kJ/mol)	$\hat{s}$ (T) (J/mol K)	C_p(T) (J/mol K)
200	−150.60	−113.42	185.87	30.20
220	−154.34	−112.82	188.73	29.78
240	−158.14	−112.23	191.31	29.50
260	−161.99	−111.64	193.66	29.32
280	−165.89	−111.06	195.83	29.20
298.15	−169.46	−110.53	197.66	29.15
300	−169.83	−110.47	197.84	29.15
320	−173.80	−109.89	199.72	29.13
340	−177.81	−109.31	201.49	29.14
360	−181.86	−108.72	203.16	29.17
380	−185.94	−108.14	204.73	29.23
400	−190.05	−107.56	206.24	29.30
420	−194.19	−106.97	207.67	29.39
440	−198.36	−106.38	209.04	29.48
460	−202.55	−105.79	210.35	29.59
480	−206.77	−105.20	211.61	29.70
500	−211.01	−104.60	212.83	29.82
520	−215.28	−104.00	214.00	29.94
540	−219.57	−103.40	215.13	30.07
560	−223.89	−102.80	216.23	30.20
580	−228.22	−102.19	217.29	30.34
600	−232.58	−101.59	218.32	30.47
620	−236.95	−100.98	219.32	30.61
640	−241.35	−100.36	220.29	30.75
660	−245.77	−99.75	221.24	30.89
680	−250.20	−99.13	222.17	31.03
700	−254.65	−98.50	223.07	31.17
720	−259.12	−97.88	223.95	31.31
740	−263.61	−97.25	224.81	31.46
760	−268.12	−96.62	225.65	31.60
780	−272.64	−95.99	226.47	31.74
800	−277.17	−95.35	227.28	31.88
820	−281.73	−94.71	228.07	32.01
840	−286.30	−94.07	228.84	32.15
860	−290.88	−93.43	229.60	32.29
880	−295.48	−92.78	230.34	32.42
900	−300.09	−92.13	231.07	32.55
920	−304.72	−91.48	231.79	32.68
940	−309.37	−90.82	232.49	32.81
960	−314.02	−90.17	233.18	32.94
980	−318.69	−89.51	233.86	33.06
1000	−323.38	−88.84	234.53	33.18

CO_2 Thermodynamic Data

T(K)	$\hat{g}$ (T) (kJ/mol)	$\hat{h}$ (T) (kJ/mol)	$\hat{s}$ (T) (J/mol K)	C_p(T) (J/mol K)
200	−436.93	−396.90	200.1	31.33
220	−440.95	396.25	203.16	32.77
240	−445.04	−395.59	206.07	34.04
260	−449.19	−394.89	208.84	35.19
280	−453.39	−394.18	211.48	36.24
300	−457.65	−393.44	214.02	37.22
320	−461.95	−392.69	216.45	38.13
340	−466.31	−391.92	218.79	39
360	−470.71	−391.13	221.04	39.81
380	−475.15	−390.33	223.21	40.59
400	−479.63	−389.51	225.31	41.34
420	−484.16	−388.67	227.35	42.05
440	−488.73	−387.83	229.32	42.73
460	−493.33	−386.96	231.23	43.38
480	−497.98	−386.09	233.09	44.01
500	−502.66	−385.20	234.9	44.61
520	−507.37	−384.31	236.66	45.2
540	512.12	−383.40	238.38	45.76
560	−516.91	−382.48	240.05	46.3
580	−521.72	−381.54	241.69	46.82
600	526.59	−380.60	243.28	47.32
620	−531.46	−379.65	244.84	47.8
640	−536.37	−378.69	246.37	48.27
660	−541.31	−377.72	247.86	48.72
680	−546.28	−376.74	249.32	49.15
700	−551.29	−375.76	250.75	49.57
720	−556.31	−374.76	252.15	49.97
740	−561.37	−373.76	253.53	50.36
760	−566.45	−372.75	254.88	50.73
780	−571.56	−371.73	256.2	51.09
800	−576.71	−370.70	257.5	51.44
820	581.86	−369.67	258.77	51.78
840	−587.05	−368.63	260.02	52.1
860	−592.26	−367.59	261.25	52.41
880	−597.50	−366.54	262.46	52.71
900	−602.76	365.48	263.65	53
920	−608.05	−364.42	264.82	53.28
940	−613.35	−363.35	265.97	53.55
960	−618.68	−362.27	267.1	53.81
980	624.04	−361.19	268.21	54.06
1000	−629.41	−360.11	269.3	54.3

CH$_4$ Thermodynamic Data

T(K)	$\hat{g}$(T) (kJ/mol)	$\hat{h}$(T) (kJ/mol)	$\hat{s}$(T) (J/mol K)	C_p(T) (J/mol K)
200	−112.69	−78.25	172.23	36.30
220	−116.17	−77.53	175.63	35.19
240	−119.71	−76.83	178.67	34.74
260	−123.32	−76.14	181.45	34.77
280	−126.97	−75.44	184.03	35.12
298.15	−130.33	−74.80	186.25	35.65
300	−130.68	−74.73	186.48	35.71
320	−134.43	−74.01	188.80	36.47
340	−138.23	−73.27	191.04	37.36
360	−142.07	−72.52	193.20	38.35
380	−145.95	−71.74	195.31	39.40
400	−149.88	−70.94	197.35	40.50
420	−153.85	−70.12	199.36	41.64
440	−157.86	−69.27	201.32	42.80
460	−161.90	−68.41	203.25	43.98
480	−165.99	−67.51	205.15	45.16
500	−170.11	−66.60	207.01	46.35
520	−174.27	−65.66	208.86	47.54
540	−178.46	−64.70	210.67	48.73
560	−182.69	−63.71	212.47	49.90
580	−186.96	−62.70	214.24	51.07
600	−191.26	−61.67	215.99	52.23
620	−195.60	−60.61	217.72	53.37
640	−199.97	−59.53	219.43	54.50
660	−204.38	−58.43	221.13	55.61
680	−208.82	−57.31	222.80	56.71
700	−213.29	−56.16	224.46	57.79
720	−217.79	−55.00	226.10	58.85
740	−222.33	−53.81	227.73	59.90
760	−226.90	−52.60	229.34	60.93
780	−231.51	−51.37	230.94	61.94
800	−236.14	−50.13	232.52	62.93
820	−240.81	−48.86	234.08	63.90
840	−245.50	−47.57	235.64	64.85
860	−250.23	−46.26	237.17	65.79
880	−254.99	−44.94	238.70	66.70
900	−259.78	−43.60	240.20	67.60
920	−264.60	−42.23	241.70	68.47
940	−269.45	−40.86	243.18	69.33
960	−274.33	−39.46	244.65	70.17
980	−279.23	−38.05	246.11	70.99
1000	−284.17	−36.62	247.55	71.79

N_2 Thermodynamic Data

T(K)	$\hat{g}$ (T) (kJ/mol)	$\hat{h}$ (T) (kJ/mol)	$\hat{s}$ (T) (J/mol K)	C_p(T) (J/mol K)
200	−38.85	−2.83	180.08	28.77
220	−42.48	−2.26	182.82	28.72
240	−46.16	−1.68	185.31	28.72
260	−49.89	−1.11	187.61	28.76
280	−53.66	−0.53	189.75	28.81
298.15	−57.11	0	191.56	28.87
300	−57.48	0.04	191.74	28.88
320	−61.33	0.62	193.6	28.96
340	−65.22	1.2	195.36	29.05
360	−69.15	1.78	197.02	29.14
380	−73.10	2.37	198.6	29.25
400	−77.09	2.95	200.11	29.35
420	−81.11	3.54	201.54	29.46
440	−85.15	4.13	202.91	29.57
460	−89.22	4.72	204.23	29.68
480	−93.32	5.32	205.5	29.79
500	−97.44	5.92	206.71	29.91
520	−101.59	6.51	207.89	30.02
540	−105.76	7.12	209.02	30.13
560	−109.95	7.72	210.12	30.24
580	−114.16	8.33	211.19	30.36
600	−118.40	8.93	212.22	30.47
620	−122.65	9.54	213.22	30.58
640	−126.92	10.16	214.19	30.69
660	−131.22	10.77	215.14	30.8
680	−135.53	11.39	216.06	30.91
700	−139.86	12.01	216.96	31.02
720	−144.21	12.63	217.83	31.13
740	−148.57	13.25	218.69	31.24
760	−152.96	13.88	219.52	31.34
780	−157.35	14.51	220.34	31.45
800	−161.77	15.14	221.13	31.55
820	−[66.20	15.77	221.91	31.66
840	−[70.64	16.4	222.68	31.76
860	−175.11	17.04	223.43	31.86
880	−179.58	17.68	224.16	31.96
900	−184.07	18.32	224.88	32.06
920	−188.58	18.96	225.58	32.16
940	−193.10	19.61	226.28	32.25
960	−19.63	20.25	226.96	32.35
980	−202.1	20.9	227.63	32.44
1000	−206.3	21.55	228.28	32.54

Appendix G

Binary Diffusion Coefficients for Selected Fuel Cell Substances

The following table lists binary diffusion coefficients for relevant fuel cell substances at one atmosphere.

Substance A	Substance B	T(K)	D_{AB} (m²/s)
Acetone	Air	298	0.11×10^{-4}
Acetone	H_2O	298	0.13×10^{-8}
Ar	N_2	293	0.19×10^{-4}
Benzene	Air	298	0.88×10^{-5}
CO_2	Air	298	0.16×10^{-4}
CO_2	N_2	293	0.16×10^{-4}
CO_2	O_2	273	0.14×10^{-4}
CO_2	H_2O	298	0.20×10^{-8}
Ethanol	H_2O	298	0.12×10^{-8}
G	H_2O	298	0.94×10^{-9}
Glucose	H_2O	298	0.69×10^{-9}
H_2	Air	273	0.41×10^{-4}
H_2	O_2	273	0.70×10^{-4}
H_2	N_2	273	0.68×10^{-4}
H_2	CO_2	273	0.55×10^{-4}
H_2	H_2O	298	0.63×10^{-8}
H_2O	Air	298	0.26×10^{-4}
N_2	H_2O	298	0.26×10^{-8}
Naphthalene	Air	300	0.62×10^{-5}
NH_3	Air	298	0.28×10^{-4}
O_2	Air	298	0.21×10^{-4}
O_2	N_2	273	0.18×10^{-4}
O_2	H_2O	298	0.24×10^{-8}

Appendix

Product Design Specifications

The following table lists preliminary product design specifications that need to be considered before beginning a fuel cell design. Many of these specifications require in-depth reports or studies (or entire departments) depending upon the scope of the project or company size.

Product Design Specifications

Product Title	
Purpose	
New or Special Features	
Competition	
Intended Market	
Need for Product	
Relationship to Existing Product Lines	
Market Demand	
Product Differentiation	
Costs	
Functional Performance	
Physical Requirements	
Uses/Environmental Conditions	
Issues to Be Resolved	

Appendix I

Fuel Cell Design Requirements and Parameters

The following tables list many of the important parameters and fuel cell stack components and materials that should be considered when designing a fuel cell. The requirements must first be thought about and estimated as shown in the "Stack Performance Parameters" table. From the initial requirements, estimates can be made about the stack weight and volume. There are also tables that display additional performance requirements, operating characteristics, and a detailed list of components. These tables are included as a starting point for the fuel cell design, and should be altered according to the fuel cell type, requirements, and design.

Initial Fuel Cell Design Requirements

Stack Performance Parameters

	Abbreviation	Value	Unit
Number of Cells	N_{cell}		N/A
Cell Voltage	V_{cell}		V
Active Area	A_{active}		cm^2
Stack Voltage	V_{stack}		V
Current	I		A
Current Density	i		mA/cm^2
Specific Power			(W/kg)
Power Density			(W/L)
Stack Power	P		(Watts)
Peak Power			(Watts)

Fuel Cell Weight, Volume, and Dimensions

Approximated Weight/Thickness of Fuel Cell Stack Repeating Units

Component	Thickness	Weight
Bipolar (Flowfield) Plate		
Anode Gasket		
Assembled Fuel Cell (MEA)		
Cathode Gasket		
Total Repeat Thickness/Weight		

Approximated Weight/Thickness for Non-Repeat Stack Components

Fuel Cell Part	Thickness	Weight
One Stack Housing		
Two Endplates		
Two Current Collectors		
Rods for Stack Clamping	N/A	
Total Non-Repeat Thickness/Weight		

Stack Dimensions

Assembled Fuel Cell (MEA) Active Area	
Active Assembled Fuel Cell (MEA) Dimensions	
Total Assembled Fuel Cell (MEA) Dimensions (including inactive margin)	
Gasket Dimensions	
Bipolar (Flowfield) Plate Dimensions	
End Plate Dimensions	
Total Dimensions of Stack Face	
Total Volume	
Total Dimensions of Fuel Cell	

Summary of Stack Weight and Volume

Fuel Cell Active Area	
Non-Repeat Mass	
Repeat Mass	
Total Mass	
Non-Repeat Thickness	
Repeat Thickness	
Total Thickness	
Total Volume	
Total Dimensions of Fuel Cell	
Overall Dimensions with Fuel Cell Housing	

Fuel Cell Performance

Additional Stack Perforance Parameters

	Abbreviation	Value	Unit
Stack Temperature	T_{stack}		°C
Stack Pressure	P_{stack}		Pa
Anode/Cathode Gas Mass Inlet Flow Ratio			
Stack (Voltage) Efficiency			
Fuel Utilization Efficiency			
Fuel Conversion Efficiency			
Response Speed			
Power Stability			
Start-Up Time			
Durability			
Operation Life			
Shelf Life			
Recyclability			
Vibration Tolerance			

Stack Components Performance Parameters

	Abbreviation	Value	Unit
Clamp Load			
Active Area Load			Pa
Gasket Load per Length			
Cell Operational Resistance per Active Area	$R_{cell\ operation}$		Ohm-cm^2
Membrane Ionic Resistance per Active Area	$R_{membrane}$		Ohm-cm^2
Electrode Ohmic Resistance per Active Area	$R_{electrode}$		Ohm-cm^2
GDL Ohmic Resistance per Active Area	R_{GDL}		Ohm-cm^2
MEA/Plate Contact Ohmic Resistance per Active Area	$R_{contact}$		Ohm-cm^2
Plate Bulk Ohmic Resistivity			Ohm-cm
Plate Bulk Ohmic Resistance per Active Area	R_{plate}		Ohm-cm^2

Appendix I

Anode Performance Parameters

	Abbreviation	Value	Unit
Pressure	P_{anode}		Pa
Mass Flow Rate	m_{anode}		kg/sec
Volumetric Flow Rate	V_{anode}		cm^3/sec
Flow Speed	v_{anode}		cm/sec
Temperature	T_{anode}		°C
Relative Humidity	RH_{anode}		%
Gas Mass Utilization Ratio	TBD		TBD

Cathode Performance Parameters

	Abbreviation	Value	Unit
Pressure	$P_{cathode}$		Pa
Mass Flow Rate	$m_{cathode}$		kg/sec
Volumetric Flow Rate	$V_{cathode}$		cm^3/sec
Flow Speed	$v_{cathode}$		cm/sec
Temperature	$T_{cathode}$		°C
Relative Humidity	$RH_{cathode}$		%
Gas Mass Utilization Ratio	TBD		TBD

Appendix I

Fuel Cell Stack Components

Item	Description	Dimensions
Bipolar Plates		
Gas Manifolds		
Clamping Bolts		
Gaskets		
Catalyst		
Electrolyte		
Gas Diffusion Backings		
Current Collectors		
End Plates		
Plastic Housing		
Blower/Fan/Pumps		
Thermocouples		
Hydrogen/Oxygen Sensors		
Pressure Transducers		
Carbon Monoxide Sensors		
Mass Flow Controllers		
Altitude Sensor		
Particulate Filters		
High-Flow Check Valve		
Humidifier		
Tubing		
Dot-Matrix LCD and Key Pad		
Microprocessor		
Heaters		

Index

Absolute enthalpy, 87
Absolute entropy, 397
AC (*see* Alternating current)
AC chargers, 54
AC resistance method, 382
Accumetrics Corporation, 75
Acetone, 413
Acetylene, 397
AC-Impedance measurement, 383
"A-Cracker," 332
Activation losses, 181–182
Activation overvoltage, 138
Activation polarization, 114
Adiabatic system, 279
AEP (annual electricity produced), 81
Aerovironment, 70
AFCs (*see* Alkaline fuel cells; Annual fuel costs)
Air, heat at constant pressure for, 399, 401
Air cooling, 163–166
Air mass flow rate, 365
Air out, 275
Alanate, 327
Alcoholysis, 330
Algae, 344
Alkaline fuel cells (AFCs), 11–12, 40
  aqueous electrolytes in, 126
  catalysts for, 196, 207
  cell interconnection in, 264
  comparison of, with other fuel types, 37–38
  direct borohydride fuel cells, 36
  gaskets for, 238–239
  history of, 8
  materials for, 191
  operating conditions for, 354
  polarization curves for, 253
  stack of, 222
Allis-Chalmers farm tractor, 8, 11
Alternating current (AC), 297, 383
Alternative energy sources, 1, 6
Alternative membranes, hydrocarbon, 194
Alternative polymer membranes, 195
Aluminum:
  for bipolar plates, 236
  conductivity of, 158
Aluminum borohydride, 331
Aluminum hydride, 331
AMC (annual maintenance cost), 83
American Fuel Cell Corp., 75
American Gas Association, 11

Ammonia, 331, 332, 397
Anabaena Cylindrica, 345
Analytic Power, 332
Annual electricity produced (AEP), 81
Annual fuel costs (AFCs), 82
Annual maintenance cost (AMC), 83
Anodes, 3, 12
  electrochemical reaction in, 158
  humidification of, 267
  performance parameters for, 420
Anodic Gibbs free energy, 158
Anodization, chemical, 226
Ansaldo Fuel Cells Spa, 75
Anuvu, Inc., 70
APFCT (Asia Pacific Fuel Cell Technologies), 73
Apollo (*see* Project Apollo)
Apollo Energy Systems Inc., 75
Applications, 53–84
  backup power, 55–57
  portable power, 53–55
  stationary power, 72, 74–83
  transportation, 57–72
APUs (auxiliary power units), 42
Aqueous electrolytes:
  charge transport in, 126–127
  ionic conductivity in, 126–127
  ions in, 127
Argon, 399
Army (*see* U.S. Army)
Asahi Chemical K-101, 193
Asahi Glass CMV, 193
Asahi Glass DMV, 193
Asia Pacific Fuel Cell Technologies (APFCT), 73
Assembly, verification of, 376
Astris Energi:
  stationary power systems, 75
  utility vehicles, 70
Atmospheric pressure system, 357
Audi, 61
Australia, hydrogen economy and, 31
Automobiles, with fuel cells, 7, 58–66
  buses vs., 66
  calculations for, 63–66
  issues for, 62–63
  manufactured, 59–61
  storage issues of, 63
  voltage requirements for, 250

Auxiliary power units (APUs), 42
Average annual efficiency, 81–82
Avista Labs (*see* ReliOn)
Avogadro's number, 395
Axial fans, 281
Axial flow compressors, 285

Back diffusion, 129, 266, 267
Backup power:
  electrolyzers, 56–57
  manufactured, 54
  voltage requirements for, 250
Backup power applications, 55–57
Backward oxidation, 111
Backwards reaction, 109–110
Bacon, Francis Thomas, 8, 11
Bacteria cells, 49
Ballard Mark V cells, 206
Ballard Power Systems, 56
Baseline conditions, 377–378
Batteries:
  "carbon battery," 9
  comparison chart, 6
  fuel cells vs., 4–5
  "gas battery," 8
  rechargeable, 53–54
  secondary, 53–54
Baur, Emil, 8–10
Benzene, 397, 413
Bernoulli equation, 261, 262
Beryllium hydride, 331
Besel S. A.:
  scooters and bicycles, 73
  utility vehicles, 70
BET (Branauer-Emmett-Teller) method, 389
BFCs (*see* Biological fuel cells)
Bharat Heavy Electricals Ltd. (BHEL), 75
BHEL (Bharat Heavy Electricals Ltd), 75
Bicycles, 69, 72–74
Binary catalysts, 203
Binary diffusion coefficients, 135, 413
Biofuels, 334
Biogases, 334
Biological fuel cells (BFCs), 49–50
  comparison of, with other fuel types, 37–38
  subcategorizes of, 36
  (*See also* Biological fuel cells, *e.g.*: Alkaline fuel cells)
Biomass, 23, 334

423

424 Index

Bioproduction, of hydrogen, 344–345
  digestion processes, 345
  photosynthesis, 344–345
Bipolar plates, 221–227
  channels, 144
  coated metallic, 224–226
  coated metallic fabrication, 236–237
  composite, 226–227
  composite fabrication, 237
  design of, 223–224, 242
  DMFC, 222
  for low-temperature fuel processing, 221–227
  manufacturing of, 235–237
  materials for, 223–224
  nonporous graphite fabrication, 235–236
  PEMFCs, 222
Bleeding, hydrogen peroxide, 203
Blowers, 283–284
BMW, 61
Boil-off, 320
Boltzmann's constant, 111, 395
Bomb, hydrogen, 18
Bombardier Recreational Products, 73
Bonded seals, 239
Boost converters:
  design of, 306–307
  single-switch, 302
Borohydrides, 331
Branauer-Emmett-Teller (BET) method, 389
Brazil, hydrogen economy and, 32
Brennstoffzellentechnik (ZBT), 75
British Royal Navy, 10
Broers, G. H. J., 8, 10
Bubbling humidification, 276
Buck converters, 306–307
Bureau of Ships (U.S. Navy), 9
Bus Manufacturing U.S.A., Inc., 67
Buses, with fuel cells, 66–68
  automobiles vs., 66
  manufactured, 67–57:16
  refueling stations for, 66
Butane, 399
n-Butane, 397
Butler-Volmer equation, 112, 138, 172

Cailleteon, Louis Paul, 8, 9
Calcium hydride, 331
Calendaring, tape (tape casting), 212
California Fuel Cell Partnership, 33
California zero emission vehicle (ZEV) mandate, 7
Canada, hydrogen economy and, 31
Capacity factor (CF), 82
Capital recovery cost (CRC), 83
Capital recovery factor (CRF), 82–83
Carbon, 397

Carbon battery, 9
Carbon bonding interactions, 374
Carbon cloth, 158
Carbon composite plates, 236
Carbon dioxide, 397
  heat at constant pressure for, 399, 401
  thermodynamic data for, 410
Carbon formation, 339
Carbon monoxide, 202, 397
  heat at constant pressure for, 399, 401
  thermodynamic data for, 409
Carbon nanofibers:
  hydrogen and, 322–323
  liquid hydrogen and, 322–323
Carbon papers, 205, 206
Carlisle, Anthony, 8
Carnot's law, 5–6
Casting, tape (tape calendaring), 212
Catalysis, 203
Catalyst(s):
  for AFCs, 196, 207
  binary, 203
  for DMFCs, 192–195, 202
  in fuel cell electrode layer, 201–208
  for MCFCs, 196–198, 208
  for Nafion membranes, 192–195
  for PAFCs, 195–196
  for PEMFCs, 192–195, 202
  single metal, 203
  for SOFCs, 198–199
  tertiary, 203, 204
Catalyst decaling, 211
Catalyst powder deposition, 210
Catalyst preparation, 210–212
Catalytic combustion, 24
Cathodes, 3, 12
  in PEMFCs, 36
  performance parameters for, 421
  water management for, 266
CCM (continuous conduction mode), 302, 303
Cell(s):
  bacteria, 49
  fuel (see Fuel cell[s])
Cell interconnection:
  in alkaline fuel cells, 264
  in SOFCs, 263–264
Cell phones, 7
  design for, 305–306
  power electronics for, 305–306
  voltage requirements for, 250
Cellex Power Products, 70
Central Technical Institute, The Hague, Netherlands, 10
Centrifugal compressors, 285
Centrifugal fans, 283
Ceramic electrolytes, 130–131
Ceramic Fuel Cells Ltd. (CFCL), 75
Ceramic interconnects, in SOFCs, 232

Ceramic seals, glass, 239
CF (capacity factor), 82
CFCL (Ceramic Fuel Cells Ltd.), 75
CFD (computational fluid dynamics) modeling, 139
Change, conservation of, 178, 183–184
Characterization, 373–392
  baseline conditions, 377–378
  conditioning, fuel cell, 376–377
  current density mapping, 388
  MEA characterization, 388–391
  neutron imaging, 388
  operating parameters for, 377–378
  polarization curves, 378–380
  resistance, fuel cell, 380–388
  setup for, 373–376
  verification of assembly, 376
Charge transport, 121–131
  aqueous electrolytes, 126–127
  in ceramic electrolytes, 130–131
  electron conductivity, 126
  ionic conductivity, 126–131
  in metals, 126
  Ohm's law and, 183, 184
  in polymer electrolytes, 127–130
  voltage loss and, 121–126
Chemical anodization, 226
Chemical hydrides, 328–331
  hydrogenation/dehydrogenation, 330
  hydrolysis reactions, 329–330
  new approaches for, 330–331
  storage of, 314
Chemical vapor deposition (CVD), 226
China, hydrogen economy and, 32
Chlamydomonas, 345
Chromia (stable oxide), 233
Circuits:
  H-bridge inverter, 303
  inverter, 303
  three-phase, 303
Clamping, 264–265
Clean Fuel Cell Energy, LLC, 213
"Clean fuels," 386
Clean Urban Transport for Europe (CUTE), 66
Clostridium butyricum, 345
CNG (compressed natural gas), 333
CO poisoning, 203, 335
CO_2 out, 274, 275
Coal gasification, 23
Coated metallic bipolar plates, 224–226
Coated metallic plate fabrication, 236–237
COE (see Cost of electricity)

Index 425

Colardeau, Louis Joseph, 8, 9
Collapse, voltage, 306
Collectors, current, 243
Combustibles, enthalpy of, 155
Combustion:
　catalytic, 24
　direct steam generation by, 24
　hydrogen, 24
　of hydrogen, 24
　in internal combustion engines, 24
Commercial energy use, 20
Commercialization:
　of DMFCs, 325
　of hydrogen, 313
　of SOFCs, 43
Composite bipolar plates, 226–227, 237
　graphite plates vs., 237
　graphite-carbon, 223
Composite fabrication, 237
Composite membranes, non-fluorinated, 195
Compressed hydrogen storage, 314, 315, 317–320
Compressed natural gas (CNG), 333
Compression force, 378
Compressive seals, 240
Compressors:
　axial flow, 285
　centrifugal, 285
　design of, 284–289
　efficiency of, 285, 386
　Roots, 285
Computational fluid dynamics (CFD) modeling, 139
Concentration:
　loss of, 139
　performance affected by, 137, 138
Condensers:
　PEMFCs and, 162
　for stack cooling, 162
Conditioning, 376–377
Conduction and conductivity, 158–159
　electron, 126
　electronic, 123
　of fuel cell materials, 158
　heat, 158–159
　ionic, 124, 126–131
　temperature affecting, 356
Conservation:
　of change, 178, 183–184
　of mass, 175
　of momentum, 175–176
　with Navier-Stokes Equation, 175–176
　of species, 177–178, 186
Consolidation Coal Company, Pennsylvania, 10
Continuous conduction mode (CCM), 302, 303
Continuous flow channels, 228
Convective flow, 172, 174
Convective mass transport, 134–135, 139–149
　from flow channels to electrode, 134–135
　in flow structures, 139–149
　layers with, 134

Convective thermal resistance, 159–160
Converters:
　boost, 302, 306–307
　buck, 306–307
　DC-to-DC converters, 302–303
　multilevel, 307
　push-pull DC-DC, 302, 303
　single-switch, 302
Coolant pumps, 161
Cooling, 160
　air, 163–166
　with condensers, 162
　with cooling plates, 162
　edge, 166–167
　with free convection, 161–162
　heat spreaders for, 162
　passive, 160
　stack, 161–162
Cooling fins, 160
Cooling plates, 162
Cooling stack, 161–162
Corrosion, 160, 356
Cost of electricity (COE), 81, 83
CRC (capital recovery cost), 83
CRF (capital recovery factor), 82–83
Crossover, hydrogen, 117
Crossover currents, 116–117
Current(s):
　conservation of charge for, 183–184
　crossover, 116–117
　internal, 116–117
Current collectors, 243
Current density:
　mapping, 388
　over electrode surface, 142
Current interrupt (iR), 381–382
CUTE (Clean Urban Transport for Europe), 66
CVD (chemical vapor deposition), 226
Cyclic voltammetry, 387–388

Daihatsu, 61
DaimlerChrysler, 58, 60, 67
Dais Analytic DCh Technologies, 75
Darcy's law, 147, 172, 174
Davtyan, O. K., 8–10
DC (direct current), 297
DC chargers, 54
DCM (discontinuous-conduction mode), 302, 303
DC-to-DC converters, 302–303
Decaling, catalyst, 211
Deere & Company, 70
Dehydrogenation, 330
Department of Energy (DOE) (see U.S. Department of Energy)
Deposition:
　electro-deposition, 210
　evaporative, 211
Design:
　of bipolar plates, 223–224, 242
　of flow fields, 227–234
　fuel cell (see Fuel cell design)
　of stacks, 247–268

Desulfurization, 336–337
Dewpoint humidification, 276
Diffusion:
　back (see Back diffusion)
　Knudsen, 180–181
Diffusive mass transport, 133, 135–139
　in fuel cell electrodes, 135–139
　layers with, 134
Digestion processes, 345
Diodes, 297
　Schottky, 298
　slow recovery, 297–298
Direct borohydride fuel cells, 36
Direct current (DC), 297
Direct hydrocarbon oxidation, 341
Direct internal reforming, 340, 341
Direct Methanol Fuel Cell Corporation, 54
Direct methanol fuel cells (DMFCs), 36, 44–46, 275
　bipolar plates, 222
　catalysts for, 192–195, 202
　cell stacks for, 248
　chemical reactions for, 45
　commercialization and, 325
　comparison of, with other fuel types, 37–38
　electrokinetic permeability for, 176
　electroosmotic flow in, 174
　flow rates of reactants in, 366
　flux in, 185
　fuel cell electrode layer of, 201–204
　gas diffusion layer of, 205–207
　gaskets for, 238–239
　materials for, 191
　modeling, 174
　operating conditions for, 354
　outlet flow rates for, 366
　PEMFCs vs., 35, 45
　polarization curves for, 252
　porous silicon-based, 258
　stack of, 222
Direct steam, 24
Direct water/steam injection, 279, 280
"Dirty fuels," 386
Discontinuous-conduction mode (DCM), 302, 303
Dissipation, heat (see Heat dissipation)
DMFCs (see Direct methanol fuel cells)
n-Dodecane, 397
DOE (see U.S. Department of Energy)
Dry gases, enthalpy of, 155, 362
Dry spraying, 211
DuPont electrolytes (see Nafion)
DuPont Nafion–117, 193, 194
DuPont Nafion–901, 193
Durability testing, 374
Dusty Gas Model, 180

Index

ECI Laboratories, 204
ECN, 73
Edge cooling, 166–167
EDS (energy dispersive spectroscopy), 390–391
EFC (Enzymatic Fuel Cell), 49
Efficiency:
  average annual, 81–82
  of compressors, 285, 386
  design for, 308–310
  of electrolyzers, 347
  energy, 100–101
  fuel, 309
  of stationary fuel systems, 81–82
  thermodynamics, 99–101
  voltage, 309–310
EIS (electrochemical) impedance spectroscopy, 383–385
EIVD, 70
Electric Power Research Institute (EPRI), 43
Electrical subsystem design, 296–307
  boost converters, 306–307
  buck converters, 306–307
  for cell phones, 305–306
  inverters, 303–304
  for larger applications, 306–307
  multilevel converters, 307
  power diodes, 297–298
  supercapacitors, 304–305
  switching devices, 298–300
  switching regulators, 300–303
Electricity generation, 2
Electrochemical (EIS) impedance spectroscopy, 383–385
Electrochemical surface area, 374
Electrochemical vapor deposition (EVD), 212
Electrochemistry, 107–118
  crossover currents, 116–117
  electrode kinetics, 110–112
  internal currents, 116–117
  kinetic performance, 117–118
  voltage losses, 112–116
Electrode(s), 178–185
  convective mass transport from flow channels, 134–135
  current density over, 142
  diffusive mass transport in, 135–139
  electrochemical behavior, 181–183
  flow channels and, 134–135
  heat source in, 184
  heat transfer, 184–185
  ion/electron transport, 183–184
  mass transport, 179–181
  modeling, 178–185
  porous nature of, 184
Electrode kinetics, 110–112
Electro-deposition, 210
Electrodes, 207

Electrokinetic permeability, 176
Electrolysis, water, 22–24
Electrolyte(s):
  aqueous, 126–127
  ceramic, 130–131
  charge transport in, 126–131
  conduction and, 116, 124
  from DuPont, 39 (See also Nafion)
  ionic conductivity of, 126–131
  ionic resistance in, 124, 125
  liquid, 124
  mass transport in, 186
  MCFCs, 197, 198
  modeling, 185–186
  polymer, 124, 127–130, 213–214
  protanic, 48
  solid, 124
  YSZ, 190
  zirconia, 49, 199, 200
Electrolyte layer, 190–199
  of AFCs, 196
  of DMFCs, 192–195
  of MCFCs, 196–198
  of Nafion membranes, 192–195
  of PAFCs, 195–196
  of PEMFCs, 192–195
  of SOFCs, 198–199
Electrolyte matrix, 197
Electrolyzers, 36, 346–348
  backup power and, 56–57
  efficiency of, 347
  high pressure in, 347–348
Electron charge, 395
Electron conductivity, 126
Electron mass, 395
Electron transport, 183–184
Electronic conduction, 123, 124
Electronic vehicles (EVs), 47
Electroosmotic drag, 129, 265–267
Electroosmotic flow, 174
Elemental analysis, 374
Elmore, G. V., 11
Emissions, environmental, 7
  of hydrogen, 313
  of natural gas, 333
End plates, 240–241, 243
Energy:
  alternative sources of, 1, 6
  conservation of, 176–177
  demand for, 19–21
  efficiency, 100–101
  hydrogen economy and, 19–20
Energy balance, 151
  for fuel cell layers, 155–157
  of fuel cell stack, 153
  for fuel cells, 154–155
  in fuel cells, 152–157
  in gaseous phase, 155–156
  general, 152–155
  procedure for, 152–153
  of solid structure, 156–157
Energy conversion, 4

Energy dispersive spectroscopy (EDS), 390–391
Energy Policy Act of 1992 (EPACT), 333
Engines:
  heat engine, 5–6
  internal combustion engines, 24, 64
Enova Systems, 67
Enthalpy:
  absolute, 87
  of combustibles, 155
  for dry gases, 155
  of dry gases, 362
  of formation, 87, 88, 397
  of liquid water, 155
  of water, 362
Entropy, absolute, 397
Environmental emissions, 7
Environmental issues, 1–2
  of fossil fuels, 1–2, 21
  of hydrogen, 2, 21
Enzymatic Fuel Cell (EFC), 49
Enzymatic fuel cells, 36
EPACT (Energy Policy Act of 1992), 333
EPDM rubber, 238
EPRI (Electric Power Research Institute), 43
ESORO, 61
Ethane, 397, 399
Ethanol, 325
  binary diffusion coefficients for, 413
  biofuels from, 334
  comparison of, with other fuel types, 16, 316
  uses of, 325
Ethyl alcohol, 397
Ethylene, 397, 399
European Fuel Cell, 75
European Union, 30, 31
Evaporative deposition, 211
EVD (electrochemical vapor deposition), 212
EvoBus, 67
EVs (electronic vehicles), 47
Exchange current density, 182
External manifolding, 260

FAAM, 73
Fabrication, 235–237
  coated metallic, 236–237
  composite, 237
  nonporous graphite, 235–236
Fans:
  axial, 281
  centrifugal, 283
  design of, 281–284
  pressure and, 283
Faraday's constant, 95, 136, 359, 395
Faraday's law, 108, 141, 359
FASTec, 70
FCT (Fuel Cell Technologies), 75
Fiat, 58, 61
Fick diffusion, 172
Fick's law, 135–136, 180

Index   **427**

Flammability, 317
Flow:
  convective, 172, 174
  in DMFCs, 174
  electroosmotic, 174
  of hydrogen, 273
  of methanol, 274
  oxidant air flow, 273, 274
Flow channels:
  continuous, 228
  convective mass transport to electrodes, 134–135
  dimensions of, 235
  electrodes and, 134–135
  heat transfer in, 184
  mass transport in, 139–144
  pressure drop in, 144–149
  shape of, 235
  spacing of, 235
Flow fields:
  design of, 227–234
  inter-digitated, 230
  pin-type, 231
  porous, 147
  serpentine, 227
  spiral inter-digitated, 230, 231
Flow rates, 358, 378
  air mass, 365
  for DMFCs, 366
  hydrogen mass, 364
  inlet, 364–365
  mass, 364–366
  nitrogen, 365, 366
  outlet, 366
  oxygen, 365, 366
  for PEMFCs, 366
  of reactants, 357–361
  stoichiometric, 358–361
Fluorinated membranes, partial, 195
Flux, 109–110
  in DMFCs, 185
  heat transfer and, 158
  in Nafion, 186
  in PEMFCs, 185
  proton, 185
  water, 185, 186
Ford, 58, 59
Formation, enthalpy of, 87, 88, 397
Formic acid fuel cells, 36
Forward reaction, 109–110
Fossil fuels, 1, 2
  in automotive market, 7
  electricity from, 2
  in energy economy, 15
  for hydrogen generation, 2, 22
  problems of, 1
Fourier's law, 167
France, hydrogen economy and, 31–32
Fraunhofer Institute for Solar Energy Systems, 24
Free convection, 161–162
Free energy (*see* Gibbs free energy)
FreedomCAR and Fuel Partnership, 33
Friction factor, 144, 145

Fuel(s), 313–348
  ammonia, 331, 332
  biofuels, 334
  chemical hydrides, 328–331
  "clean," 386
  "dirty," 386
  distribution of, 133, 260–262
  electrolyzers and, 346–348
  ethanol, 325
  fossil, 1, 2, 7
  gasoline, 333–334
  hydrogen, 18–19, 315–323, 344–345
  metal hydrides, 325–328
  methanol, 324–325
  natural gas, 332, 333
  petroleum-based, 333–334
  processing, 334–444
  propane, 333
  sources of, 2
Fuel cell(s), 1–13, 183–184
  advantages of, 3, 4
  batteries vs., 4–5
  components of, 273–275
  defining, 3–6
  fuel distributed to, 133, 260–262
  heat engine vs., 5–6
  history of, 7–12
  interconnection of, 262–264
  limitations of, 4
  mechanics of, 258–259
  need for, 6–7
  number of, in stacks, 254–255
  oxidants distributed to, 133, 260–262
  types of, 35–51
  (*See also* specific types, e.g.: PEM fuel cells)
Fuel cell concentration loss (mass transport loss), 139
Fuel cell design, 273–310
  for efficiency, 308–310
  electrical subsystem, 296–307
  fuel subsystem, 276–296
  hybrid power systems, 307–308
  limitations on, 247
  parameters for, 249
  for performances, 419–421
  specifications for, 415
  for stacks, 417–419
Fuel cell electrode layer, 199–209
  of AFCs, 207
  catalysts in, 201–208
  of DMFCs, 201–207
  gas diffusion layer, 205–207
  of MCFCs, 208
  of PAFCs, 201–207
  of PEMFCs, 201–207
  of SOFCs, 198–199, 208, 209
Fuel cell layers (*see* Layers)
Fuel cell pumps, 291–296
Fuel Cell Technologies (FCT), 75
Fuel efficiency, 309

Fuel processing, 334–344
  carbon formation, 339
  desulfurization, 336–337
  direct hydrocarbon oxidation, 341
  internal reforming, 339–341
  low-temperature, 210–216
  methanol reforming, 343–344
  partial oxidation, 341–343
  pyrolysis, 343
  steam reforming, 337–339
Fuel subsystem design, 276–296
  blowers, 283–284
  compressors, 284–289
  fans, 281–284
  humidification systems, 276–281
  pumps, 291–296
  turbines, 289–291
FuelCell Energy:
  MCFCs and, 43
  stationary power systems, 76
Fuji Electric, 76
Full bridge DC-to-DC converters, 303
FutureGen, 33

Gas(es):
  biogases, 334
  compressed natural gas, 333
  dry, 155, 362
  heat and, 399, 401
  humidifying, 276
  hydrogen, 18, 315–320, 401
  hydrogen gas, 18, 315–320
  liquefied natural gas, 333
  natural gas, 332–333
  reactant, 260
  temperature and, 401
  "town gas," 23
  (*See also* specific types, e.g.: Oxygen)
"Gas battery," 8
Gas diffusion electrode (GDE), 11, 46
Gas diffusion layer (GDL), 205–207
Gas flows, 155–156
Gas storage cylinders, 318
  compressed, 313, 314, 317
  natural, 317
Gas streams, 147
Gaseous phase, 155–156
Gasification, coal, 23
Gaskets, 237–240
  for AFCs, 238–239
  for DMFCs, 238–239
  for PEMFCs, 238–239
  PTFE for, 238
  selection of, 243
  for SOFCs, 239–240
Gasoline, 333–334
  comparison of, with other fuel types, 16, 316
  hydrogen vs., 16, 18
  internal combustion engines with, 64
  propane vs., 333
  sulfur in, 337
Gaz de France, 25

428 Index

GDE (*see* Gas diffusion electrode)
GDL (gas diffusion layer), 205–207
GE (General Electric), 8–10
GenCell, 76
General Electric (GE), 8–10
General energy balance, 152–155
General Motors (GM):
 automobiles, 58, 59
 stationary power systems, 76
German Aerospace Research Establishment, 24
Germany:
 hydrogen economy and, 32
 hydrogen refueling stations in, 25, 26
Gibbs free energy, 92–93
 anodic, 158
 in electrochemical processes, 107
 electrode kinetics, 110–111
 formation and, 397
 in modeling, 174
 pressure and, 102
Gillig Corporation, 67
Glass seals, 239
Global warming, 1
Glucose, 413
GM (*see* General Motors)
Gold coating, 226
Government support, 27–28
Graphite:
 for bipolar plates, 223
 conductivity of, 158
Graphite foil layer, 226
Graphite plates, 223, 237
Graphite topcoat layer, 226
Graphite-carbon composite plates, 223
Graphitic nanofibers, 322
Grove, William, 8, 11
"Grove cell," 8
Grubb, Thomas, 8, 9

H2 Logic Ap2, 70
HaveBlue LLC., 71
H-bridge inverter circuits, 303
HDW (Howaldts werke-Deutsche WerftAG), 71
Heat, 356 (*See also* Heat transfer; Temperature)
Heat conduction, 158–159
Heat dissipation:
 free convection and, 161, 162
 through natural convection and radiation, 159–160
Heat engines, 5–6
Heat exchanger model, 162–163
Heat generation, 158, 159, 184
Heat management, 160–167
 air cooling, 163–166
 edge cooling, 166–167
 exchanger model, 162–163
 with free convection, 161–162
 for PEMFCs, 160
 of portable fuel cells, 160
 stack cooling, 161–162
Heat sinks, 160, 185
Heat spreaders, 162

Heat transfer, 151–167
 conduction, 158–159
 dissipation, 159–162
 electrodes, 184–185
 energy balance in fuel cells, 152–157
 in flow channels, 184
 flux and, 158
 in fuel cell layers, 155–158
 generation, 158
 management of, 160–167
 ohmic, 184
 one-dimensional, 166
Heise, G. W., 11
Helinsky University of Toronto, 49
Helium, 399
HFR (high-frequency resistance) method, 382–383
HHV (higher heating value), 90–91
High-chrome interconnects, 232
Higher heating value (HHV), 90–91
High-frequency resistance (HFR) method, 382–383
High-frequency switching, 307
Hindenburg incident, 18
Hino Motors Ltd., 68
History, of fuel cells, 7–12
Honda:
 automobiles, 58, 59
 scooters and bicycles, 73
Hot-pressing, 211, 215–216
Howaldts werke-Deutsche WerftAG (HDW), 71
H-Power Corp., 76
Humidification:
 of anodes, 267
 bubbling, 276
 dewpoint, 276
Humidification systems, 276–281
Humidity, relative, 362
Humidity (psychrometric) chart, 279, 280
Humidity of reactants, 361–363
Hybrid power systems, 307–308
Hydrides:
 chemical, 328–331
 metal, 24, 25, 325–328
Hydrocarbon(s), 17, 48
Hydrocarbon alternative membranes, 194
Hydrogen, 15–33, 397
 biomass and, 23
 bioproduction of, 344–345
 carbon nanofibers and, 322–323
 catalytic combustion of, 24
 characteristics of, 16–19
 coal gasification of, 23
 combustion of, 24
 commercialization of, 313
 comparison of, with other fuel types, 16, 316
 compressed storage of, 318–320
 designing systems for, 18–19

Hydrogen (*Cont.*):
 distribution of, 22
 environmental issues of, 2, 17, 21
 in flow-through mode, 360
 as fuel, 18–19, 315–323, 344–345
 as gas, 18, 315–320
 gasoline vs., 16, 18
 heat at constant pressure for, 399
 in internal combustion engines, 24
 as liquid, 18, 320–323
 long-term use of, 28–29
 metal hydrides applications, 24
 novel, 25
 partial oxidation of, 23
 pipeline systems for, 22
 pollution and, 17, 317
 production, 22–24, 316
 reforming, 22–23
 refueling stations, 25–26
 safety with, 18–19, 317–318
 steam reforming, 22–23
 storage, 18, 22, 24–25, 63
 thermodynamic data for, 406
 water electrolysis for, 23–24
Hydrogen bomb, 18
Hydrogen combustion, 24
Hydrogen crossover, 117
Hydrogen economy, 15–33
 development of, 21
 European Union and, 30, 31
 government support for, 27–28
 hydrogen characteristics, 16–19
 infrastructure investment, 27–33
 Japan and, 29–31
 key players in, 29–33
 long-term use projections, 28–29
 in private industry, 32–33
 research and development, 29–33
 United States and, 29, 30
 world energy demand and, 19–20
Hydrogen flow, 273
Hydrogen fuel cells, 2
Hydrogen gas, 18, 315–320
 compressed storage of, 318–320
 heat at constant pressure for, 401
 safety with, 317–318
Hydrogen mass flow rate, 364
Hydrogen out, 273
Hydrogen peroxide, 397
Hydrogen peroxide bleeding, 203
Hydrogen refueling stations, 25–26
Hydrogenation, 330
Hydrogenics, 73
Hydrogenics Corp., 76
Hydrolysis, 63, 329–330
Hydrophilic electrodes, 207
Hydrophobic electrodes, 207

Index

Hydroxyl, 397
Hysteresis, 379
Hyundai, 58, 60

ICP-MS (inductively coupled plasma mass spectroscopy), 391
IdaTech, 76
Ideal gas law, 318
IGBTs (see Insulated gate bipolar transistors)
IGCTs (see Integrated gate-commutated thyristors)
IHI (Ishikawajima-Harima Heavy Industries), 77
Imaging, neutron, 388
Impedance analyzers, 375
Imperial Chemical Industries, Great Britain, 25
Impregnation, 210–211
Impregnation reduction, 210–211
India, hydrogen economy and, 32
Indirect internal reforming, 340
Indium tin oxide, 226
Inductively coupled plasma mass spectroscopy (ICP-MS), 391
Industrial Research laboratory (IRL), 76
Inlet flow rates, 364–365
Insulated gate bipolar transistors (IGBTs), 297–299
Integrated gate-commutated thyristors (IGCTs), 297, 300
Intelligent Energy:
  scooters and bicycles, 73
  stationary power systems, 77
Interconnects:
  of fuel cells, 262–264
  in SOFCs, 232
Inter-digitated flow field, 230
Interfacial contact resistance, 224
Internal combustion engines:
  combustion in, 24
  with gasoline, 64
  hydrogen in, 24
Internal currents, 116–117
Internal manifolding, 260, 261
Internal reforming, 339–341
  direct, 340, 341
  indirect, 340
International Energy Agency, 32–33
Inverters, 303–304
Ion(s), 127
Ion transport, 183–184
Ionac membranes, 193
Ionic conduction, 123, 126–131
  in aqueous electrolytes, 126–127
  of ceramic electrolytes, 130–131
  electronic conduction vs., 124
  metal conductors vs., 124
  of polymer electrolytes, 127–130

Ionic crystals, 124
Ionic resistance, 124, 125
Ionomer impregnation, 210
iR (current interrupt), 381–382
IrisBus, 67
IRL (Industrial Research laboratory), 76
Ishikawajima-Harima Heavy Industries (IHI), 77
Ishikawajima-Shibaura Machinery Co. (ISm), 77
ISm (Ishikawajima-Shibaura Machinery Co.), 77
Italy, hydrogen economy and, 32

Jacques, William W., 8, 9
Japan:
  hydrogen economy and, 29–31
  SOFCs in, 42
Justi, Edward, 12

Ketelaar, J. A. A., 8, 10
Kia, 61
Kinetic energy, 110–112
Kinetic performance, 117–118
Knudsen diffusion, 180–181
KOH (see Potassium hydroxide)
Kordesch, Karl, 8, 11
Korea, hydrogen economy and, 32
Kureha E-715, 206
Kurimoto, 71

Langer, Carl, 8, 9
LANL (see Los Alamos National Laboratory)
Laptop computers, 250
Layers:
  with convective mass transport, 134
  with diffusive mass transport, 134
  energy balance in, 155–157
  in gaseous phase, 155–156
  heat transfer in, 155–158
  with mass transport, 134
  solid structure, 156–157
Lead oxide, 226
Leakage, 5
LHV (lower heating value), 91
Limiting current(s), 386–387
Limiting current density, 137
Liquefied natural gas (LNG), 333
Liquid electrolytes, 124
Liquid hydrogen, 18, 320–323
  carbon nanofibers and, 322–323
  safety with, 18
  storage of, 25, 314, 320–321
  use of, 314
Liquid molten carbonate fuel cells (see Molten carbonate fuel cells)
Lithium borohydride, 331
Lithium hydride, 331
LNG (liquefied natural gas), 333

Los Alamos National Laboratory (LANL), 71, 226
Lower heating value (LHV), 91
Low-temperature fuel processing, 210–216
  bipolar plates for, 221–227
  catalyst decaling, 211
  catalyst powder deposition, 210
  catalyst preparation, 210–212
  dry spraying, 211
  electro-deposition, 210
  evaporative deposition, 211
  impregnation reduction, 210–211
  ionomer impregnation, 210
  painting, 211
  for PEMFC MEA, 213
  SOFC manufacturing, 212–216
  spraying, 210–211
  spreading, 210
Lynnetech, 256, 257
Lysholm (screw) compressors, 285

Macchi-Ansaldo, 68
MAN (bus manufacturer), 68
Manhattan Scientifics:
  scooters and bicycles, 73
  utility vehicles, 71
Manifolding:
  external, 260
  internal, 260, 261
Manufacturing fins, 162
Mapping, current density, 388
Mass, conservation of, 175
Mass balance, 363–369
Mass flow rates:
  air, 364–366
  hydrogen, 364
  nitrogen, 365, 366
  oxygen, 365, 366
Mass transport, 133–149
  convective, 134–135, 139–149
  diffusive, 133, 135–139
  electrodes, 179–181
  in electrolytes, 186
  in flow channels, 139–144
  layers with, 134
  modeling, 174
Mass transport loss (fuel cell concentration loss), 139
Masterflex AG, 73
Material balances, 364
Matthey, Johnson, 343
Maxwell-Stefan equation, 174, 186
Mazda, 58, 60
M-C Power Corporation:
  MCFCs and, 43
  stationary power systems, 77
MCFCs (see Molten carbonate fuel cells)
MEA (see Membrane electrode assembly)
Medis Technologies, 54

Membrane(s):
  alternative, 194, 195
  alternative polymer, 195
  composite, 195
  hydrocarbon, 194
  hydrocarbon alternative, 194
  Ionac, 193
  non-fluorinated, 195
  non-fluorinated composite, 195
  Pall, 193
  partially fluorinated, 195
  perfluorinated, 195
  PFSA, 194
  polymer, 195
  polymer alternative, 195
Membrane electrode assembly (MEA), 388–391
  Branauer-Emmett-Teller method, 389
  characterization, 388–391
  energy dispersive spectroscopy, 390–391
  hot-pressing, 215–216
  inductively coupled plasma mass spectroscopy, 391
  limiting current, 386
  of PEMFCs, 213
  porosity determination, 389
  scanning electron microscopy, 390
  transmission electron microscopy, 390
  X-ray diffraction, 390
  X-ray fluorescence, 391
MEMs (mico-electro-mechanical) fuel cells, 257
Mercury porosimetry, 389
MERDC (U.S. Army Mobility Equipment Research and Development Center), 10–11
MES-DEA, 73
Metal(s):
  charge transport in, 126
  as conductors, 124
  electron conductivity of, 126
Metal brazes, 239
Metal conductors, 124
Metal hydrides, 325–328
  applications of, 24
  hydrogen and, 24
  problems of using, 326
  storage, 25, 314, 315
Metal hydrides applications, 24
Metal oxide semiconductor field effect transistors (power MOSFETs), 297, 298
Metallic bipolar plates, coated, 224–226
Metering pumps, 280
Methane, 397
  comparison of, with other fuel types, 16
  heat at constant pressure for, 399
  thermodynamic data for, 411
Methanol, 324–325
  biofuels from, 334
  comparison of, with other fuel types, 16, 316
  reforming, 343–344

Methanol flows, 274
Methanol out, 274, 275
Methyl alcohol, 397
Mico-electro-mechanical (MEMs) fuel cells, 257
Microbial fuel cells, 36, 259
Microcracks, 225
Micro-fuel cells, 305
Micro-fuel stack, 259
Micropores, 225
Microscopy, 390
Military use, 7, 55
Milliohms (ohmic resistance), 380
Mitsubishi, 60
Modeling, 171–186
  conservation of change, 178
  conservation of energy, 176–177
  conservation of mass, 175
  conservation of momentum, 175–176
  conservation of species, 177–178
  DMFC, 174
  electrodes, 178–185
  electrolytes, 185–186
  mass transport, 174
  Navier-Stokes Equation, 175–176
  one-dimensional, 171
  PEMFC, 174
  two-dimensional, 171
Modeling software, 384
Molar fractions, 96
Molten carbonate fuel cells (MCFCs), 10–11, 43–44
  aqueous electrolytes in, 126
  catalysts for, 196–198, 208
  comparison of, with other fuel types, 37–38
  electrode material of, 208
  electrolyte, 197, 198
  fuel processing for, 335
  history of, 8
  materials for, 191
  operating conditions for, 354
  plates, 233–234
  polarization curves for, 253
  polarization in, 208
  SOFCs vs., 43
  stack of, 222
  stationary fuel cell systems with, 43
  turbines in, 289
Momentum, conservation of (Navier-Stokes Equation), 175–176
Mond, Ludwig, 8, 9
Mosaic Energy, 77
MOSFETs (see Metal oxide semiconductor field effect transistors)
MTI MicroFuel Cells, 54
MTU CFC Solutions, 77
MTU Friedrichshafen, 71
Multilevel converters, 307
Multi-walled carbon nanotubes (MWNT), 322, 323
MWNT (multi-walled carbon nanotubes), 322, 323

NADH (nicotinamide adenine dinucleotide), 49
NADHP (nicotinamide adenine dinucleotide phosphate), 49
Nafion, 39, 127–129
  catalysts for, 192–195
  characteristics of, 194
  membranes, 186, 192–195
  901 type, 193
  117 type, 193, 194
  water flux in, 186
Nanofibers:
  carbon, 322–323
  graphitic, 322
Naphthalene, 413
National Aeronautics and Space Administration (NASA):
  AFCs and, 8, 40
  PEM fuel cells, 9–10
  Project Apollo, 8, 9, 12
  research by, 8, 9
National Renewable Energy Laboratory (NREL), 345
Natural convection, 159–160
Natural gas, 332, 333
  compressed, 333
  liquefied, 333
  for steam reforming, 23
  steam reforming of, 22
Natural gas cylinders, 317
Navier-Stokes Equation (conservation of momentum), 175–176
Navy (see U.S. Navy)
Neon, 399
Neoplan, 68
Nernst equation, 103, 135, 137, 181
Nernst potential, 103–604
Nernst-Planck equation, 172, 186
Net output voltage, 92–99
Neutron imaging, 388
New Flyer Industries, Ltd., 67
New Hydrogen Project (NHP), 30
NHP (New Hydrogen Project), 30
Nicholson, William, 8
Nickel:
  for bipolar plates, 236
  conductivity of, 158
Nickel powder, sintered, 207
Nicotinamide adenine dinucleotide (NADH), 49
Nicotinamide adenine dinucleotide phosphate (NADHP), 49
Niedrach, Leonard, 8, 9
Nissan, 59
Nitrogen, 397
  heat at constant pressure for, 399, 401
  thermodynamic data for, 412
Nitrogen mass flow rate, 365, 366
n-Octane, 397
Non-fluorinated composite membranes, 195

Index

Nonporous graphite fabrication, 235–236
Nonrepeat components, 418
NovaBus Corporation, 67
Novel hydrogen storage methods, 25
NREL (National Renewable Energy Laboratory), 345
Nusselt number, 156, 163, 165
Nuvera:
  backup power fuel cells, 56
  stationary power systems, 77
Nyquist plots, 385

Octane, 399
OCV (see Open-circuit voltage)
Ohmic heat, 184
Ohmic losses, 121–124, 256
Ohmic polarization, 121
Ohmic resistance (milliohms), 197, 380
Ohm's law:
  charge transport and, 183, 184
  voltage drop, 96
One-dimensional heat transfer, 166
One-dimensional models, 171
ONSI (see United Technologies Company)
Open-circuit voltage (OCV), 376, 379
Operating conditions, 353–369
  flow rates of reactants, 357–361
  fuel cell mass balance, 363–369
  humidity of reactants, 361–363
  pressure, 353–355
  temperature, 356–357
ORR (oxygen reduction reaction), 45
Ostwald, Friedrich Wilhelm, 8
Outlet flow rates, 366
Oxidant(s), 3
  distributed to fuel cells, 133, 260–262
  in fuel cells, 260–262
Oxidant air flow, 273, 274
Oxidation:
  backward, 111
  direct hydrocarbon oxidation, 341
  forward, 111
  of hydrogen, 23
  overcoating, 226
  partial, 23, 336, 341–343
  selective, 203
Oxygen, 397
  heat at constant pressure for, 399
  thermodynamic data for, 407
Oxygen Generating Plant (U.S. Navy), 10
Oxygen mass flow rate, 365, 366
Oxygen reduction reaction (ORR), 45

PAFCs (see Phosphoric acid fuel cells)
Painting, 211
Palcan Fuel Cells, Ltd., 74
Pall membranes, 193
Palladium hydride, 331
Parallel (Z-shape) flow, 261, 262
Partial oxidation, 341–343
  of hydrogen, 23
  steam reforming vs., 336
Partially fluorinated membranes, 195
Particle size distribution, 374
Passive cooling, 160
PCFCs (see Protonic ceramic fuel cells)
PEM Technologies, Inc., 74
PEMFCs (see Polymer electrolyte membrane fuel cells)
PEMs (see Polymer electrolyte membrane fuel cells)
Perfluorinated membranes, 195
Perfluorosulfonic acid (PFSA), 192, 194
Performance:
  concentration affecting, 137, 138
  of stack, 417, 419–420
Permeability, electrokinetic, 176
Persulfonated polytetrafluoroethylene (PTFE), 127, 192, 264
  for gaskets, 238
  treatment with, 206
  water management with, 205
Petroleum-based fuels, 333–334
Peugeot:
  automobiles, 61
  scooters and bicycles, 74
PFSA (see Perfluorosulfonic acid)
Phosphoric acid fuel cells (PAFCs), 11, 40–41
  advantages/disadvantages of, 41
  aqueous electrolytes in, 126
  catalysts for, 195–196
  chemical reactions for, 41
  comparison of, with other fuel types, 37–38
  electrolyte resistance in, 380
  fuel cell electrode layer of, 201–204
  fuel processing for, 334
  gas diffusion layer of, 205–207
  materials for, 191
  operating conditions for, 354
  plates, 234
  polarization curves for, 252
Photosynthesis, 344–345
Physical vapor deposition (PVD), 226
Pin-type flow field design, 231
Pipeline systems, 22
Planck's constant, 111, 395

Plate(s):
  composite, 237
  cooling, 162
  end, 240–241, 243
  graphite, 237
  graphite-carbon composite, 223
  MCFC, 233–234
  PAFC, 234
  SOFC, 232–233
Plate fabrication (see Fabrication)
Platinum, conductivity of, 158
Platinum hydroxide layer, 377
Platinum ink, 211
Plug Power Corp., 56, 77
Poisson equation, 159
Polarization:
  activation, 114
  in MFCFs, 208
  ohmic, 121
Polarization curves, 251, 378–380
  for AFCs, 253
  calculating, 251
  defined, 378
  for DMFCs, 252
  for MCFCs, 253
  for PAFCs, 252
  for PCFCs, 253
  for PEMFCs, 251
  regions of, 379
  for SOFCs, 252
Pollution, 1, 17, 315, 317
PolyFuel, 54
Polymer degradation, 374
Polymer electrolyte(s), 124
  charge transport in, 127–130
  ionic conductivity of, 127–130
  preparing, 213–214
Polymer electrolyte membrane fuel cells (PEMFCs), 36–39, 274
  bipolar plates, 222
  catalysts for, 192–195, 202
  comparison of, with other fuel types, 37–38
  condensers and, 162
  configuration example, 3
  cooling plates in, 162
  DMFCs vs., 35, 45
  electrokinetic permeability for, 176
  electrolyzer, 346
  electroosmotic flow in, 174
  flow rates of reactants in, 358, 366
  flux in, 185
  fuel cell electrode layer of, 201–204
  fuel processing for, 335
  gas diffusion layer of, 205–207
  gaskets for, 238–239
  heat management for, 160
  history of, 8, 9–10
  humidification system, 276
  humidity of reactants in, 361
  materials for, 191, 222

432 Index

Polymer electrolyte membrane
  fuel cells (PEMFCs)
  (*Cont.*):
  MEA in, 155
  MEA of, 213
  modeling, 174
  operating conditions for,
    354, 356
  other fuel cells and, 36
  outlet flow rates for, 366
  polarization curves for, 251
  "reversible," 36
  stack cooling for, 162
  stack of, 222, 268, 361
  temperatures for, 356
  total flow rate at stack
    entrance, 145
  water management for, 205,
    265–268
Polymer membranes,
  alternative, 195
Porosity:
  determination of, 389
  of electrodes, 184
Porous electrolyte matrix, 197
Porous flow fields, 147
Porous silicon-based DMFCs,
  258
Portable fuel cells, 53–55
  heat management of, 160
  manufactured, 54
  problems with, 1
  ranges for, 53
Portable sector:
  military use in, 7, 55
  power for, 7
Positive displacement
  compressors, 285
Potassium hydride, 331
Potassium hydroxide (KOH),
  196, 345
Potentiostat/galvanostat, 375
Powder deposition, catalyst,
  210
Power cycling, 374
Power diodes, 297–298 (*See
  also* Diodes)
Power MOSFETs (*see* Metal
  oxide semiconductor field
  effect transistors)
Power plants, fossil-fueled, 2
Power systems, hybrid,
  307–308
Powerzinc Electric, 74
Prandtl number, 156
Pratt & Whitney, 11, 12
PRDs (pressure release
  devices), 317
Preferential oxidizer (PROX),
  335
Preis, H., 10
Pressure, 378
  atmospheric, 357
  clamping and, 264
  in electrolyzers, 347–348
  fans affecting, 283
  in flow channels, 144–149
  high, 347–348
  operating, 353–355
  thermodynamics, 102–104
Pressure release devices
  (PRDs), 317
Private industry, 32–33

Processing (*see* Fuel
  processing)
Processing fuels, 334–444
Project Apollo, 8, 9, 12
Propane, 333, 397
  comparison of, with other
    fuel types, 16, 316
  gasoline vs., 333
  heat at constant pressure
    for, 399
Propylene, 397
Protanic electrolytes, 48
Proton Energy Systems, 78
Proton exchange membrane
  fuel cells (*see* Polymer
  electrolyte membrane fuel
  cells)
Proton flux, 185
Proton Motor Fuel Cell, 71
Protonic ceramic fuel cells
  (PCFCs), 36, 47–49
  comparison of, with other
    fuel types, 37–38
  flow rates of reactants in,
    358
  polarization curves for,
    253
  SOFCs vs., 48
PROX (preferential oxidizer),
  335
PSA Peugeot Citron, 61
Psychrometric (humidity)
  chart, 279, 280
PTFE (*see* Persulfonated
  polytetrafluoroethylene)
Pumps, 291–296
Push-pull DC-DC converters,
  302, 303
PVD (physical vapor
  deposition), 226
Pyrolysis, 343

Quantum Fuel Systems
  Technologies, 71
Quick Start initiative, 31

Radial blowers, 283
Radiation, 159–160
Raney metals, 207
R&D (research and
  development), 29–33
Reactant gases, 260
Reactants:
  flow rates of, 357–361
  humidity of, 361–363
Reaction kinetics, 138
Reaction sites, 117–118
Real-Time Radiography, 388
Rechargeable (secondary)
  batteries, 53–54
Rectangular channels, 146
Reforming, 314
  direct internal, 340, 341
  fuel processing, 337–344
  heat sink, 185
  hydrogen, 22–23
  indirect internal, 340
  internal, 339–341
  methanol, 343–344
  with natural gas, 22, 23
  performance, 343
  steam, 22–23, 336,
    337–339

Refueling stations:
  for buses, 66
  in Germany, 25, 26
  for hydrogen, 25–26
  in United States, 26
Regenerative fuel cells, 36, 47
  (*See also specific types,
    e.g.*: Zinc air fuel cells)
Regulators, switching,
  300–303
Relative humidity, 362
Relich-Kwong equation of
  state, 318
ReliOn:
  MCFCs and, 56
  stationary power systems,
    78
Renault:
  automobiles, 61
  IrisBus, 67
Renewable Energy Sixth
  Framework Programme,
  30
Repeating units, 418
Research and development
  (R&D), 29–33
Research Centre Jülich,
  71–72
Residential energy use, 20
Resistance:
  AC resistance method, 382
  bulk, 224
  convective thermal, 159–160
  current interrupt, 381–382
  cyclic voltammetry,
    387–388
  EIS impedance spectroscopy,
    383–385
  in electrolytes, 124, 125
  fuel cell, 380–388
  HFR method, 382–383
  interfacial contact, 224
  ionic, 124, 125
  limiting current, 386–387
  stoichiometry sweeps,
    385–386
  utilization sweeps,
    385–386
Reverse (U-shape) flow, 261
Reversible condition, 92–99
"Reversible" PEMFC, 36
Reynolds number, 145–146,
  156, 164
Roots compressors, 285
Russia, hydrogen economy
  and, 32

Safety, with hydrogen, 18–19,
  317–318
  in gas form, 317–318
  in liquid form, 18
Sanyo Electric Company, 78
S/C (steam-to-carbon ratio),
  336
Scanning electron microscopy
  (SEM), 390
Scarr, R. F., 11
Scenedesmus, 344
Schatz Energy Research
  Center, 78
Schlogls formulation, 174
Schottky diodes, 298
Schumacher, E. A., 11

Index 433

Scooters, with fuel cells, 69, 72–74
  manufactured, 73–74
  voltage requirements for, 250
Screw (Lysholm) compressors, 285
SCRs (*see* Silicon controlled rectifiers)
Seals:
  bonded, 239
  compressive, 240
Secondary (rechargeable) batteries, 53–54
Selective oxidation, 203
SEM (scanning electron microscopy), 390
Semiconductors, 124, 297
Sensible thermal energy, 87, 88
Serpentine flow field, 227
Sherwood number, 134–135
Siemens Power Generation, Inc., 78, 263
Silicon, 238
Silicon controlled rectifiers (SCRs), 297, 299–300
Single metal catalysts, 203
Single-switch boost converters, 302
Single-walled carbon nanotubes (SWNT), 322, 323
Sintered nickel powder, 207
Slow recovery diodes, 297–298
Smart Fuel Cell:
  portable fuel cells, 54
  stationary power systems, 78
Sodium borohydride, 331
Sodium hydride, 331
SOFCs (*see* Solid oxide fuel cells)
Solid electrolytes, 124
Solid oxide fuel cells (SOFCs), 42–43
  bonded seals, 239
  catalysts for, 198–199
  cell interconnection in, 263–264
  ceramic interconnects, 232
  commercial viability of, 43
  comparison of, with other fuel types, 37–38
  compressive seals, 240
  configuration of, 257
  electrode material of, 208, 209
  fabrication of, 264
  fuel processing for, 335
  gaskets for, 239–240
  high-chrome interconnects in, 232
  history of, 10
  interconnects, 232
  manufacturing, 212–216
  materials for, 191
  MCFCs vs., 43
  operating conditions for, 354, 357
  PCFCs vs., 48
  plates, 232–233
  polarization curves for, 252

Solid oxide fuel cells (SOFCs) (*Cont.*):
  stack of, 222
  temperatures for, 357
  tubular, 263
  turbines in, 289
  yttria stabilized zirconia in, 130
Solid structure layers, 156–157
Spacers, 237
Species, conservation of, 177–178, 186
Spectracarb 2050A-1041, 206
Spectroscopy:
  EDS, 390–391
  EIS, 383–385
  ICP-MS, 391
Spiral inter-digitated flow field, 230, 231
Spraying:
  catalyst decaling, 211
  dry spraying, 211
  evaporative deposition, 211
  impregnation reduction, 210–211
  painting, 211
Stable oxide (chromia), 233
Stack(s), 247–268
  AFC, 222
  cell interconnection in, 262–264
  clamping, 264–265
  configuration of, 255–260
  constructing, 267–268
  cooling, 161–162
  design for, 417–419
  design of, 247–268
  dimensions, 418
  DMFC, 222, 248
  energy balance of, 153
  fuel distribution in, 260–262
  for micro-fuel, 259
  non-repeat components, 418
  number of cells in, 254–255
  oxidants distributed in, 260–262
  parts of, 268
  PEMFC, 222, 268, 361
  performance parameters for, 417, 419–420
  repeating units, 418
  sizing, 249–254
  SOFC, 212, 222
  temperature of, 163
  thickness of, 418
  two-cell, 256
  volume of, 419
  water management and, 265–268
  weight of, 418, 419
Stack cooling, 161–162
  with condensers, 162
  with cooling plates, 162
  with free convection, 161–162
  heat spreaders for, 162
  for PEMFCs, 162

Stainless steel:
  for bipolar plates, 236
  for coating, 226
  conductivity of, 158
  type 316, 158
Stationary fuel cell systems:
  calculations for, 74, 80–81
  disadvantages of, 43, 44
  economics of, 81–83
  manufactured, 75–79
  with MCFCs, 43
Stationary power applications, 72, 74–83, 250
Stationary sector, 7
Steady-state heat conduction, 159
Steady-state testing, 374
Steam, 399
Steam reforming, 337–339
  hydrogen, 22–23
  of natural gas, 22
  partial oxidation vs., 336
Steam-to-carbon ratio (S/C), 336
Stefan-Boltzmann constant, 160
Stefan-Maxwell model, 180
Step-down switching, 301
Step-up switching, 301
Stoichiometric flow rate, 358–361
Stoichiometry (utilization) sweeps, 385–386
Storage:
  compressed, 318–320
  hydrogen, 18, 22, 24–25, 318–320
  of hydrogen gas, 318–320
  for liquid hydrogen, 25
  for metal hydrides, 25
  novel hydrogen, 25
  underground, 25
Structural analysis, 374
Subsystem design:
  electrical, 296–307
  fuel, 276–296
Sulfur, 337
Sulzer Hexis, 78
Supercapacitors, 304–305
Suzuki, 61
Switching:
  devices for, 298–300
  high-frequency, 307
  regulators for, 300–303
  step-down/up, 301
SWNT (single-walled carbon nanotubes), 322, 323

Tafel-type expressions, 171, 172
Tanner, H. A., 11
Tape casting (tape calendaring), 212
TARGET (partnership), 11
Teflon:
  conductivity of, 158
  electrolytes from (*see* Nafion)
Teledyne Energy Systems Inc., 79
TEM (transmission electron microscopy), 390

434 Index

Temperature, 101–102, 378
  conductivity affected by, 356
  gas and, 401
  in gas flows, 155–156
  kinetic performance affected by, 117
  operating, 356–357
  for PEMFCs, 356
  for SOFCs, 357
  of stack, 163
  viscosity variance with, 146
  voltage and, 94
  wet bulb, 279
Tertiary catalysts, 203, 204
Test stations, 374, 375
Testing (*see* Characterization)
Texas Instruments, 8, 11
Thermal energy, 87, 88
Thermal resistance, convective, 159–160
Thermocouples, 161
Thermodynamic data:
  for carbon dioxide, 410
  for carbon monoxide, 409
  for hydrogen, 406
  for methane, 411
  for nitrogen, 412
  for oxygen, 407
  for water, 408
Thermodynamics, 87–104
  basic concepts, 87–92
  defined, 87
  efficiency, 99–101
  net output voltage, 92–99
  pressure, 102–104
  reversible condition, 92–99
  temperature, 101–102
Thin-film approach, 202
Thompson, C., 8, 9
Thor Industries, 68
Three-phase inverter circuits, 303
Thyristors (*see* Silicon controlled rectifiers)
Titanium:
  for bipolar plates, 236
  conductivity of, 158
Tokyo Gas, 79
Toray TGPH 090, 206
Toshiba, 79
"Town gas," 23
Toyota:
  automobiles, 58, 59
  buses, 68
  Hino Motors Ltd., 68
  stationary power systems, 79
  utility vehicles, 72
Transient testing, 374
Transition State Theory, 110
Transmission electron microscopy (TEM), 390
Transportation applications, 57–72
  automobiles, 58–66
  bicycles, 69, 72–74
  buses, 66–68
  energy demand for, 20
  scooters, 69, 72–74
  utility vehicles, 66, 69–72

Transportation market, 7
Tubular SOFCs, 263
Turbines, 289–291
Turbocompressors, 286
Two-cell stack, 256
Two-dimensional models, 171

Underground storage, 25
Union Carbide, 11
United States:
  hydrogen economy and, 29, 30
  hydrogen refueling stations in, 26
United Technologies Company (UTC), 41
  Fuel Cells, 41
  stationary power systems, 79
U.S. Army:
  Mobility Equipment Research and Development Center (MERDC), 10–11
  PAFCS research by, 11
  Signal Corps, 9
U.S. Department of Energy (DOE), 22, 29
  carbon nanofibers and, 322
  PAFCs and, 41
U.S. Navy Bureau of Ships, 9
U.S. Navy Oxygen Generating Plant, 10
Universal gas constant, 395
University of Tasmania, 74
U-shape (reverse) flow, 261
UTC (*see* United Technologies Company)
Utility vehicles, with fuel cells, 66, 69–72
Utilization (stoichiometry) sweeps, 385–386

Van Hool, 68
Vapor, water, 365
Vapor deposition:
  chemical, 226
  electrochemical, 212
Vectrix, 74
Vehicular applications (*see* Transportation applications)
Vehicular pressurized hydrogen tanks, 25
Velocity, 144–145
Verification of assembly, 376
Viscosity, 146
Volkswagen, 58, 60
Voltage:
  requirements for fuel cell applications, 250
  temperature and, 94
Voltage collapse, 306
Voltage efficiency, 309–310
Voltage loss:
  charge transport and, 121–126
  electrochemistry and, 112–116

Voltage loss (*Cont.*):
  from reactant depletion, 137–138
Volvo, 67

Water, 397
  enthalpy of, 155, 362
  heat at constant pressure for, 401, 403
  saturated, 403
  thermodynamic data for, 408
Water electrolysis, 22–24
Water flux, 185, 186
Water management:
  for cathodes, 266
  methods for, 266–268
  for PEMFCs, 205, 265–268
  pressure and, 353
  with PTFE, 205
  stacks and, 265–268
Water out, 273–275
"Water splitting" organisms, 344
Water transport, 176
Water vapor, 365
Wet bulb temperature, 279
Wright, Charles R. Alder, 8, 9

X-ray diffraction (XRD), 390
X-ray fluorescence (XRF), 391
XRD (X-ray diffraction), 390
XRF (X-ray fluorescence), 391

Yamaha Motor Company, 74
Yttria stabilized zirconia (YSZ), 130, 263
  electrolytes of, 190
  structure of, 189

ZAFCs (*see* Zinc air fuel cells)
ZBT (Brennstoffzellentechnik), 75
Zentrum für Sonnenenergie-und Wasserstoff-Forschung (ZSW), 79
Zero emission vehicle (ZEV) mandate, 7
Zinc air fuel cells (ZAFCs), 6, 46–47
Zinc regenerative fuel system (ZRFC), 47
Zinc-air technology, 46, 47
Zirconia electrolytes, 49, 199, 200
ZRFC (zinc regenerative fuel system), 47
Z-shape (parallel) flow, 261, 262
ZSW (Zentrum für Sonnenenergie-und Wasserstoff-Forschung), 79
ZTEK Corp., 79